The Earth
and
Land Use Planning

GARY B. GRIGGS

JOHN A. GILCHRIST

UNIVERSITY OF CALIFORNIA, SANTA CRUZ

The Earth and Land Use Planning

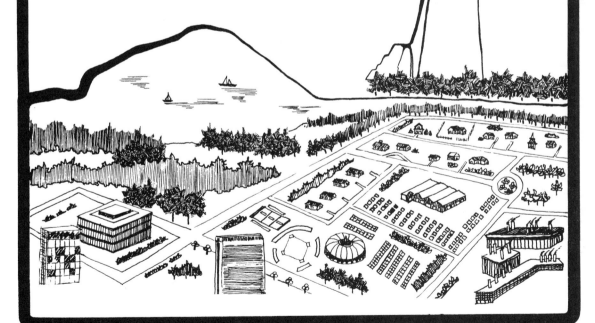

DUXBURY PRESS   NORTH SCITUATE, MASSACHUSETTS

Library of Congress Cataloging in Publication Data
Griggs, Gary B
   The earth and land use planning.
   Includes bibliographical references and index.
   1. Earth sciences. 2. Land use — Environmental aspects.   I. Gilchrist, John
A., joint author.   II. Title.
QE26.2.G74   550   77-5609
ISBN 0-87872-127-4

DUXBURY PRESS
A Division of Wadsworth Publishing Company, Inc.

The Earth and Land Use Planning was edited and prepared for composition by
Sheila Steinberg. Interior design by Dorothy Booth. Cover design and chapter
opening illustrations by Cindy Daniels.

L.C. Cat. Card No.: 77-5609
ISBN 0-87872-127-4
Printed in the United States of America
1 2 3 4 5 6 7 8 9 — 81 80 79 78 77

Contents

Preface

THE initial outline and organization for *The Earth and Land Use Planning* arose from a course in environmental geology that one of us had taught at the University of California, Santa Cruz, for several years. It became clear that certain geologic and hydrologic processes were affecting society to a greater and greater degree and that these processes needed to be studied and understood. That understanding alone was inadequate quickly became apparent when we, as a geologist and an environmental planner, began working together in planning an environmental impact assessment within local government. Recognizing a landslide or a rapidly retreating sea cliff was one thing, but planning for land use in such areas and then implementing land use decisions was quite a different thing. Thus land use planning became a joint focus for the book as it relates to geologic and hydrologic considerations.

We believe this is a major departure from the "first-generation" texts published on "environmental geology" in recent years. These early books and collections of readings presented an instructional overview of the issues. We have tried to look deeper and see environmental geology as a focal point where both earth scientists and environmental planners can converge to solve important problems and make important decisions involving land usage and planning. Having two completely different backgrounds, earth science and environmental planning, but having converged in our

work, we feel capable of presenting such a viewpoint.

Although many geologic hazards and environmental problems are regional in scope, such as volcanic activity on the West Coast, and strip mining in the Appalachians, every effort has been made to present environmental geology and planning in a broad geographic perspective. Examples are drawn not only from many places in the United States, but, where appropriate, from other regions of the world. We feel this should make the book particularly useful throughout this country. The land use and environmental planning orientation of the book streamlines its coverage to a degree and, therefore, the book avoids the comprehensive but unfocused coverage of issue-oriented texts. The basic processes and terminology appropriate to the subject are explained and covered in detail.

It did not seem appropriate to us, however, to offer an entire course in physical geology within this text. A glossary and appendices are included to define all necessary terms, and boxed essays are utilized to explain certain concepts and examples that amplify the text. For example, a boxed discussion of soils appears in chapter 7 within a discussion of erosion and sedimentation.

We feel the book is entirely self-contained and although some previous earth science would make it more meaningful to the reader, no such knowledge is required. Should the course instructor or the reader desire to cover material beyond the scope of this book we suggest the text be supplemented with one of the brief paperbacks on physical geology, energy, or more general environmental problems as is appropriate.

In addition to the book's land use planning orientation and its broad geographic coverage, we feel it is timely and useful for some other reasons.

The metric system is used throughout with English equivalents given where the authors felt appropriate. The coverage is up to date and includes very recent examples and information. A great deal of effort has been spent on illustrations to make the dynamic nature of the subject as visual and clear as possible. An entire chapter is included on basic concepts of land use planning and environmental impact assessment, both areas in which earth scientists are becoming more and more involved. We feel that the book should be useful as a text or reference for people in earth sciences, environmental sciences, geography, and environmental planning.

Acknowledgements

The authors are grateful to Donald O. Doehring, Colorado State University; Thomas Dunne, University of Washington; Donald Eschman, University of Michigan; Ira Furlong, Bridgewater State College; Ralph Gram, San Jose State University; Henry T. Hall, University of Minnesota; Robert M. Hordon, Rutgers University; and David M. Mickelson, University of Wisconsin, who reviewed the manuscript at various stages, and to Sheila Steinberg for her careful editing. We also want to give thanks to Venetia Bradfield, Cindy Daniels and Judith Bateman for their work in illustrating the book.

The Earth
and
Land Use Planning

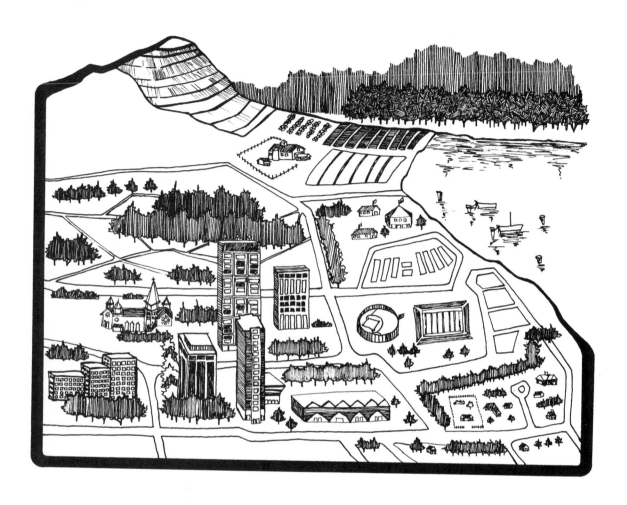

CHAPTER 1

Introduction

ALMOST every day newspapers carry stories on accounts of earthquakes, floods, hurricanes, or some other geological hazard or resource problem. It is difficult to visualize the human suffering, property damage, and economic disruption that result from inevitable natural disasters whose timing is completely uncertain. Despite this uncertainty, the magnitude of these events and their potential impact can be evaluated on the basis of past occurrences and existing knowledge (White and Haas, 1975).

The San Francisco Bay area of California will eventually be struck by a great earthquake, which may kill hundreds of people and injure thousands. Communications will be severed, utilities will be put out of commission, reservoirs may fail, hundreds of old buildings will collapse while many newer ones, including public buildings, will suffer major structural damage and become useless or unsafe.

A "dormant" volcano in the Cascades of the Pacific Northwest may erupt and cover hundreds of square kilometers with volcanic ash, ruining crops and contaminating water supplies. Mudflows generated by such an eruption could completely bury resort areas and entire towns in the valleys below.

A large hurricane will again strike the low-lying and heavily populated Miami area of the Florida coast. Storm tides, high winds and waves, and serious flooding will trap thousands of people, making evacuation impossible. Death and destruction will be even greater than during previous hurricanes because of increased population density and lack of adequate routes for evacuation.

Each of these events is possible, in fact very probable. The chapters that follow will deal with these hazards among others and bring out a number of points: prediction and forecasting will not in themselves prevent catastrophe; local awareness and responsibility are essential to prevent further encroachment into hazardous areas; and the failure to plan land use will only increase the potential for major loss and catastrophe.

We must understand that the land and water are both dynamic systems and limited resources which need to be considered in almost every action we take. The demands of more and more people and their concentration in urban areas require that we both safeguard these resources and give explicit consideration to natural hazards and processes in making land use decisions. In the past, land was treated strictly as a commodity — to be bought and sold in the real estate market. Natural obstacles such as hills or wetlands were either avoided where adjacent land was cheap or flattened and filled where land values were higher. Today regions can no longer be viewed as blank pieces of paper on which to write the story of irresponsible development; instead, the earth's surface everywhere has to be seen as a dynamic system, a combination of existing physical conditions that is constantly acted upon by natural processes, producing certain limitations or constraints on land use. To protect ourselves from geologic and hydrologic disasters, to guard against property damage and loss of life, and to conserve irreplaceable resources, the earth scientist and planner need to work together, each forging practical and constructive solutions from their knowledge and experience.

The successful interaction of human activity and nature requires the knowledge of the resources to be developed and conserved and also an awareness of hazards to be avoided or mitigated. Earth science information is essential to this integration and includes basic data from geology, soils, hydrology, and related scientific and engineering disciplines, along with the interpretation of those data.

Earth scientists are still working toward a basic understanding of many geological and hydrological processes acting at the earth's surface. Investigation and monitoring of the events that precede earthquakes or volcanic eruptions are examples of these research efforts. A clear understanding of a process that can be hazardous (excess runoff, leading to overbank flooding, for instance) or the recognition of an environmental constraint or limitation (such as the yield of a given groundwater reservoir or **aquifer**) is necessary before any rational planning or land use assessment can be made. The integration of geologic knowledge and the planning process is clearly exemplified by the land uses that might be appropriate in a 100-year floodplain (the area along a river that would be inundated by a flood likely to occur once every 100 years on the average). To calculate how much of a floodplain will be covered by water during the 100-year flood we need to analyze the existing discharge data for the stream in question. It is then necessary to make some calculations utilizing this data that can then be applied to accurate topographic maps of the floodplain. The ultimate goal in this case is the zoning or delineation of risk zones in the floodplain, which would then establish controls or limitations on all land usage or development.

What happens if we choose to disregard or ignore these natural hazards and limitations and simply let everyone use the land as they wish? What is the danger or cost of failing to respond to the natural limitations of the physical environment? When there were fewer people on the earth and they were more widely dispersed, the problems were relatively insignificant. Now, however, depending upon the severity of the hazard and its recurrence interval, costs in terms of property damage and human life can be extremely high. The earthquake of February 4,

1976, in Guatemala, for example, resulted in the death of about 20,000 persons. A recent study by the California Division of Mines and Geology (Bulletin 198, 1973) entitled "The Nature, Magnitude, and Cost of Geologic Hazards in California and Recommendations for Their Mitigation" is a good analysis of the problem of hazard assessment. Given a continuation of present practices, it is estimated that property damage and the dollar equivalent of loss of life di-

rectly attributable to geologic processes and conditions, and the loss of mineral resources due to urbanization, will amount to more than $55 billion in California alone between 1970 and the year 2000 (see figure 1.1). Average annual flood losses in the United States by the year 2000 are estimated to be about $3.5 billion.

To reduce or minimize these losses we can work toward extending the application of existing techniques and knowledge or to-

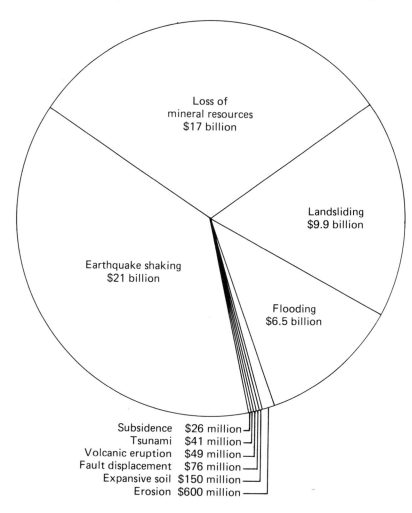

Figure 1.1. *Geologic Hazards in California to the Year 2000: A $55 Billion Problem.* REDRAWN FROM: *J. T. Alfors, J. E. Burnett, and T. E. Gay,* Urban Geology Master Plan for California: The Nature, Magnitude and Costs of Geologic Hazards in California and Recommendations for Their Mitigation *(California Div. of Mines and Geology Bulletin 198, 1973).*

ward improving our hazard reduction capabilities. Both seem logical and necessary. Losses from earthquake shaking, for example, can and should be reduced through a combined effort involving geologic and seismologic research, engineering practices, building codes, urban planning and zoning, fiscal and taxation policy, and preparedness planning. Priority efforts, such as strengthening older hazardous buildings, demolishing them, or reducing their use, need to be applied to reduce the loss of life. Adequate implementation of measures that can reduce losses is dependent upon the level of enforcement provided by local government. If all presently feasible loss reduction measures were applied and all current practices were upgraded to the current state of the art throughout California alone, an estimated $38 billion reduction of the projected $55 billion losses in that state could be achieved. The total costs of applying the loss reduction measures are estimated to be about $6 billion (see figure 1.2). In many areas legislation is requiring local governments to recognize and come to grips with these or other hazard reduction problems. The need exists for earth scientists and planners to work together to identify problem areas and then to plan for appropriate future land uses.

For any earth scientist who has been concerned with the problems of geologic hazards and land use, or for any environmental planner who has attempted to utilize geologic information in his or her work, there are still some obvious communication and information gaps. Geologists, in the minds of most planners, lack an understanding of the planning process and how their work might best fit into it. Geologists also need to learn how to communicate the re-

sults of their work to people other than their own colleagues. In the minds of many geologists, the common desire of planners is for a brief report unsupported by technically sound data. Geology in the past has only been very incidental to planning. All this is beginning to change, however. Many earth scientists are now very much involved in the applied aspects of geology and are keenly interested in putting existing data into a usable form and then into the hands of planners who will use it. New geologic maps depicting landslide distribution, slope stability, soil thickness, and depth to water table are being generated and used. The publicized failures and disasters where urbanization or development has encroached on unstable land, combined with the availability and demand for geologic data, have begun to bring the earth sciences into the planning process. Geology and hydrology are as essential to planning as any other components or inputs, and planners have to realize this and become familiar with these disciplines.

Within this book we have discussed geologic hazards in the first five chapters, including descriptions of the basic physical processes, how, why, and where they occur, how to recognize hazardous areas, and how hazards can be mitigated or avoided through the planning process. The latter sections of the book cover waste treatment and disposal, geologic and hydrologic resources, and planning environmental impact and land use control. These chapters should acquaint environmental planners with the physical processes and limitations of the earth and inform scientists how geologic and hydrologic information fit into the planning process.

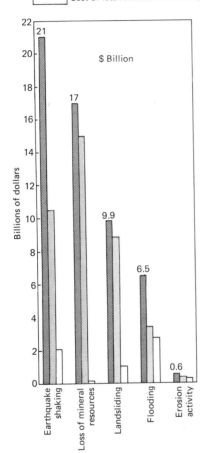
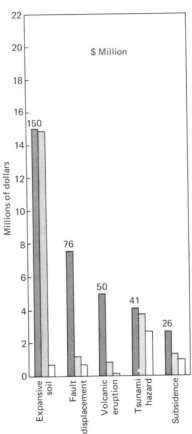

Explanation

Total losses, 1970-2000, under current practices
Loss-reduction possible, 1970-2000
Cost of loss-reduction measures, 1970-2000

Figure 1.2. *Estimated Total Losses Due to Each of Ten Geologic Problems in California for the Period 1970 to 2000.* REDRAWN FROM: *J. T. Alfors, J. E. Burnett, and T. E. Gay,* Urban Geology Master Plan for California: The Nature, Magnitude and Costs of Geologic Hazards in California and Recommendations for Their Mitigation (California Div. of Mines and Geology Bulletin 198, 1973).

REFERENCES

Alfors, J. T., Burnett, J. E., and Gay, T. E. Jr. *Urban Geology Master Plan for California: The Nature, Magnitude and Costs of Geologic Hazards in California and Recom-* mendations for Their Mitigation. California Div. of Mines and Geology Bulletin 198, 1973.

White, G. F. and Haas, J. E. *Assessment of Research on Natural Hazards.* Cambridge, Mass.: The MIT Press, 1975.

CHAPTER 2

Earthquakes and Faulting

Contents

INTRODUCTION

THE earth is continually evolving and undergoing change. From our viewpoint at the surface of the earth, we observe and are affected by both surface processes and the external expression of activity occurring deep within the earth. This activity may be very slow and take place over millions of years, as in the case of mountain building. It may also occur very quickly without warning, as in the case of a major earthquake. Much of the large scale earth movement of concern to us is concentrated along faults, or breaks in the earth's crust. When movement occurs suddenly along a fault, energy is released in the form of an earthquake.

Major earthquakes in the past have destroyed everything from primitive villages to modern cities. Since 1900 alone, there have been twenty-six major earthquakes throughout the world that have resulted in serious loss of life (see table 2.1). Most people in the United States tend to think of the 1906 San Francisco earthquake, which resulted in the death of about 700 people, as a major disaster. Although property loss was very high due to the fires that followed the shock, the

Table 2.1. *Major Historic Earthquakes and Their Effects*

Location of Earthquake	Date	Worldwide Magnitude	Deaths	Cost of Damage
Shensi, China	1556		830,000	
Lisbon, Portugal	1755		60,000	
Calabria, Italy	1783–1786		50,000	
Sanriku, Japan	1896		28,000	
Kangra, India	1905	8.7	19,000	
Messina, Italy	1908		82,000	
Avezzano, Italy	1915		30,000	
Kansu, China	1920	8.6	180,000	
Kwanto, Japan	1923	8.2	143,000	
Concepcion, Chile	1939	8.3	25,000	
Turkey	1939	7.9	25,000	
Agadir, Morocco	1960		10,000	
Chile	1960	8.9	10,000	
Peru	1970	7.7	50,000	
Nicaragua	1972	6.2	5,000	
Guatemala	1976	7.5	20,000	
United States				
Charleston, South Carolina	1886		60	$ 23 million
San Francisco, California	1906	8.3	700	$ 524 million
Santa Barbara, California	1925	6.3	13	$ 8 million
Long Beach, California	1933	6.3	115	$ 50 million
Bakersfield, California	1952	7.7	14	$ 60 million
Hebgen Lake, Montana	1959	7.1	28	$ 11 million
Alaska	1964	8.4	131	$ 500 million
Puget Sound, Washington	1965		7	$ 125 million
San Fernando, California	1971	6.5	65	$1000 million

SOURCES: B. A. Bolt; W. L. Horn; G. A. Macdonald; and R. F. Scott, *Geological Hazards* (New York: Springer-Verlag, 1975).

E. A. Keller, *Environmental Geology* (Columbus, Ohio: Charles E. Merrill Publishing Company, 1976).

H. W. Menard, *Geology, Resources, Society* (San Francisco: W. H. Freeman and Company, 1974).

fatalities from that event were relatively low on a worldwide scale. In 1923 a major earthquake near Tokyo in Japan killed 143,000 people. In 1556 a large earthquake collapsed thousands of cave dwellings in the loose, windblown deposits, or **loess**, of China's Shensi Province, killing an estimated 830,000 people. Another damaging earthquake will certainly strike the San Francisco Bay area, with its high rise structures and shore development, in the future. It is imperative that we understand why and where earthquakes occur and do everything feasible to minimize their effects.

MEASURING EARTHQUAKES

When the earth moves suddenly along a fault line, shock waves radiate outward in all directions. Three different kinds of waves, which move at different velocities (or speeds), are generated. As these waves radiate outward, and pass through different material, they can be reflected (or bounced back), refracted (or bent), and can change velocity. Ground motion is further complicated by the occurrence of **foreshocks** and **aftershocks**, which precede and follow major earthquakes.

Seismographs record the arrival time of shock waves. If we measure the time interval between various arrivals and plot the information on a graph (see figure 2.1), the distance to the **epicenter** of the earthquake can be reasonably well determined. (The location within the earth where the earthquake actually occurs is known as the **focal point**, and the point directly above this at the earth's surface is the epicenter.) If arcs are swung from each of three seismographic sta-

tions with radii equal to the distance to the epicenter, the location of the epicenter can be determined from the intersection of the arcs (see figure 2.2).

Earthquakes can be measured in terms of either energy released (**magnitude**) or actual effects (**intensity**). The first is based on the record of the earthquake on a **seismograph** and the second on the observations of the people involved. They are completely separate in intent but are often confused by the general public.

Richter Magnitude Scale

The Richter Magnitude Scale was developed by Charles F. Richter of the California Institute of Technology and utilized initially in 1935. It is a measurement of the energy of an earthquake at its source. The maximum **amplitude** of the P or **compressional wave** is the basis for the determination of the Richter Magnitude Scale. Seismographs from several different stations are usually used in the computations. The calculations involved require comparisons and some conversions to obtain the value at a standard distance of 100 kilometers (60 miles) from the epicenter. The amplitude is then transformed to a numerical value by means of a logarithmic scale. Thus an increase of 1.0 on the Richter Scale (for instance, from 6.0 to 7.0) represents a tenfold increase in the measured wave amplitude, and an increase of 31 times in the amount of energy released by the earthquake (see figure 2.3). The amplitude of an 8.2-magnitude earthquake is not twice as large as a shock of a 4.1-magnitude earthquake, but 10,000 times as large. Correspondingly an 8.2-magnitude earthquake releases nearly 1 million times more energy than one of magnitude 4.1. The comparisons that are

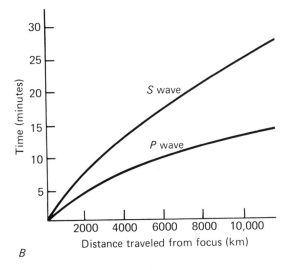

B

Figure 2.1. *Travel Time Graph for Shock Waves from an Earthquake. The increasing time lag between arrival of P (compressional) wave and S (transverse) wave can be used to determine the distance to an earthquake.*

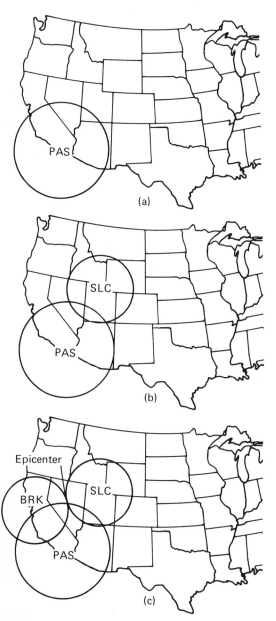

Figure 2.2. *Locating an Earthquake. Determination of the epicenter for an earthquake by the construction of arcs from three seismographic stations.* REDRAWN FROM: *D. N. Cargo and B. F. Mallory, Man and His Geologic Environment, 1974, Addison-Wesley, Reading, Mass.*

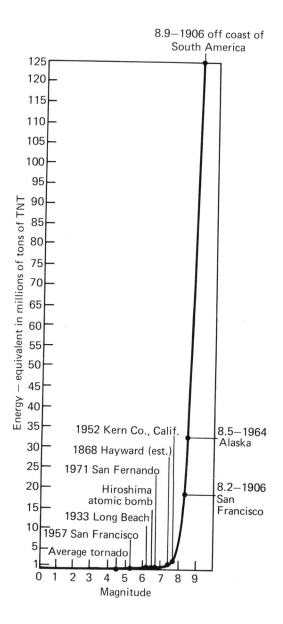

Figure 2.3. *Earthquake Energy Release. Energy released by different magnitude earthquakes expressed in millions of tons of TNT.*

made in newspapers between earthquakes of varying magnitudes are often meaningless because the logarithmic nature of the Richter Magnitude Scale is not mentioned. The occurrence of earthquakes of such varying size has, therefore, made it necessary to install and maintain seismographs of differing sensitivities.

Earthquake Intensity Scales

Several slightly different scales of earthquake intensity have been utilized in the past. Intensity is a measure of the destructive capacity of an earthquake or the effects of the shock as observed by people. It is not an absolute scale. Intensity can vary as a function of the distance from the epicenter, the nature of the underlying material, and even the type of construction in the area affected.

The first scale of earthquake intensity was developed by De Rossi and Forel in the 1880s in Europe. This scale, with values ranging from I to X, was widely used for a number of years to compare the effects of various shocks throughout the world. Its main disadvantage was the lumping of a great deal of major damage under classification X, which as seismology progressed became inadequate. The Italian seismologist Mercalli set up a new scale in 1902 based on a range of I to XII and included a more refined analysis of major damage. This Mercalli scale was modified in 1931 by two American seismologists to take into account modern features such as tall buildings and automobiles and trucks (Iacopi, 1973). It is the Modified Mercalli (MM) Scale that is still used today (see table 2.2). The varying intensity grades of an earthquake have usually been expressed on a map with lines drawn through areas of equal intensity. Although

an earthquake has a single magnitude, the intensity value can vary considerably and these should therefore not be confused.

EARTHQUAKES AND SEISMIC AREAS

The relatively new concept of global or **plate tectonics** has given geologists a framework from which they can more clearly understand the interrelationships between the large scale processes shaping the earth's surface and the worldwide distribution of earthquake activity. The earth's surface appears to consist of a number of large plates that are moving slowly relative to one another. The boundaries of these plates are marked by characteristic geological features as well as by earthquake activity (see figure 2.4). Where these plates pull apart or move away from each other, oceanic ridges and rises occur which are sites of active vulcanism and shallow-focus seismic activity. Where two plates collide or come together, two different types of features can exist depending upon the nature of the plates. If two thick, low density continental plates collide, compressional mountain ranges and shallow earthquakes result. If a continental and oceanic plate or two oceanic plates collide, underthrusting occurs and creates a deep trench. A zone of earthquakes dips landward from the floor of the trench to depths as great as 700 kilometers (420 miles). In addition to those locations where plates are moving away from each other, a third case involves two plates moving past one another, without creating or consuming crust. This gives rise to **transform** or **strike-slip faulting** and shallow seismic activity. A plot of worldwide earthquake activity clearly outlines the

Table 2.2. *Modified Mercalli (MM) Intensity Scale of 1931*

I	Not felt except by a very few under especially favorable circumstances.
II	Felt only by a few persons at rest, especially on upper floors of buildings. Delicately suspended objects may swing.
III	Felt quite noticeably indoors, especially on upper floors of buildings, but many people do not recognize it as an earthquake. Standing motor cars may rock slightly. Vibration like passing of truck. Duration estimated.
IV	During the day felt indoors by many, outdoors by few. At night some awakened. Dishes, windows, doors disturbed; walls make cracking sound. Sensation like heavy truck striking building. Standing motor cars rocked noticeably.
V	Felt by nearly everyone, many awakened. Some dishes, windows, etc., broken; a few instances of cracked plaster; unstable objects overturned. Disturbances of trees, poles, and other tall objects sometimes noticed. Pendulum clocks may stop.
VI	Felt by all, many frightened and run outdoors. Some heavy furniture moved; a few instances of fallen plaster or damaged chimneys. Damage slight.
VII	Everybody runs outdoors. Damage negligible in buildings of good design and construction; slight to moderate in well-built ordinary structures; considerable in poorly built or badly designed structures; some chimneys broken. Noticed by persons driving motor cars.
VIII	Damage slight in specially designed structures; considerable in ordinary substantial buildings, with partial collapse; great in poorly built structures. Panel walls thrown out of frame structures. Fall of chimneys, factory stacks, columns, monuments, walls. Heavy furniture overturned. Sand and mud ejected in small amounts. Changes in well water. Persons driving motor cars disturbed.
IX	Damage considerable in specially designed structures; well-designed frame structures thrown out of plumb; great in substantial buildings, with partial collapse. Buildings shifted off foundations. Ground cracked conspicuously. Underground pipes broken.
X	Some well-built wooden structures destroyed; most masonry and frame structures destroyed with foundations; ground badly cracked. Rails bent. Landslides considerable from river banks and steep slopes. Shifted sand and mud. Water splashed (slopped) over banks.
XI	Few, if any, (masonry) structures remain standing. Bridges destroyed. Broad fissures in ground. Underground pipelines completely out of service. Earth slumps and land slips in soft ground. Rails bent greatly.
XII	Damage total. Practically all works of construction are damaged greatly or destroyed. Waves seen on ground surface. Lines of sight and level are distorted. Objects are thrown upward into the air.

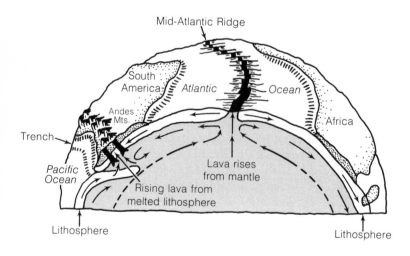

Figure 2.4. *Cross-Section Through the Earth Showing Plate Motion. The rise, lateral spreading, and sinking of the lithosphere (the upper 200km of the earth's surface including the crust and upper mantle) by convection currents in the earth's mantle.* REDRAWN FROM: *P. J. Wyllie, "Earthquakes and Continental Drift," University of Chicago Magazine (Jan./Feb. 1972): 16.*

major plate boundaries or margins (*see* figure 2.5).

The active faulting and seismic activity along western North America seem to be the surface expressions of the plate boundary along which the Pacific Ocean is moving northwest relative to North America. The San Andreas Fault system, which includes the main **trace** (principal segment of the fault) and its branch faults, extends for over 1000 kilometers (600 miles) from Cape Mendocino on the northern California coast nearly to the Gulf of California to the south. This is the zone along which much of this large scale plate motion is occurring. Driving across the southern part of the San Francisco peninsula, we cross from the North American plate to the Pacific plate. With this concept we can begin to understand the fault

system in California and its true magnitude. The fault is not simply a narrow fracture in the ground that we can fence off, designate as open space, and easily control or avoid. This zone of faulting and deformation created between two huge moving plates is many kilometers wide and is very complex.

California is characterized, therefore, by many active faults and is an area prone to high earthquake activity (*see* figure 2.6). Small shocks occur daily along the San Andreas Fault system, and major damaging earthquakes are common but less frequent. The rather diffuse seismic pattern that characterizes the western United States outside of California (*see* figure 2.7) can be interpreted as the expression of a very wide soft boundary between two rigid moving plates (Atwater, 1970). In the Northwest, the

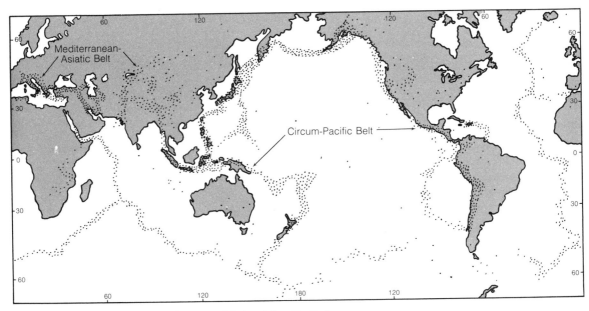

Figure 2.5. *Worldwide Distribution of Seismicity. Dotted areas are those with major earthquake activity.* REDRAWN FROM: *Barazangi and Dorman, Bull. Seismol. Soc. America, Vol. 59:1 (1969) p. 369.*

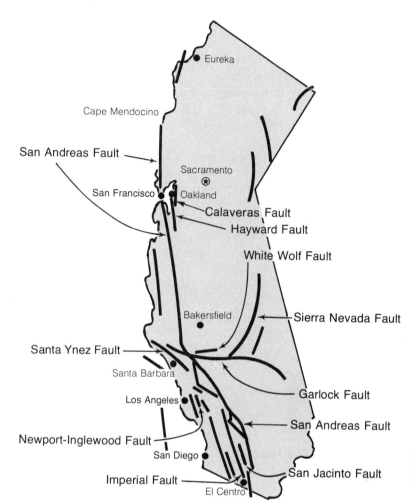

Figure 2.6. *Major Faults in California. These faults have produced significant earthquakes in historic time.*

Map labels: Eureka, Cape Mendocino, San Andreas Fault, Sacramento, San Francisco, Oakland, Calaveras Fault, Hayward Fault, White Wolf Fault, Bakersfield, Sierra Nevada Fault, Santa Ynez Fault, Santa Barbara, Los Angeles, Garlock Fault, Newport-Inglewood Fault, San Andreas Fault, San Diego, San Jacinto Fault, Imperial Fault, El Centro

underthrusting of the Pacific plate beneath the North American plate along the Aleutian Trench generates considerable activity in the state of Alaska.

Although the remainder of the United States is far from any plate boundaries, major and minor earthquakes have occurred throughout the country (see figure 2.7), perhaps because of weak points within the North American plate. Charleston, South Carolina, for example, is over 1500 kilometers (900 miles) from the eastern edge of the North American plate; yet it lies in a seismi-

cally active area, and a major earthquake that struck there in 1886 killed twenty-seven people. New Madrid, Missouri, also lies in the middle of the plate, but it experienced three great earthquakes in 1811 and 1812 that affected the course of the Mississippi River. A resident of the then sparsely populated area wrote: "The whole land was moved and waved like the waves of the sea. With the explosions and bursting of the ground, large fissures were formed, some of which closed immediately, while others were of varying widths, as much as 30 feet"

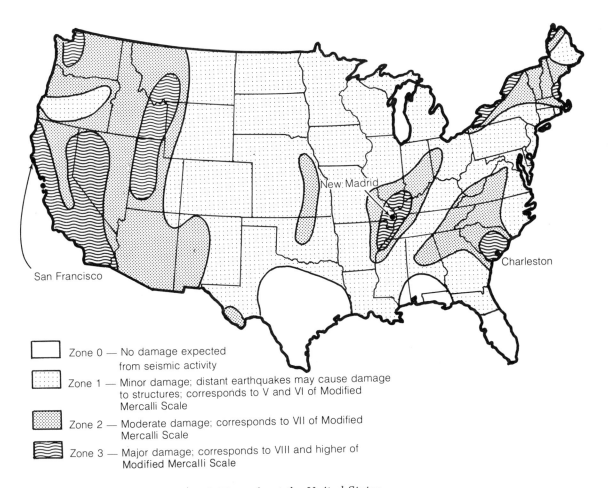

Figure 2.7. *Zones of Seismic Risk Throughout the United States. The frequency of possible earthquakes within the zones has not been considered, but rather the zones are based on historical damaging earthquakes, their intensities, evidence of strain release, and distribution of geological structures related to earthquake activity.* REDRAWN FROM: *D. N. Cargo and B. F. Mallory, Man and His Geologic Environment, 1974, Addison-Wesley, Reading, Mass.*

(Time, September 1, 1975). Some geologists feel that these were among the largest earthquakes ever to occur within the United States. In 1895 an earthquake centered there was felt by residents in twenty-three states. A lake was formed when four acres of land sunk; numerous chimneys were also demolished. In October 1965, a shock of 5.2 magnitude shook Missouri, causing minor damage over a wide area. Earthquake hazards are not restricted then to the western states, but can occur in other parts of the

United States. Future seismic activity in the midcontinent area and along portions of the East Coast is impossible to predict with our present knowledge.

RECOGNITION OF FAULT FEATURES

Because of the fear that is usually aroused whenever the word "fault" is mentioned, it is important to make an initial distinction between an active fault and an inactive fault. **Active faults** are those along which movement has occurred in historic or recent geologic time, and along which movement is likely to reoccur. **Inactive faults** are older features along which there has been no indication of motion in historic or recent geologic time and where there is no reason to expect a recurrence of movement. A very critical examination may be necessary in some cases first to recognize a fault and then to determine whether it is active or inactive. If the future stability of a hospital or school rests on the determination, then it is critical to look carefully at all the evidence.

An example of the problem of recognition and distinction between active and inactive faults occurs with the White Wolf Fault, just south of Bakersfield in central California. The fault is short and relatively insignificant, but it unexpectedly gave rise to the greatest earthquake to hit California since 1906 when movement occurred in 1952, generating a 7.7-magnitude shock. Because of the effect of surface processes and **weathering** in the area, the **scarps** (low cliffs) and cracks formed by this shock were scarcely visible ten years later. Field evidence alone, then, may not provide enough information to distinguish between an active and inactive fault.

The San Fernando earthquake in California in combination with the accompanying surface rupture of February 1971 is a similar example. This area had been one of low to moderate seismic activity; in fact, it was less active seismically than other parts of the Los Angeles, California, area. There was no clear indication of an active fault where the surface break occurred, and certainly nothing in the recent seismic history of the area to suggest that it was particularly likely to experience a 6.6-magnitude earthquake. This is an area where we could have said that the geology was well known and understood.

Recent investigations by Allen (1975) reveal that the very short historic record of earthquakes in North America should be used with extreme caution in estimating future seismic activity. Those parts of the world with the longest historic records of earthquakes — some 2000 years for Japan and the Middle East, and 3000 years for China — indicate surprisingly large long-term temporal and spatial variations. For example, the record for the last 3000 years in the Kansu and northern China regions (which comprise an area four times larger than California and Nevada combined) reveals relatively frequent seismic activity during the first and last parts of this period, but during an 800-year period from AD 200 to 1000, large shocks were almost totally lacking (see figure 2.8). This long historical record gives us a new perspective on our own relatively short 200-year earthquake record in the United States. Work in Japan has shown that major earthquakes have commonly occurred on short, relatively insignificant faults, whereas the larger and more conspicuous faults have shown considerably less seismic activity (Allen, 1975). Although most large earthquakes in California have occurred along the San Andreas Fault system, some major exceptions have also been recorded. Major earthquakes in the midcontinent region of the United States, Missouri for instance, have occurred in the midst of historically seismically inactive areas. All this information simply makes us more aware of how little we still know and of how much additional work is necessary to understand the problems of earthquakes and faulting.

A logical starting point in the study of faulting is in the field where we can begin to recognize and to identify the **geomorphic** and geologic features commonly found along fault zones. The untrained person might cross over or travel alongside a major

fault, such as the San Andreas, for miles and be unaware of its presence. Once a person becomes aware of the fault, however, there are a number of easily recognizable features that give clear evidence of its presence. It is certainly easier initially to study a section of a large conspicuous fault, such as the San Andreas, where the geomorphic features or landforms (which are more recognizable than geologic differences, such as rock types) are well developed (see figure 2.9), than to attempt to recognize and delineate similar features along faults or suspected faults that may be much less prominent.

Topographic maps and stereo aerial photographs are the principal tools needed at this stage of an investigation along with a good deal of field work. The rock types, topography, and vegetative cover are all going to affect the kinds of fault features recognizable in any given region. Areas with low relief or gentle hills covered only with grass, such as central California, provide the best exposures of fault topography.

Geomorphic Features

Fault Valley

A valley is often the most conspicuous topographic feature of a major fault zone and is very clearly seen from the air. A fault valley may have one of two origins:

1.   Fault movement may have broken up the rock in the immediate vicinity of a fault so that it was easily eroded while the bedrock in the adjacent area remained resistant. Tomales Bay, north of San Francisco, along the San Andreas Fault, is a good example of a fault valley now filled with sea water (see figure 2.10).

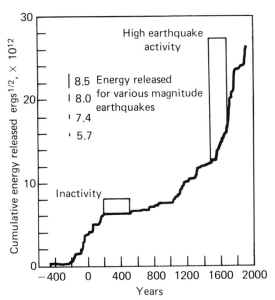

Figure 2.8. *Earthquake History in China. Historic earthquake activity for Kansu and northern China region, 466 BC to present, expressed in cumulative energy released. Notice temporal variations in release of energy.* ADAPTED FROM: *C. R. Allen, "Geological Criteria for Evaluating Seismicity," Geological Society of America Bulletin, 86 (1975): 1041-57.*

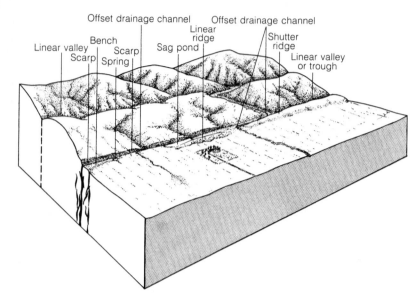

Linear valley
Scarp
Bench
Scarp
Spring
Offset drainage channel
Sag pond
Linear ridge
Offset drainage channel
Shutter ridge
Linear valley or trough

Figure 2.9. *Fault Zone Topography. Block diagram showing typical landforms found along recently active strike-slip faults.* ADAPTED FROM: *R. D. Borcherdt, ed., "Studies for Seismic Zonation of the San Francisco Bay Region," U.S. Geological Survey Professional Paper 941-A (1975).*

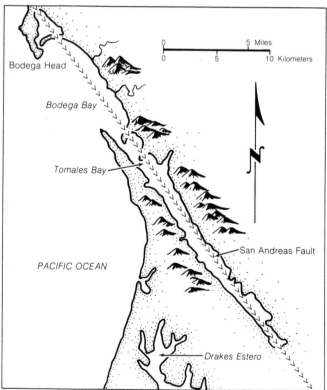

Bodega Head

Bodega Bay

Tomales Bay

PACIFIC OCEAN

San Andreas Fault

Drakes Estero

0   5 Miles
0   5   10 Kilometers

N

Figure 2.10. *Fault Valley. Tomales Bay is a fault valley that forms an embayment along the San Andreas fault north of San Francisco.*

Figure 2.11. *Sag Pond and Several Saddles in the Background Marking the Trace of the San Andreas Fault in Central California. Dashed lines indicate general locations of fault zone.* PHOTO BY: *Richard Farrington, U.S. Forest Service.*

Figure 2.12. *A Linear Ridge. Shadows in the center of the photo and dashed line mark a linear ridge along the San Andreas Fault.* PHOTO BY: *Richard Farrington, U.S. Forest Service.*

2. A block or several blocks may have dropped down between several **traces** of a fault so that parallel or subparallel series of valleys and ridges resulted. The Imperial Valley in southern California is an example of a valley formed by a depressed block.

Saddle

A saddle is similar to a fault valley but occurs on the side of a hill where the removal of crushed or sheared material has left a low area or swale (see figure 2.11).

Scarps

A sharp ridge or cliff or varying dimensions is a common feature along fault traces.

Where **thrust faulting** has occurred, scarps may be quickly modified by surficial processes as was the 3-foot scarp formed during the 1971 San Fernando earthquake. On the other hand, scarps formed as a result of lateral offset of ridges or hills may become more permanent features of the landscape. Repeated vertical offset has produced steep slopes, or scarps, in some cases thousands of feet high, such as those of the San Bernardino and San Jacinto mountains of southern California.

Linear Ridges

Ridges or low linear hills that are parallel to the trace of a fault are very common and may result from several different processes (see figure 2.12):

1. Ridges may form along hillsides where surface rupture has occurred and erosion begins along the line of weakness. The downslope portion of the break is still resistant whereas the crushed rocks along the fault line are gradually removed. As erosion proceeds the still resistant rock will stand out as a linear ridge.

2. Long slices of rock may also be squeezed or forced upward between parallel faults, leaving a linear ridge. With time, weathering rounds and smooths the feature but its elongation and trend along the fault remain clear.

Landslides

One contributing factor to the abundance of landslides in the California Coast Ranges is the San Andreas Fault system. The fracturing and weakening of rocks along the fault, and the repeated seismic shaking during earthquakes, have led to a concentration of slides along the fault. In some hillside areas, the landslides may be as prominent as any other topographic feature (see figure 2.13).

Offset Streams

The routes of streams and rivers that cross the trace of a fault can be instructive in indicating recent movement. Streams will commonly be offset and show bends in their courses as they have adjusted to the progressive movement that has occurred along the fault (see figure 2.14). Care has to be taken using such evidence, however, as it can be misleading. Streams may erode into other

Figure 2.13. *Fault Zone Topography. Hummocky landslide topography along the San Andreas Fault zone, east of Watsonville, California. The fault zone is enclosed by dashed lines.* PHOTO BY: *Richard Farrington, U.S. Forest Service.*

Figure 2.14. *Offset Streams. Streams offset right laterally along the San Andreas as it crosses the Carrizo Plain. Dashed line delineates the fault trace.* PHOTO BY: *Robert E. Wallace and Parke D. Snavely, Jr., U.S. Geological Survey.*

drainages and "capture" them, for instance, which may confuse the apparent offsets.

Sag Ponds and Impounded Groundwater

Ponds are very common features along faults, especially along the San Andreas. Due to movement along the fault, either (1) areas are elevated and surrounding drainage is cut off so that all undrained areas become ponds along the fault trace, or (2) crushed rock or **fault gouge** may be impermeable to groundwater flow so that impoundment occurs and a spring or lake is formed (see figure 2.11). Even if water does not reach the ground surface, it may be damp enough to support vegetation, such as willows, that is anomalous for the area and can be recognized as such.

More generally, the topography along an active fault is very anomalous relative to the surrounding terrain (see figure 2.15). Once this is recognized, it will become clear that the drainage is disturbed, vegetation patterns vary in regular ways, and landslides and downslope movements are common. The geomorphic features in themselves are indicative of recent or active faulting. In addition, (1) the documentation of seismic activity, (2) the measurement of **creep** (very slow, gradual movement along a fault), or (3) the offset of young surface of near-surface deposits also are clear evidence of an active fault.

Figure 2.15. *Aerial View of the San Andreas Fault in the Panorama Hills, Kern County, California. Note the fault valley, linear ridges, and offset streams.* PHOTO COURTESY: *U.S. Forest Service.*

Geologic Features

Juxtaposition of Different Rock Types and Ages

One of the earliest indications that a great deal of motion had occurred along the San Andreas Fault was the recognized offset of rocks of the same type and age. By looking at displacements of progressively older rocks, we can get an idea of the history of movement along the fault through time (see figure 2.16).

Crushed and Deformed Rocks

It is very common to find a shear zone along a fault with either a thin or wide zone of broken and crushed rocks, fault gouge, or chemically and physically altered rocks. These zones are commonly very weak and are easily eroded.

THE EFFECTS OF EARTHQUAKES

Having a clearer idea of where seismic activity occurs and why, let us consider what

happens during an earthquake and how this affects us. An earthquake affects the earth's surface, directly, through ground shaking and surface displacement, and indirectly, through landsliding and related phenomena, and **tsunamis**, or seismic sea waves.

Ground Shaking

As seismic waves pass through the earth's crust, the severity of ground motion or shaking at a particular point depends upon several factors (Albee and Smith, 1966):

1. total energy released in the form of seismic waves

2. distance from the source of the earthquake

3. nature of the surface and subsurface material

Maximum **ground acceleration** has commonly been used as a parameter indicating the severity of motion during an earthquake. By acceleration we mean the increase in velocity imparted to the ground by

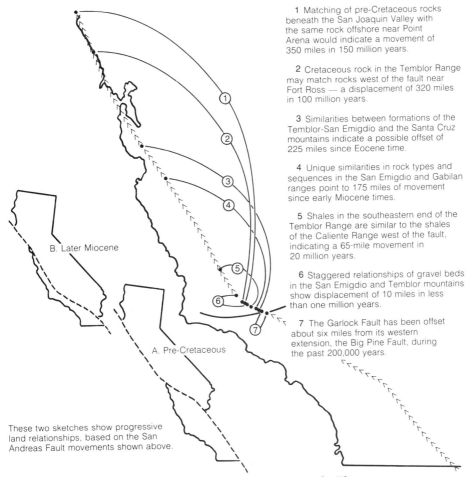

1 Matching of pre-Cretaceous rocks beneath the San Joaquin Valley with the same rock offshore near Point Arena would indicate a movement of 350 miles in 150 million years.

2 Cretaceous rock in the Temblor Range may match rocks west of the fault near Fort Ross — a displacement of 320 miles in 100 million years.

3 Similarities between formations of the Temblor-San Emigdio and the Santa Cruz mountains indicate a possible offset of 225 miles since Eocene time.

4 Unique similarities in rock types and sequences in the San Emigdio and Gabilan ranges point to 175 miles of movement since early Miocene times.

5 Shales in the southeastern end of the Temblor Range are similar to the shales of the Caliente Range west of the fault, indicating a 65-mile movement in 20 million years.

6 Staggered relationships of gravel beds in the San Emigdio and Temblor mountains show displacement of 10 miles in less than one million years.

7 The Garlock Fault has been offset about six miles from its western extension, the Big Pine Fault, during the past 200,000 years.

B. Later Miocene

A. Pre-Cretaceous

These two sketches show progressive land relationships, based on the San Andreas Fault movements shown above.

Figure 2.16. *Displacement Along the San Andreas Fault. The progressive offset of rock units of increasing age along the San Andreas Fault in California is due to continued movement.* DATA FROM: M. L. Hill and T. W. Dibblee, Jr., "San Andreas, Garlock, and Big Pine Faults, California," Geological Society of America Bulletin, Vol. 64 (April 1953): 443-458.

the passage of seismic waves. Acceleration is normally expressed in terms of gravitational force (g), where 1g = 9.8 m/sec². The frequency and cumulative effect of the sequence of pulses are also of major importance in considering the effects of shaking on structures.

In general, passing from more dense **crystalline rock** through less dense **sedimentary rock**, into unconsolidated and finally water-saturated alluvial materials, seismic waves tend to become reduced in velocity and increased in amplitude, and accelerations become greater (see figure 2.17). Ground motion lasts longer and is more damaging in unconsolidated or water-

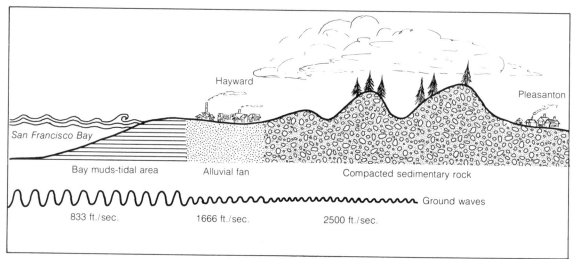

Figure 2.17. *Variations in Seismic Wave Behavior in Different Materials. The relative frequency, amplitude, and duration of seismic waves increase in passing from highly consolidated material to less consolidated material or water-saturated sediments. Ground shaking, therefore, will last longer and have a greater amplitude in bay muds than in the surrounding hills underlain by sedimentary rocks.* REDRAWN FROM: *"Hayward City Seismic Element," Paper of the Planning Department, City of Hayward, California (1973).*

saturated materials. Structures located on such materials suffer far more during earthquakes than those situated on bedrock. This has been clearly and repeatedly demonstrated in large earthquakes: San Francisco in 1906, Alaska in 1964, and Peru in 1970. Reports on the 1906 San Francisco earthquake documented the exaggerated shaking and consistently greater damage to buildings in the lower waterfront areas of the city underlain by the thickest bay mud and land fill (*see* figure 2.18). On the higher bedrock hills, intensities were considerably less.

Ground motion amplification can become particularly troublesome to **"long-period structures"** — usually, high rise buildings. Damage commonly results when the **wave period** of the building (the length of time it takes for a shock wave to pass through the building from bottom to top), as determined by building height and construction characteristics, is coincident with the natural wave period of the ground. Taller buildings have a longer wave period (two seconds or more) and are subjected to increased damage when founded on ground such as bay mud, alluvium, or loosely compacted fill, which also has a long natural period. In the 1957 Mexico earthquake, it was noted that multistory building damage was far greater in the poor ground areas of Mexico City than in firmer ground areas closer to the earthquake epicenter (Steinbrugge, 1968).

Many earthquakes have shown that poor ground is a greater hazard than close proximity to the fault and epicenter. In the 1906

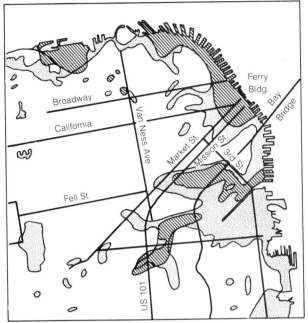

Greatest intensity 1906

Artificial fill

Figure 2.18. *Relationship Between Artificial Fill and Intensity of Seismic Shaking. In the waterfront area of San Francisco damage due to seismic shaking was far greater during the 1906 earthquake in areas underlain by artificial fill than in those underlain by bedrock.* REDRAWN FROM: *G. O. Gates, "Earthquake Hazards,"* Proceedings of the Conference on Geologic Hazards and Public Problems, R. A. Olson and M. W. Wallace, eds. (Santa Rosa, Calif.: Office of Emergency Preparedness, 1969).

earthquake again, the cities of Santa Rosa and San Jose, in California, both underlain by alluvial deposits, suffered to an extent out of proportion to their distance from fault movement. The effects of future seismic shaking on the development that has occurred during the last sixty years on San Francisco Bay fill are open to question. This will be partly determined by the nature and configuration of the fills and the nature of underlying foundation materials (*see* chapter 5).

Although empirical curves for the attenuation, or reduction, of acceleration of shock waves with distance from faulting had been published (*see* figure 2.19) prior to 1966 very little data had been recorded within 30 kilometers (18 miles) of faulting because of the shortage of instrumentation (Housner, 1965). Subsequently, the number of close-in records has increased by an order

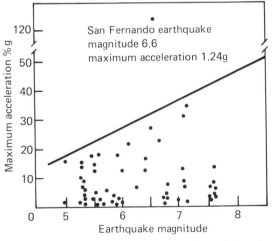

Figure 2.19. *Earthquake Magnitude and Acceleration. Comparison of maximum acceleration recorded for various magnitude earthquakes based on data collected prior to 1965.* ADAPTED FROM: *G. W. Housner, "Intensity of Earthquake Ground Shaking Near the Causative Fault,"* Proceedings of the Third World Conference on Earthquake Engineering (Auckland, N.Z.: Earthquake Research Institute, 1965).

of magnitude with particularly informative data from the 1971 San Fernando, California, earthquake. The data now available indicate that maximum acceleration increases with earthquake magnitude and existing values from empirical curves are underestimated. For example, prior to the San Fernando earthquake it was estimated that a 6.5-magnitude shock should produce an acceleration of 0.1 g on solid ground (Gutenberg and Richter, 1956). During that shock, however, several high frequency peaks over 1.0 g actually occurred (Ploessel and Slosson, 1974). In other words, maximum acceleration recorded was about ten times greater than what was expected for an earthquake of that magnitude. It seems clear that a large number of major structures in California and elsewhere have been designed using accel-

eration data that were far too low. The effects of future ground shaking from major earthquakes on such structures are at this point very poorly known or understood. Overall, damage from seismic shaking, due to the tremendous size of the areas that can be affected, will be far greater than damage from any other earthquake phenomena.

Surface Faulting or Displacement

During larger earthquakes, the fault slippage often extends to the earth's surface where sudden and abrupt ground displacement occurs (see figure 2.20). The displacement, or slip along the fault plane, may be primarily horizontal or vertical. During the 1906 San Francisco earthquake maximum horizontal

Figure 2.20. *Ground Rupture Along the Northern San Andreas Fault During the 1906 San Francisco Earthquake. Maximum horizontal offset during the earthquake was 16 feet.* Photo courtesy: *U.S. Geological Survey.*

Figure 2.21. *Offset Fence. This fence was displaced right laterally along the Northern San Andreas Fault during the 1906 earthquake.* PHOTO COURTESY: *U.S. Geological Survey.*

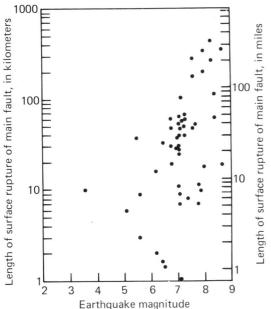

Figure 2.22. *Length of Observed Surface Rupture in Relation to Earthquake Magnitude. Observations of rupture length often underestimate the actual source dimensions of the earthquake because (1) the rupture expressed at the surface may represent only part of the total rupture, or (2) the surface rupture may be obscured by vegetation or water. Note that larger magnitude earthquakes generally occur along faults with the greatest rupture length.* REDRAWN FROM: *R. D. Borcherdt, ed., "Studies for Seismic Zonation of the San Francisco Bay Region," Geological Survey Professional Paper 941-A (1975).*

displacement was about 6 meters (see figure 2.21). At Hebgen Lake, Montana, as a result of a 7.1-magnitude earthquake in 1959, vertical displacement was also about 5 meters. In the 1971 San Fernando earthquake a combined horizontal and vertical displacement of about 1 meter occurred (see figure 2.26). As a general rule the amount of displacement that can occur during an earthquake is directly related to the total length of a fault. The longer the fault, the greater the potential for a large magnitude earthquake and the greater the amount of displacement likely (Bonilla, 1970). In addition to the relative ground displacement during an earthquake, the total length of associated surface faulting is also variable (see table 2.3 and figure 2.22). Larger magnitude earthquakes generally produce surface rupture over greater distances.

From an engineering standpoint, no structure can be designed to withstand, without some damage, a large displacement along an underlying fault. Many housing developments in California are located on major faults, and damage to houses and other structures in the event of surface displacement is inevitable (see figures 2.23 and 2.24). Along the east side of San Francisco

Table 2.3. *Length and Amount of Observed Displacement Related to Earthquakes in California, Nevada, and Baja California*

Date	Location	Magnitude	Length, Miles	Observed Displacement			Displacement By Resurveying
				Maximum Vertical Component	Maximum Horizontal Component	Maximum Total Across Zone	
1857	Fort Tejon, Calif.	7¾	40–250	?	Large		
1868	Hayward, Calif.	7	?	?	Some		
1872	Owens Valley, Calif.	7¾	ca. 50	13'	18'		
1899	San Jacinto, Calif.	ca. 6¾	ca. 2	?	?		
1906	San Francisco, Calif.	8.3	190 or 270	3'	21'	16'	
1915	Pleasant Valley, Nev.	7.6	24	15'	—	15'	
1932	Cedar Mtns., Nev.	7.3	38	24"	34"	35"	
1934	Excelsior Mtns., Nev.	6.5	0.85	5"	—	5"	
1934	Colorado River Delta, Baja Calif.	7.1	?	?	?	?	
1940	Imperial Valley, Calif.	7.1	40	—	19'	19'	ca. 9' Horiz.
1947	Manix, Calif.	6.2	1	—	3"	3"	
1950	Fort Sage Mtns., Calif.	5.6	5½	5"–8"	—	8"	
1951	Superstition Hills, Calif.	5.6	1.9	—	Slight	Slight	
1952	Kern County, Calif.	7.7	40	4'	2'–3'	3.6'	1–2' Horiz. 1–2' Vert.
1954	Rainbow Mtns., Nev.	6.6	11	12"	—	12"	
1954	Fallon, Nev.	6.8	14	30"	—	30"	
1954	Fairview Peak, Nev.	7.2	35	14'	12'	18½'	ca. 8' Horiz. ca. 4' Vert.
1954	Dixie Valley, Nev.	6.9	31	7'	7'	7'	ca. 7' Vert.
1956	San Miguel, Baja Calif.	6.8	12	36"	31"	36"	
1966	Parkfield, Calif.	5.6	ca. 25	—	4"	4"	

Source: Compilation by A. L. Albee and J. L. Smith, "Earthquake Characteristics and Fault Activity in Southern California," in *Engineering Geology in Southern California*, edited by R. Lung and R. Proctor (Los Angeles: Special Publication of Association of Engineering Geologists, 1966).

Bay, numerous public buildings, hospitals, and schools are located astride the active traces of the Hayward Fault. The relation of schools to the fault is so common that in drawing a line from school to school, one fairly well delineates the Hayward Fault zone (see figure 2.25; Gates, 1969). Such intense development close to the fault will make major damage inevitable when surface breakage occurs next. Thus, surface displacement during the San Fernando earth-quake resulted in extensive damage to many residences, destruction of a number of roads, streets, and water, sewer, and natural gas lines (see figure 2.26).

Most major faults do not consist of a single break but instead comprise many parallel and interfingering breaks that extend over a considerable width (1 kilometer or more along portions of the San Andreas Fault; see figure 2.27). The width of this fractured zone indicates that during the history

Figure 2.23. *The San Andreas Fault in the Daly City-Pacifica Area of the San Francisco Peninsula in 1956. This photograph shows the area before any development had occurred.* PHOTO COURTESY: *U.S. Geological Survey.*

Figure 2.24. *Development Along the San Andreas Fault, 1966. This photograph shows the same area as figure 2.23 ten years later, after urbanization had occurred with no consideration of the fault.* PHOTO COURTESY: *U.S. Geological Survey.*

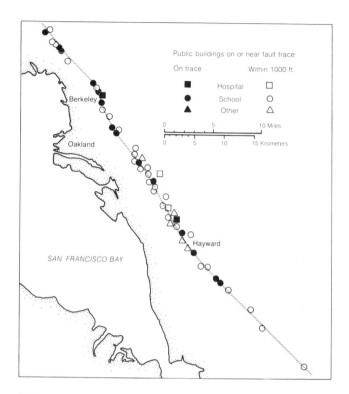

Figure 2.25. *Public Buildings and the Hayward Fault. This map delineates the locations of schools, hospitals, and other public buildings relative to the Hayward Fault along the eastern edge of San Francisco Bay.* REDRAWN FROM: *A. E. Alquist, "Seismic Hazards — A Question of Public Policy," in Environmental Planning and Geology, D. R. Nichols and C. C. Campbell, eds. (Washington, D.C.: U.S. Government Printing Office, USGS-HUD, 1969).*

Figure 2.26. *Offset of a Street Curb, Sidewalk and House During the 1971 San Fernando Earthquake. Movement during this earthquake reached a maximum of three feet in an oblique direction.* PHOTO COURTESY: *U.S. Geological Survey.*

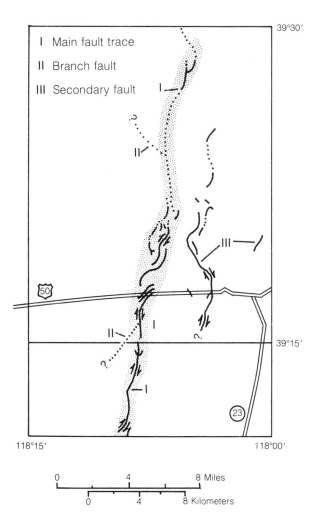

I	Main fault trace
II	Branch fault
III	Secondary fault

0   4   8 Miles

0   4   8 Kilometers

Figure 2.27. *Detailed Delineation of the Calaveras Fault. Main fault traces, branch faults, and secondary faults indicating the width of the San Andreas fault zone in an area of northern California. Dotted lines indicate uncertain location.* REDRAWN FROM: *M. G. Bonilla, "Surface Faulting and Related Effects," in Earthquake Engineering, R. L. Wiegel, ed., © 1970, fig. 3.4. Reprinted by permission of Prentice-Hall, Inc., Englewood Cliffs, New Jersey.*

of the fault displacements have not been limited to a single break, but have tended to migrate. Furthermore, future earthquakes in all probability will not take place on a single break. These uncertainties make land use planning in areas of active faulting a somewhat complicated undertaking.

Fault Creep

In addition to the sudden surface displacement that occurs during large earthquakes, there is another slower, more gradual movement known as fault creep. This may be equally important to overall fault motion but may or may not be accompanied by earthquakes. During the past sixty years or so, triangulation with very accurate instrument measurements has been utilized to document and record the extent and rate of fault creep in California. The average right lateral motion along the central San Andreas Fault is estimated to have been 32 ± 5 millimeters (1.25 ± .20 inches) per year from 1907 to 1971 (Savage and Burford, 1973). This motion has gradually been fracturing and displacing a winery at Almaden, California, and has offset houses, sidewalks, curbs, and streets in Hollister just to the north (see figure 2.28).

If recognized before construction, creep must be considered in the location, placement, and/or design of structures. Failure either to recognize the Hayward Fault zone or to realize that creep was occurring along it has led to the gradual deformation of a number of structures (Radbruch et al., 1966). This fault is a branch of the San Andreas Fault, which passes along the east side of San Francisco Bay. Creep has apparently been occurring along the Hayward Fault for the past thirty or forty years at an average rate of

Figure 2.28. *Fault Creep in Hollister, California. Creep in this photograph is slowly offsetting a sidewalk and retaining wall. Arrows indicate relative motion along the fault.* PHOTO BY: *Gary Griggs.*

6 to 10 millimeters (.24 to .40 inches) per year. Cracks have appeared in a major tunnel that supplies the East Bay with water, and railroad tracks have been deformed. Evidence of building and street deformation exists throughout the town of Hayward.

A controversy of sorts now exists over exactly what this slow creep activity and the associated **microearthquakes** or small tremors indicate. Is this activity evidence of gradual strain release, or of strain accumulation, and hence indicative of a major future earthquake?

Ground Failure

It generally takes a long period of time for most earth materials to reach equilibrium or stability. Many of the hillsides, mountains, and valley bottoms in geologically young and active areas, like California and Alaska, are often unstable. Landslides and mass movements are common on slopes, and compaction or subsidence may occur in unconsolidated valley sediments. Strong ground motion during earthquakes produces rapid changes in the condition of such unstable materials. The changes that occur, such as liquefaction and loss of strength in fine-grained materials, result in landslides, differential settlement, subsidence, ground cracking, lurching, and other ground surface changes (see figure 2.29).

Cracking, fracturing, or lurching of the ground associated with changes in surface material may occur during seismic shaking. During the 1964 Alaskan earthquake, ground cracking occurred along river flood plains and road and railroad embankments due to either **liquefaction** or lateral spreading at depth. Damage from this type of motion has been extensive in weak water-saturated materials such as those along the waterfront areas during the 1906 San Francisco earthquake and those in the irrigated alluvium and soils of the San Joaquin Valley during the 1952 Bakersfield shock.

Liquefaction

Liquefaction occurs when the strength of loose, saturated, coarse-grained material is greatly reduced. An initially stable granular material can be transformed quickly by vibration into a fluid state where the solid particles are in suspension, similar to quicksand, and all frictional resistance is lost. The Juvenile Hall landslide during the 1971 San Fernando earthquake resulted from liquefaction of a shallow sand layer and involved failure of an area over 1 kilometer long on a slope of only 2.5 percent (see figure 2.30). Structures built on thick layers of surficial granular material may gradually sink or settle differentially during a major earthquake. The well-publicized tilting and settling of large apartment buildings during the Niigata, Japan, earthquake of 1964 was a result of liquefaction of near surface sand deposits (see figure 2.31; Nichols and Buchanan-Banks, 1974).

Certain clays, because of their geological history develop the ability to liquify or become quick, like quicksand (Aune, 1966). **Quick clays** are generally confined to the far north and have caused serious damage in eastern Canada and Scandanavia (Kerr,

Figure 2.29. *Slumping in Artificial Fill. Along the shore of Lake Merced, near San Francisco, slumping of fill occurred during the 1957 Daly City earthquake on magnitude 5.3.* PHOTO BY: *M. Bonilla, U.S. Geological Survey.*

Figure 2.30. *The Juvenile Hall Landslide, 1971. This slide occurred during the San Fernando earthquake and involved an area almost 1 mile long. It resulted from liquefaction of a subsurface layer of sand and had a failure surface with a slope of only 2.5 percent.* PHOTO BY: *Robert Wallace, U.S. Geological Survey.*

Figure 2.31. *Liquefaction During Niigata Earthquake, Japan. Settlement and tilting of these buildings resulted from liquefaction of the subsurface material.* PHOTO BY: *Joseph Panzien.*

1963; Liebling and Kerr, 1965). These clays are produced when fine-grained material is deposited in marine or brackish water and subsequently raised above sea level. The sediments consist of very fine glacial flour, including individual clay minerals that initially have salt water in their pore spaces.

The ions in the salty or brackish water act to stabilize the individual clay particles through cohesion or bonding. However, as these deposits are raised above sea level, perhaps due to **glacial rebound**, the salt water may be flushed out with fresh groundwater removing the bonds and decreasing the strength of the clay. They now can quicken or liquify during a disturbance and lead to ground failure.

This type of failure was a major factor in the disastrous 1964 Alaskan earthquake, when the Turnagain Heights area of Anchorage was destroyed by slides. Failure and massive sliding was related to the physical and engineering properties of the Bootlegger Cove clay, a glacial estuarine-marine deposit that underlies much of Anchorage. This unit contains zones of low shear strength, high water content, and high sensitivity that failed under the vibrational stress of the earthquake. Even though Anchorage was 130 kilometers (78 miles) from the epicenter, it bore the brunt of the earthquake, in part because it is Alaska's largest city. Water mains, gas, sewer, telephone, and electric systems were destroyed. Roads and railroads were badly damaged. In the downtown area, entire blocks collapsed into depressions. Apartments, houses, and commercial build-

ings were completely destroyed (*see* figures 2.32 and 2.33). Modern geologic maps were available and geologists had warned in print that the clay would be unstable in the event of future earthquakes. Geologists in the future may have to do far more than study and write about problems. Agencies conducting research have to accept the responsibility of putting this information into the hands of those who are in a position to utilize it. Otherwise geologic research in these areas will only be a futile exercise.

Mass Downslope Movements

The 1970 Peru earthquake of 7.8 magnitude generated a destructive type of mass downslope movement that had never before been observed. A cataclysmic avalanche of about 25 million cubic meters of ice and rock fell from a glacier-covered peak at an elevation of 5500 to 6500 meters (18,000 to 21,000 feet). The melting of ice and the incorpora-

Figure 2.32. *Ground Failure in Anchorage. During the 1964 Alaska earthquake extensive damage occurred in Anchorage, 130 kilometers away from the epicenter, due to the failure of weak material in the subsurface.* PHOTO COURTESY: *U.S. Army.*

Figure 2.33. *Wreckage of the Government Hill School, Anchorage, Alaska. Although Anchorage was 130 kilometer's from the epicenter of the 1964 earthquake, extensive damage occurred. The graben in the foreground is about 4 meters deep.* PHOTO COURTESY: *U.S. Geological Survey.*

tion of water and water-saturated sediments into the avalanche converted it to a highly fluid mudflow. The flow of debris swept downslope about 14 kilometers (8.5 miles) to several villages at an average velocity of well over 160 kilometers per hour (100 mph). In the Rio Santa Valley below, an area of 15 square kilometers (6 square miles) was devastated. Yungay and its 20,000 inhabitants was completely buried by mud and rock as a lobe of the mudflow swept over a ridge and covered the prosperous and picturesque city (see figures 2.34 and 2.35; Ericksen et al., 1970).

A number of large landslides and mud or debris flows occurred during the 1906 San Francisco earthquake in the steep hills surrounding the San Francisco Bay area. The San Andreas, Hayward, and Calaveras fault zones in this area are marked by a number of geologically recent slides that in many places define the fault traces quite closely. The steep hills on either side of the bay are particularly vulnerable to sliding.

The 7.1-magnitude earthquake at Hebgen Lake, Montana, in 1959 generated a major rock slide that carried some 33,000,000 cubic meters of loose rock across the Madison River. This slide buried a number of campers in the area and also dammed the river and created a lake (see figure 2.36).

More recently, during the 1971 San Fernando earthquake, aerial photographic interpretation indicated that more than 1000 seismically triggered landslides occurred over a 250 square kilometer portion of the hilly and mountainous terrain above the San Fernando Valley. Landslides were common

Figure 2.34. *Nevado Huascaran, Peru. Oblique view of highest peak in Peru, which was the source of debris avalanche triggered by 1970 earthquake. The flow buried the towns of Yungay and Ranrahirca, killing over 20,000 people almost instantly.* Photo courtesy: *U.S. Geological Survey.*

Figure 2.35. *Material Carried by 1970 Peru Debris Avalanche. A block of grandodiorite estimated to weigh 700 tons was transported downslope by the debris avalanche during the 1970 Peru earthquake.* Photo courtesy: *U.S. Geological Survey.*

Figure 2.36. *Madison Canyon Slide, Montana. Aerial view of the slide that occurred during the 1959 Montana earthquake and created Earthquake Lake.* PHOTO BY: *J. R. Stacy, U.S. Geological Survey.*

in excavations and numerous roads were blocked by the sliding of cut slopes (Morton, 1971). Considering the kinds of landslides and mass movements that have accompanied major recent earthquakes, the areas over which they have occurred, and their distances from epicenters, this phenomenon is extremely difficult to plan for or to predict.

Seismic Sea Waves (Tsunamis)

Tsunamis or seismic sea waves (often incorrectly called "tidal waves") commonly accompany large submarine earthquakes and explosive volcanic eruptions. The coastal areas surrounding the Pacific Ocean Basin are subject to far more of these waves than other coastal areas of the world due to the

seismic activity that occurs in the vicinity of the trenches that encircle the Pacific Ocean.

In the open sea these waves are usually unnoticeable as they are often less than 1 meter high. They travel at velocities of 600 to 800 kilometers per hour (360 to 480 mph), however, and can reach distant coastlines in a matter of hours. On reaching the coast the water from tsunamis has been driven inland for several kilometers often destroying everything in its path. Waves reportedly have reached heights of as much as 40 meters, with a rise of 15 meters being common. Japan, Hawaii, and Alaska have been seriously affected by tsunamis from past earthquakes in and around the Pacific Basin (see chapter 5 for further discussion).

In the past, very little attention has been given to planning for tsunami hazards. It is imperative that we recognize this hazard and begin to apply stringent land use controls to all low-lying coastal areas susceptible to tsunami inundation. Controls could include the restriction of all land uses that are not coastally dependent (examples of coastally dependent uses are marinas and fishing piers), exclusion of all critical facilities from coastal areas (hospitals, police, and fire stations), installation of warning systems, and development of evacuation plans. Fortunately for the United States, the National Oceanic and Atmospheric Administration (NOAA) operates a reasonably effective tsunami warning system. Several hours notice can allow evacuation of threatened areas, preventing injuries and loss of life.

EARTHQUAKES AND FAULTING: CASE HISTORIES AND PLANNING IMPLICATIONS

St. Francis Dam Failure of 1928

The failure of the St. Francis Dam was an important event that focused public atten-

tion, at least for a short time, on the need for geologic considerations in the location of dam sites. The dam itself was built about 70 kilometers northwest of Los Angeles along the Santa Clara River in the 1920s. It was part of the then rapidly expanding network of pipelines, canals, dams, and reservoirs being hastily constructed to bring water to southern California. Constructed of solid concrete, about 62 meters high and 210 meters long, the dam at capacity held 46 million cubic meters of water (see figure 2.37). It failed completely the night of 12 March 1928, less than one year after the reservoir was filled, and when the water was within 30 centimeters (12 inches) of the top of the dam.

Different but equally unstable rock types occurred on either side of the dam (see figure 2.38). Both ancient and recent landslide scars occurred in **schist** on one side. Sedimentary rocks on the opposite abutment were dropped into the water after the dam failure and were found to disintegrate rapidly (Clements, 1966). A fault contact between the two units, which was marked by a wide zone of sheared and **brecciated rock**

Figure 2.37. *St. Francis Dam and Reservoir, California. This dam along the Santa Clara River in Southern California was one of those quickly built in the 1920s to serve the rapidly growing Los Angeles area.* PHOTO COURTESY: *Los Angeles Dept. of Water and Power.*

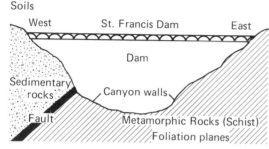

Figure 2.38. *Idealized Geology of the St. Francis Dam Site. Note the fault contact between the two types of rock and the foliation planes of the schist parallel to the canyon walls.* REDRAWN FROM: *W. H. Keller, Environmental Geology (Columbus, Ohio: Merrill Publishing Company, 1976).*

and fault gouge, passed under the dam. This fault also appeared on the existing fault map of California.

Failure of the dam could have been due to any of three causes: (1) slippage of the schists, (2) slumping of the sediments as a result of weakening from water saturation, or (3) seeping of water under pressure along the fault and removal of the gouge. When failure did occur, chunks of the westerly section were carried several kilometers downstream, the easterly section (resting on schist) collapsed, and the central section remained standing (see figure 2.39). Evidence suggested that failure occurred as a result of

Figure 2.39. *St. Francis Dam After Collapse, 13 March 1928. Note massive pieces of concrete dam carried downstream and large landslide at left abutment.* PHOTO COURTESY: *Los Angeles Dept. of Water and Power.*

seepage along the fault plane. Enough of the soft gouge was eventually washed away so that a stream of water poured through the opening. This quickly enlarged the flow, which removed the weak water-soaked sediments and eventually led to the collapse of the entire western abutment. The sudden discharge of water then probably swirled across the canyon, undercutting the opposite bank. The schist soon gave way and without support this side of the dam slid down into the canyon as well. Failure was total and as the water raced downstream to the ocean it took the lives of more than 500 people and destroyed approximately $10 million worth of property.

A number of inquiries and studies followed the dam failure and many reports have been published discussing the event (Clements, 1966). The consensus of opinion of all geologists and engineers was that the dam failed because of adverse geological conditions at the site that were either unrecognized or ignored.

The San Fernando Earthquake

At 6:01 in the morning on 9 February 1971, the San Fernando area was struck by one of the most devastating earthquakes in California history (see figure 2.40). Although the shock had a Richter Scale magnitude of only 6.6, it shook a wide and heavily populated area, leaving death and destruction in its wake. The sixty-five lives lost made it the

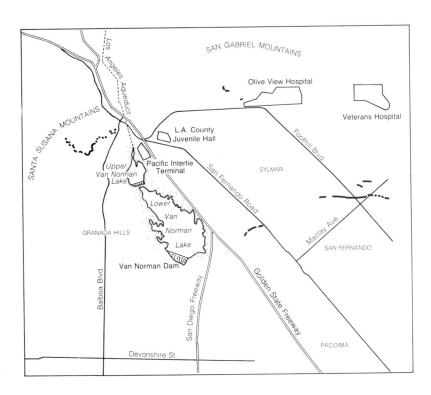

Figure 2.40. *San Fernando Earthquake, 1971. This index map of geographic area strongly affected by San Fernando earthquake shows locations of major structures damaged. Heavy lines indicate surface faulting.*

third worst earthquake in California's history, and the property damage of over a half-billion dollars was exceeded only by the 1906 San Francisco shock. Although it was an expensive disaster to California, something has been learned as a result. The ground shook early in the morning while highways were relatively free of traffic and before most people had gone to work in offices and public buildings. Schools were also empty. These factors minimized the loss of life.

The particular location of the shock had not been previously suspect anymore than the heart of Los Angeles, where the damage would have been more catastrophic. Earthquakes of this size are not uncommon in California, but this one struck the edge of a large metropolitan area. It is certain that earthquakes of this size, and larger, will rock other places in the United States, both rural and urban, in the future. Local, state, and federal geologists have collected a vast amount of data on the San Fernando earthquake. This needs to be interpreted and converted to information that the appropriate people and agencies can utilize in future planning for earthquakes. Highway departments, public utilities, building departments, and planning staffs need to use this information in the planning of future structures.

Seismic activity in the San Fernando area had been low in comparison with the rest of southern California during recent years. From 1934 to 1963 this area had seismic activity equivalent to less than four 3.0-magnitude shocks per 100 square kilometers (40 square miles). In historic time only one strong destructive earthquake is known to have occurred in the immediate area. The San Fernando earthquake, therefore, was not caused by movement on a major mapped fault, although such a fault may have been inferred from the abrupt linear scarp along the southern base of the San Gabriel Mountains (Greensfelder, 1971).

A number of other destructive earthquakes in California have occurred on faults that were either unmapped or not known to be seismically active. The towns of Vacaville, Dixon, and Winters in the Sacramento Valley were severely damaged by an earthquake in 1892 on an unmapped fault. The Imperial Valley earthquake of 7.1 magnitude, which occurred in 1940 and caused moderate to severe damage in several towns, took place on an unknown fault. More recently, in 1954, Eureka, in northern California, suffered moderate damage from a 6.6-magnitude shock, also on an unmapped fault. It is evident from these examples that major destructive earthquakes can occur on faults whose seismic activity or even existence is unknown. There is a real need to continue our efforts in mapping known faults, and also those whose existence we only suspect, for they seem just as capable of generating destructive earthquakes as those that have been clearly delineated.

Structural Damage and Lessons Learned

The San Fernando earthquake offers a unique opportunity to assess many of the important scientific, engineering, and human concerns associated with earthquakes in a modern urban environment. Two recent references (Greensfelder, 1971, and National Academy of Sciences-National Research Council [NAS-NRC], 1971) have been utilized in the following discussion of the damage that occurred during the earthquake and the lessons that have been learned.

The collapse or partial collapse of a number of structures due to seismic shaking during the earthquake led to the death of 65 people and serious injuries to over 1000 more. This emphasizes the hazards associated with urbanization near active fault zones, whether recognized or not. Were there unrecognized geological clues that might have revealed that this area and these faults were particularly hazardous? Are there geologically similar areas in which comparable earthquakes might occur?

Forty-four persons were killed at the Sylmar Veterans Hospital (see figure 2.40) when two unreinforced masonry buildings collapsed; the structures of the complex were over forty years old and were built before any modern building codes involving earthquake engineering were enacted. The

newly dedicated Olive View Hospital, which had been designed to withstand earthquakes and had been built according to the most-up-to-date building codes, was also seriously damaged. One two-story building collapsed so that the second floor came to rest at ground level. The first floor contained offices and examination rooms fortunately unoccupied at the early hour of the earthquake. Four five-story staircases pulled away from the main building and three of these collapsed outward (see figure 2.41). These buildings, in addition to two other hospitals in the area, were so severely damaged that they were no longer operative when they were most needed. Certain critical public structures need to be designed so that they will remain functional even after experiencing the most severe ground shaking.

More than 1300 buildings and 1700 mobile homes suffered major damage. Old

Figure 2.41. *Olive View Hospital After the 1971 San Fernando Earthquake. This building had just been completed according to the newest building code provisions for earthquakes. Seismic shaking of the loose subsurface alluvium led to extensive damage. Note wheel chair hanging from the roof of the collapsed stairwell.* PHOTO BY: *James Kahle, California Div. of Mines and Geology.*

unreinforced masonry walls, new chimneys, and concrete block walls were cracked and knocked down over a large portion of the San Fernando Valley and as far away as downtown Los Angeles. There are many thousands of such old buildings in California that will collapse if subjected to future strong ground shaking. Programs should be undertaken to either raze such buildings or render them safe, wherever possible, within some reasonable time period.

School buildings in the area of strong shaking that were designed and constructed since the passage of the **Field Act** in 1933 in California did not suffer structural damage severe enough to have been dangerous to the occupants had the schools been in session.

Older school buildings, however, which did not meet Field Act provisions, suffered potentially hazardous damage and some were condemned as a result. As recently as 1974, almost 180,000 California children were still attending school in buildings that did not meet the 1933 earthquake safety standards. Building replacement simply has not been considered critical to either appropriate officials or to the public when voting on school bonds.

The old earthfill dam on Lower Van Norman Lake, built in 1916 on an emergency basis, suffered major failure and sliding on its upstream face (see figure 2.42). It posed the threat of total failure (which may well have occurred with several more seconds of shak-

Figure 2.42. *Van Norman Dam. This old earth fill dam was heavily damaged by the 1971 San Fernando earthquake. Note the failure of the concrete facing. Eighty thousand people live directly below the dam.* PHOTO BY: *Kathy Sullivan.*

ing) so that 80,000 people living in the area immediately below the dam were evacuated for four days while the lake was drained. There are 228 dams and reservoirs with capacities of more than 80,000 cubic meters (50 **acre-feet**) of water in the nine-county San Francisco Bay area. Many are terminal storage and distribution reservoirs. They are, therefore, located on high ground in the midst of densely populated areas, which are also seismically active. A number of these dams were constructed prior to the passage

of the state's dam safety laws in 1929, which are no doubt in themselves seriously out of date. An improved plan to bring older dams in earthquake prone areas up to the best modern safety standards, or else to replace them as quickly as possible, is imperative.

Modern freeways and overpasses in the Sylmar area were severely damaged during the earthquake (see figure 2.43). Many bridge spans collapsed — one killing two men parked beneath it. Concrete pavement slabs, 20 centimeters thick (8 inches), were com-

Figure 2.43. *Freeway Damage, 1971 San Fernando Earthquake. Photo shows the concrete overpass connecting the Golden State Freeway with Foothill Boulevard, which collapsed during the San Fernando earthquake. Nearly two dozen overpasses failed during the intense shaking of the 6.6-magnitude shock.* Photo courtesy: *U.S. Geological Survey.*

pressed into ridges and thrust over one another, apparently as a result of land compression in a north-south direction. Lateral spreading affected areas of the freeway underlain by fill. The disruption of transportation resulting from damage of this type can greatly magnify the disastrous effects of an earthquake. Given similar conditions, the proximity of the shock and the kind of earth movement, the freeways of the San Francisco Bay area would suffer the same devastation. Can we build structures designed to withstand such forces now that we have a better idea of their magnitudes? If so, how much are we willing to spend? If not, how much risk are we willing to take in construction of this sort?

Networks for the distribution of electrical power, water, and gas, for sewage disposal, and for transportation of food and other necessities continue to grow in size and complexity with urbanization. However, the vulnerability of such systems became clear with the San Fernando earthquake when components of each of these systems were damaged and disrupted. Water lines, for example, were broken in 1400 places. The area is served by two major aqueducts that cross the San Andreas Fault on the way from Owens Valley. If breakage had occurred, only a four-hour reserve water supply would have existed. The need for careful planning in earthquake prone areas was clearly pointed out in the report by the NAS-NRC (1971).

The unexpected occurrence of an earthquake in this location, and the concentration of the most severe damage in zones of ground breakage forcefully illustrate both the importance and the difficulty of responsible and practicable seismic zoning. No evidence from pre-viously completed geological or seismological studies had been generally interpreted as indicating that the region was a more likely place for a damaging earthquake than many other parts of the southern California seismic region. This experience points out again that the short-term local seismic history is not in itself an adequate base for estimating earthquake risk. Until we gain a better understanding of earthquake processes and probabilities, due regard for public safety demands that seismic hazards be considered high throughout wide areas, and seismic zoning maps must reflect this. Many agencies and groups are working constructively on the problem of recognizing seismic hazards, but this effort is so important that it deserves more support.

THE INDUCEMENT, CONTROL, AND PREDICTION OF EARTHQUAKES

We have now arrived at the point where we are able to upset equilibrium within the earth and generate earthquakes, some planned, some unplanned. The detonation of large nuclear explosions, the injection or entry of fluids into the subsurface environment, or surface loading by water impoundment have been the principal generating mechanisms of inducing earthquakes. Realizing this capacity to generate tremors has led to additional research, as well as the usual speculation, on the employment of these techniques to release stored seismic energy in the form of small earthquakes. Thus the detonation of underground blasts or the injection of fluids may in the future lessen the potential for large earthquakes in critical areas. Along with this interest and research on earthquake control, scientists in the United States, Japan, China, and the Soviet Union have been at work on a number of techniques directed toward eventual earthquake prediction.

Nuclear Testing

Since the limited test ban treaty of 1963, both the United States and the Soviet Union have carried out numerous underground nuclear tests. By 1969 the United States had reportedly conducted nearly 200 underground tests, primarily at the Nevada test site, while the Soviet Union had conducted about 50 underground tests (Committee for Environmental Information, 1969).

Underground explosions can be assigned magnitudes just as earthquakes, according to the seismic energy released. A 1 megaton explosion (equivalent to 1 million tons of TNT) would be slightly less than 7.0 magnitude. The energy from such an explosion produces three types of effects: (1) seismic waves that radiate outward; (2) displacement at the site, either within the earth or at the surface; and (3) additional earthquakes triggered by the explosion.

Underground explosions are similar to earthquakes in that they are shocks that cause rock displacement and generate seismic waves. Due to the number of tests and their close monitoring, reasonably accurate predictions can now be made regarding energy transmission and resultant ground motion. The degree of movement and damage to be expected of tall buildings in Las Vegas due to bomb tests of various magnitudes in central Nevada are now being estimated. A number of complexities, how-

ever, make these effects difficult to calculate or to predict accurately.

Displacement of the earth's surface or subsurface as a result of a nuclear explosion is now well documented. Displacements seem to be most common where faults exist at the surface. Motion along 600 meters of the Yucca Fault in central Nevada has been observed as a result of testing, with maximum displacement of a meter having occurred. The Nevada test site is criss-crossed with surface faults and the effects have become more dramatic as the test sizes have increased (see figure 2.44). The Fault-less blast of 1968 showed a long fault opening within seconds of the explosion with maximum vertical displacement of 4.5 meters.

A considerable amount of research has been directed toward the possibility of triggering destructive earthquakes by nuclear explosions. As tests grow in size, approaching levels that may damage buildings or dams, this becomes a very serious consideration. It has been clearly established that relatively small earthquakes are triggered close to the sites of some underground nuclear explosions (Healy and Marshall, 1970). This post-shot activity may last for several weeks afterwards and seems to be a result of movement on previously recognized faults in the immediate vicinity. Energy released by these "after-shocks" may be in fact equivalent to or greater than that released by the explosion itself.

The concept that nuclear blasts have triggered additional earthquakes more than a few tens of kilometers from test sites has been investigated, but is still uncertain. The work of Healy and Marshall (1970) indicates that for certain intervals, greater numbers of earthquakes were recorded in the period immediately after a blast, relative to the

Figure 2.44. *Nuclear Testing, Nevada. The sequence of photographs shows the effect on the ground surface of nuclear testing at the Nevada test site.* PHOTO COURTESY: *U.S. Energy Research and Development Administration.*

period immediately preceding the blast, but for other intervals, the opposite is true. More research no doubt will be done in this area because of the implications of large future tests.

Injection or Entry of Fluids into the Subsurface

Perhaps the best known and documented case of artificial and accidental earthquake generation is the incident that occurred near Denver, Colorado, at the site of the Rocky Mountain Arsenal beginning in 1962 (Healy et al., 1970). The arsenal, operated by the Chemical Corps of the United States Army, has manufactured chemical weapons since 1942. The contaminated waste water resulting from this manufacture was put out to evaporate in open air reservoirs until 1961, when it began to seep into the ground, contaminating the groundwater and endangering crops. The army's solution to this problem was to drill an injection well to dispose of the wastes. Normally all space below the water table is filled with fluids, but depending upon rock permeability existing fluids can be moved out by high pressure injection. Various oil field brines and chemical and radioactive wastes have been disposed of in this way.

The army's injection began in March 1962 into a 4000-meter well that ended in fractured **gneiss**. Earthquakes began to occur the following month and continued to occur at a rate of four to eighty-five per month thereafter. Injection initially occurred at an average rate of 16 million liters per month for a year and one-half followed by nearly a year of no injections. There was a subsequent period of gravity flow and then injection under pressure. After four years, the program of disposal was halted because of a suggested correlation between the well and earthquakes in the Denver area (see figure 2.45; Healy et al., 1970).

The proximity of the earthquakes to the Denver metropolitan area created considerable public interest and concern. Many earthquakes of Richter Scale magnitude between 3.0 and 4.0 were felt over a large area and caused minor damage near the epicenters. The sudden appearance of seismic ac-

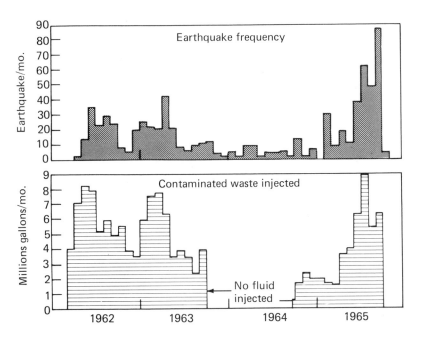

Figure 2.45. *Correlation Between Earthquakes and Fluid Injection. The upper half shows the number of earthquakes recorded in the Denver, Colorado, area each month; the lower half shows the monthly volumes of contaminated waste water injected into the arsenal well.* REDRAWN FROM: *J. H. Healy, W. E. Robey, D. T. Griggs, and C. B. Raleigh, "The Denver Earthquakes," Science 161 (1968): 1301–10. Copyright 1968 by the American Association for the Advancement of Science.*

tivity close to a large city raised serious questions and the possibility of earthquake generation by operations at the arsenal had to be evaluated quickly. The first question was that of seismic activity in the area prior to injection. A review of seismographic data indicated that the continued earthquake activity was not normal. Four earthquakes in the Denver area with magnitudes greater than 5.0 during 1967 were further substantiation of this abnormal activity. Additional research indicated that the probability of an earthquake swarm occurring randomly in this particular part of Colorado at this precise time was extremely low, about 1:2,500,000 (Healy et al., 1970).

Earthquake mechanisms were then analyzed. Radiation of seismic energy was consistent with right lateral strike-slip motion on vertical fault planes aligned with the trend of the seismic zone. The elongation of the epicentral zone in a WNW direction parallel to one of the two possible fault planes (see figure 2.46) strongly suggested that a zone of vertical fractures existed along the trend prior to fluid injection. An examination of basement rock cores confirmed the presence of the vertical fractures.

These observations suggested that a regional stress field of **tectonic** origin had produced the earthquakes. The release of the stored tectonic strain was triggered by the fluid injection into the basement rock. The fluid served both to increase the pore pressure and to reduce the frictional resistance to faulting. Prior to 1967, the possibility of a really destructive earthquake occurring in Denver could be reasonably considered remote. The four earthquakes greater than 5.0 magnitude that occurred during that year, however, suggested there was no longer any assurance that such a destructive

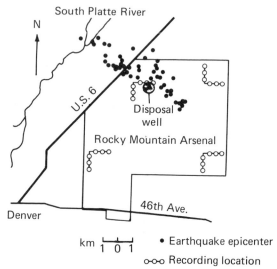

Figure 2.46. *Denver Earthquakes. Epicenters of earthquakes located in January and February of 1966 in the Denver area were plotted using a dense network of seismic stations.* REDRAWN FROM: *J. H. Healy, W. B. Robey, D. T. Griggs, and C. B. Raleigh, "The Denver Earthquakes,"* Science 161 (1968): 1301–10. Copyright 1968 by the American Association for the Advancement of Science.

earthquake would not occur. Those studying the problem felt confronted with the question of alleviating the earthquake hazard. At that time there was strong opinion that it might be possible to reduce the size and number of earthquakes by removing substantial quantities of fluid from the reservoir. Although theoretically feasible, the engineering difficulties and costs involved in such an operation may be prohibitively expensive. On the other hand, what may be the extent of damage to Denver in the event of a major earthquake? These kinds of comparisons are, unfortunately, rarely made.

There is now additional evidence that earthquakes are associated with fluid injection or flow of liquid into the subsurface in other areas. It is suspected by some that the pumping of fluid into the Inglewood oil field to raise the pressure and increase oil recovery was responsible for the earth movement that ruptured the Baldwin Hills Reservoir in southern California in 1963.

There have been a number of incidents in which earthquakes have apparently been triggered by the construction of dams and the impounding of large volumes of water (see table 2.4). The first known correlation between reservoir filling and earthquake occurrence was in Greece. In 1929, Marathon Lake, north of Athens, began filling and in 1931 the first earthquakes were felt. Two damaging earthquakes occurred in 1938 and shocks have continued to occur. At Lake Kremasta, in western Greece, waterloading of a regional stress field evidently led to energy release along a fault culminating in a 5.9-magnitude shock in addition to numerous fore- and aftershocks.

A well-documented case of earthquakes induced by reservoir filling in the United States is at Lake Mead behind Hoover Dam on the Colorado River. Although little seismographic information was available prior to dam construction, the area was thought to be free of earthquakes. Filling of the lake began in 1935 and the maximum water level was reached by 1941. The first shocks were felt in 1936, and the largest, a 5.0-magnitude shock, was recorded in 1939 when the lake had reached 80 percent capacity. At least 6000 minor earthquakes have occurred in the area since dam construction. The load of 42 cubic kilometers of water apparently began to depress blocks along faults that had been quiescent during Pleistocene and recent times. And "the tectonic equilibrium was sufficiently delicate to be disturbed by this comparatively minor additional load" (Richter, 1958).

Another well-known case of reservoir filling followed by earthquakes is the Koyna Dam in India. The reservoir formed by the

Table 2.4. *Reservoirs Where Earthquakes Have Occurred During and After Filling*

Reservoir	Year Filling Began	First Earthquake Reported	Damage
Marathon, Greece	1929	1931	Not known.
Oued Fodda, Algeria	1932	1933	Not known.
Lake Mead, United States	1935	1936	None known.
Lake Kariba, Rhodesia and Zambia	1958	1961	None known.
Lake Grandval, France	1959	1961	Not known.
Koyna, India	1962	1963	177 dead, 2,300 injured, extensive damage.
Vogorno, Switzerland	1964	1965	Not known.
Lake Kremasta, Greece	1965	1965	1 dead, 60 injured, 1,680 houses damaged.

SOURCE: "Earthquakes Related to Reservoir Filling," a report by the Joint Panel on Problems Concerning Seismology and Rock Mechanics, Division of Earth Sciences, National Academy of Sciences/Engineering.

dam is located on the Indian Precambrian shield, one of the least seismic regions of the world. The first earthquakes were felt in 1963 within a few months of when the lake began to form. Maximum lake level was reached in 1965; two severe shocks occurred in September 1967 and a 6.4-magnitude event occurred in December of that year resulting in 177 deaths and 2300 injuries in the area. Most of the buildings in the village of Koyna Naga were either destroyed or heavily damaged. Similar examples of seismicity apparently induced by reservoir filling have occurred in Algeria, Rhodesia and Zambia, France, and Switzerland.

The question has recently been raised of whether the 1975 series of earthquakes at Oroville, California, might be related to the recently filled Oroville Reservoir (see figure 2.47). Although some investigations have already been made, it is too early to be certain of any connection and additional seismic and geologic work is necessary. The dam is the nation's largest earth fill dam, being 236 meters high and impounding 4,365 billion cubic meters of water. Seismic networks in the area showed no earthquake activity from 1963 to 1967 when the reservoir began to fill. Approximately seven years passed before this series of earthquakes, with a maximum magnitude of 6.0, occurred.

The weight of the water itself in some locations, such as Lake Mead, may be responsible for triggering movement on preexisting faults. Another explanation involves the increased pore water pressure in the rocks underlying reservoirs. This increase will decrease the shear strength of the rock and may result in the release of tectonic strain along fractures or zones of weakness. In the case of existing dams that have induced seismic activity, only the lowering of the water level can bring about a decrease in the hazard that has been created. Prior to dam construction geologic and seismic studies certainly need to be carried out to try better to understand the relationship between the tectonic setting and the effects of the future impoundment of water.

Earthquake Control and Prediction

The first phase of a field experiment to control earthquakes by fluid injection and withdrawal has been successfully completed by scientists of the United States Geological Survey (USGS; Rayleigh, 1973). Field experiments have been carried out in the Rangely oil fields in northwest Colorado with the purpose of "taming" earthquakes. Water has been deliberately pumped and withdrawn through wells that penetrate a fault zone deep beneath the surface of the field. The experiments began in 1970, following the Denver earthquakes, and were undertaken to determine if stresses in buried strata could be relieved by injection of fluid. A series of minor earthquakes appear to have been generated at this locality in the past by high pressure injection, known as "water flooding," a common process applied to increase oil field production.

The experiments carried out by the USGS have shown that tremors can be triggered by the injection of water and numerous small tremors can be halted by the withdrawal of water (see figure 2.48). These results, according to Rayleigh of the National Center for Earthquake Research, indicate that

stresses in rocks deep in the earth (the Rangely earthquakes are at 1800 meters) can be relieved artificially. This gives us

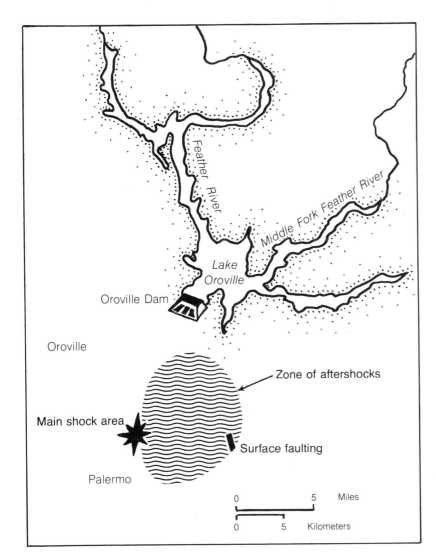

Figure 2.47. *Oroville Dam and Earthquake. The Oroville Dam and lake area in northern California showing areas of shocks and surface faulting during the August 1, 1975 earthquake. The dam is the nation's largest earth fill dam and the reservoir was filled in 1967.*

hope that eventually it may be possible to prevent catastrophic earthquakes, such as might occur on California's San Andreas fault, by inducing gradual fault movement. This would release energy in small doses, reducing the chances for the occurrence of a single major quake. It should be emphasized, however, that such an undertaking is many years away, and that much research needs to be carried out.

Although no one is routinely or consistently predicting earthquakes yet, a number of new discoveries are beginning to show considerable promise for the near future.

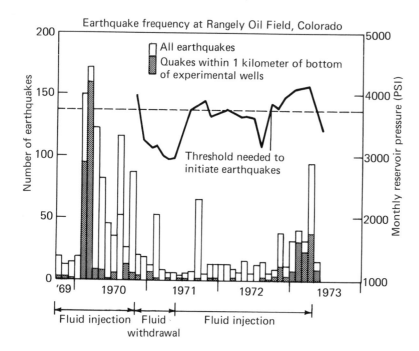

Figure 2.48. *Earthquake Control in Colorado. The level of earthquake activity at Rangely Oil Field, Colorado, was controlled by water pumping to maintain reservoir pressure either above or below a critical threshold level.* REDRAWN FROM: *R. E. Wallace, "Goals, Strategy, and Tasks of the Earthquake Hazard Reduction Program,"* Geological Survey Circular 701 *(1974).*

Seismologists in Japan have forecasted intermediate-sized earthquakes in one region with some accuracy. In the Soviet Union, three relatively large earthquakes have been predicted on an experimental basis. Phenomena as diverse as slow creep, variations in the tilt of the land surface, the distribution of microearthquakes, variations in electrical conductivity, changes in the chemistry of groundwater, and differences in the velocities of seismic waves are all being studied worldwide.

For some time now seismologists in this country have been establishing and monitoring creep meters along the San Andreas Fault system in the Hollister area of central California. These meters are sophisticated devices that usually consist of a wire attached to pillars on opposite sides of the fault (*see* figure 2.49). Depending upon their orientation across the fault trace, they are capable of measuring creep by lengthening or shortening of the distance between the pillars at the opposite ends of the wire. On wider segments of the fault, precision **theodolites** or lasers and fixed targets are being utilized to very accurately measure distances across the fault trace at oblique angles. Stations can then be reoccupied and distances checked for amounts and rates of motion. As mentioned previously, seismologists disagree somewhat on the importance of creep; some believe it relieves strain, reducing the possibility of a major earthquake. Others think creep is an indication of the energy stored and a signal of a greater seismic event in the area.

The measurement of the tilt of the earth's surface, a technique used mostly by the Japanese, is another clue to impending

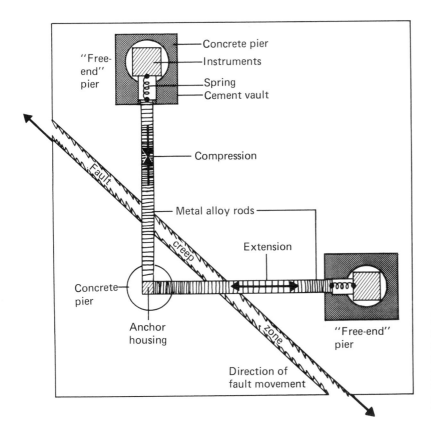

Figure 2.49. *The Components and Operation of a Creep Meter. Notice that the right lateral movement will produce compression along the North-South line and extension along the East-West line.*

earthquakes. According to field work and laboratory experiments, changes in rock volume along a fault due to the buildup of stress before an earthquake result in a slight tilt in the ground surface. By utilizing very delicate levels to record tilt changes, Japanese scientists have carefully predicted a number of earthquakes. Microearthquake activity commonly increases markedly prior to large tremors, so a careful monitoring of low level seismic activity may be another valuable clue in the forecasting of larger events. An increase in the electrical conductivity of the ground was used successfully to predict a strong earthquake in the Kamchatka region by the Soviets. Another phe-

nomenon, a change in the chemical composition of groundwater from deep wells, in particular an increase in radon emitted, gave warning before two different earthquakes in Tashkent, USSR.

The most significant recent discovery is an anomaly observed in seismic wave velocities prior to an earthquake. The ratio of the compressional wave velocity (V_p) to that of the **shear** or **transverse wave** (V_s) has been found to decrease from its normal value some weeks or months before a shock, then gradually to increase and reach approximately its normal value just before an event (*see* figure 2.50). The timing of the change appears closely related to the earthquake

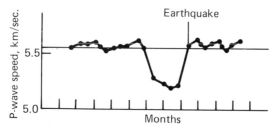

Figure 2.50. *The Utilization of Changes in Speed of Seismic Waves in Earthquake Prediction. The speed of P wave may decrease for months, then rise to normal just before an earthquake.* REDRAWN FROM: *R. E. Wallace, "Goals, Strategy, and Tasks of the Earthquake Hazard Reduction Program," Geological Survey Circular 701 (1974).*

magnitude with earlier anomalous signals observed for larger earthquakes. Seismic velocity variations occurred prior to the San Fernando earthquake and several others recently studied in California and in the Appalachians (Whitcomb et al., 1973). Scientists at Stanford University, the California Institute of Technology, and Lamont Doherty Geological Observatory have presented hypotheses, in somewhat differing forms, to explain this and other earthquake-warning signals (Hammond, 1973). The phenomenon of **dilatancy** (dilation), or the increase in rock volume and appearance of microcracks and voids, is now believed largely responsible for the various earthquake precursors. The hypotheses suggest that when rocks along a fault are stressed almost to their breaking point, they dilate and cracks open up. As groundwater flows in from surrounding regions and fluid pressure in the rocks increases, the shear strength of the rocks gradually decreases to the point where they fail, and an earthquake occurs.

This dilatancy model offers reasonable explanations for the other seemingly unre-

lated preearthquake observations. As fluid enters a region, increased pore pressure may be responsible for the increased microearthquake activity, and the volume increase could produce upward movements of the crust, or tilting. Since the electrical conductivity of the rocks depends upon the amount of water they contain, the increase in conductivity could be explained by greater water content. The increased flow rate of water into the zone would also lead to more active transport and the release of short lived isotopes such as radon.

Thus the model appears to fit well with data from the Soviet Union, Japan, and the United States and, in part because of this work, the United States and the Soviet Union have begun formal collaboration on earthquake prediction.

PLANNING IN SEISMIC AREAS

The question that logically may follow this discussion of earthquake prediction, especially in light of the development and urbanization that has occurred along faults around the Pacific Basin, is what would a city do if it were warned about an impending earthquake? It is not too difficult to imagine the chaos that would erupt if a major earthquake were pinpointed for a specific day. Perhaps worse yet, what if an earthquake were predicted for some thirty- or sixty-day period? Could everyone in San Francisco, Anchorage, Managua, or Tokyo retreat to the surrounding countryside for a month or two?

According to a report entitled *Earthquake Prediction and Public Policy* released in 1975 by the NAS-NRC, an earthquake prediction system could be very useful in saving human life. Many steps could be taken to safeguard the population of an area: the evacuation of vulnerable buildings and dangerous locations, the lowering of water levels behind dams, the shutdown of nuclear power plants, the closing of vulnerable gas lines, roadways, and other key transportation routes, restricted traffic flow in certain areas, increased precaution against fire, and the like. Some major recommendations of the report include the following:

1. Priority should be assigned to the saving of human lives, with secondary emphasis placed on minimizing property losses and economic disruptions.

2. A predictive system should supplement a vigorous earthquake hazard reduction program and not substitute for it.

3. Scientists, rather than politicians, should develop, assess, and issue the predictions of earthquakes.

4. Responsibility for planning and responding to earthquake predictions should be given to existing federal, state, local, and private agencies involved in disaster preparedness and community problems.

Another approach, somewhat more immediate, is that of updating and improving engineering standards and fault zone planning for earthquake prone areas. Planning for earthquakes involves the organization of people and their activities in such a way that personal and material damage during an earthquake are minimized. The unpredictability of earthquakes obviously makes this difficult. Although it can be said with near

certainty that California will experience a great earthquake within the next century, knowledge of just when or where it will ,occur has somehow eluded both geologists and planners. The extent to which any precautionary measures should be undertaken can easily become arbitrary. Although the first question most often asked is what is planning for earthquakes going to cost and can we afford it, we should perhaps instead be asking what risks are we willing, and not willing, to take? Government, at each level, with appropriate scientific input, must decide to what degree it wants to regulate itself and thereby avoid extensive damage to life and property. Earthquake information must be evaluated and the risks involved determined to see what specifically can be done and what expenditure of time, energy, and money is necessary (see table 2.5).

The earthquake-planning effort at this point can be expended in two directions:

1. Existing substandard structures must be recognized and methods and timetables must be devised to bring them up to acceptable standards or to eliminate them.

2. Future land use in seismic areas must be well planned, utilizing our most up-to-date knowledge and experience.

A preliminary necessity to either is the merging of economic and scientific schools of thought into a planning framework. Geologists have remained too long unheeded, often stranded in the shadows of "progress." The tremendous need for scientific opinion and input was long ago made manifest, but the roles of geologists and agencies like the USGS have only recently been altered. No longer are these agencies

Table 2.5. *A Scale of Acceptable Risks*

Level of Acceptable Risk	Kinds of Structures	Extra Project Cost Probably Required to Reduce Risk to An Acceptable Level
1. Extremely low[1]	Structures whose continued functioning is critical, or whose failure might be catastrophic: nuclear reactors, large dams, power intertie systems, plants manufacturing or storing explosives or toxic materials	No set percentage (whatever is required for maximum attainable safety)
2. Slightly higher than under level 1[1]	Structures whose use is critically needed after a disaster: important utility centers; hospitals; fire, police, and emergency communication facilities; fire stations; and critical transportation elements such as bridges and overpasses; also smaller dams	5 to 25 percent of project cost[2]
3. Lowest possible risk to occupants of the structure[3]	Structures of high occupancy, or whose use after a disaster would be particularly convenient: schools, churches, theaters, large hotels, and other high-rise buildings housing large numbers of people, other places normally attracting large concentrations of people, civic buildings such as fire stations, secondary utility structures, extremely large commercial enterprises, most roads, alternative or noncritical bridges and overpasses	5 to 15 percent of project cost[4]
4. An "ordinary" level of risk to occupants of the structure[3,5]	The vast majority of structures: most commercial and industrial buildings, small hotels and apartment buildings, and single family residences	1 to 2 percent of project cost, in most cases (2 to 10 percent of project cost in a minority of cases)[4]

1. Failure of a single structure may affect substantial populations.

2. These additional percentages are based on the assumption that the base cost is the total cost of the building or other facility when ready for occupancy. In addition, it is assumed that the structure would have been designed and built in accordance with current California practice. Moreover, the estimated additional cost presumes that structures in this acceptable-risk category are to embody sufficient safety to remain functional following an earthquake.

3. Failure of a single structure would affect primarily only the occupants.

4. These additional percentages are based on the assumption that the base cost is the total cost of the building or facility when ready for occupancy. In addition, it is assumed that the structures would have been designed and built in accordance with current California practice. Moreover the estimated additional cost presumes that structures in this acceptable-risk category are to be sufficiently safe to give reasonable assurance of preventing injury or loss of life during and following an earthquake, but otherwise not necessarily to remain functional.

5. "Ordinary risk": Resist minor earthquakes without damage; resist moderate earthquakes without structural damage, but with some nonstructural damage; resist major earthquakes of the intensity or severity of the strongest experienced in California, without collapse, but with some structural as well as nonstructural damage. In most structures, it is expected that structural damage, even in a major earthquake, could be limited to repairable damage. (Structural Engineers Association of California).

Source: *Meeting the Earthquake Challenge*, Joint Committee on Seismic Safety of the California Legislature, January 1974, p. 9.

and individuals needed merely in an "advisory" capacity. Instead, some equal degree of influence is required to balance the economically based decisions of nonscientifically oriented planning boards. Several factors have led to a perpetuation of past practices and have placed us in our present predicament. Planning commissions and similar groups are usually made up of political appointees and therefore commonly have little, if any, scientific background. Secondly, few geologists, or scientists of any kind, desire or even consider politics or public service as a parallel to their scientific careers. Although some progress has been made, the fact remains that geologists are sadly needed in many strategic decision-making positions. Planners and others involved in making land use decisions also need to acquire some basic geologic knowledge.

Earthquake Planning for Existing Structures

The problem of dealing with existing substandard structures is not a new one and is one where shortsighted economic considerations have prevailed to the present day. A major hazard in the event of any large future earthquake is that of the failure of older buildings and structures such as dams to withstand the shock. Cities are full of structures that were built before adequate earthquake building regulations were adopted. Many of these, especially older unreinforced brick and masonry buildings, are hazardous. A case in point is the 1971 San Fernando earthquake and the failure of the Sylmar Veterans Hospital.

Through the years the California legislature has appropriated funds to bring some schools up to more recent earthquake resistant standards. Yet many substandard schools and other public buildings still stand and are still being utilized. In 1960, 1500 of California's schools failed to meet the standards of the Field Act of 1933. In San Francisco alone, more than half the existing schools are considered poor earthquake risks (Alquist, 1969). The most logical approach toward the elimination of this problem seems to be that taken by the Joint Committee on Seismic Safety of the California legislature. The committee is conducting a detailed study of earthquake-related problems confronting California, developing seismic safety plans and policies, and recommending legislation to minimize the effects of a major earthquake on people, property, and the economy. The appropriation of funds or passage of bond issues for the replacement, or reinforcement where possible, of the most hazardous public buildings or structures would be a first step in the right direction.

Hazardous Structure Abatement Programs

Some farsighted local communities have recently adopted hazardous building abatement ordinances, including provisions for the identification and abatement of buildings susceptible to earthquake damage. The nonconforming building ordinance is a legal zoning tool that can provide a mechanism for removal of hazardous structures. Building abatement should be a long range program to avoid undue economic hardship and dislocation problems, but, at the same time, it should provide for the early removal of unsound structures that endanger the greatest number of lives — dams, schools, hospitals, theaters, and large office buildings. Evaluation of structural competence for all buildings suspected of being substandard should be based either on a preliminary survey or the date of adoption and enforcement of building codes. A community might allow buildings to remain for longer periods if their use or occupancy is greatly reduced.

In establishing a program of hazardous structures abatement, the following buildings should be given priority:

1.   Unreinforced masonry structures

2.   Buildings constructed prior to a specific date determined by the history of adoption and enforcement of the jurisdiction's building codes

3.   Critical facilities

   a. essential facilities (buildings whose use is necessary during an emergency)

   b. buildings whose occupancy is involuntary

   c. high occupancy buildings

Single family dwellings should be given lowest priority in an abatement program since they are predominantly wood frame construction and should, therefore, perform relatively well during seismic shaking. Older wood frame houses, however, would greatly benefit from inexpensive repairs to strengthen their earthquake resistance, and home owners should be encouraged to make these repairs.

Parapets are a particular problem that should be included in any structural abatement ordinance. These types of overhanging decor readily collapse in an earthquake and can cause death or injury to persons on well-traveled streets below. Fortunately parapet corrective work is relatively inexpensive and easy to perform. Therefore

many communities have found it politically more expedient to adopt a parapet ordinance before launching into a total building abatement program.

Seismic Hazard Planning

In planning for seismic hazards, planners and geologists must differentiate between the various types of hazards and design their solutions accordingly. Planning to prevent hazards involves knowledge of the hazard, formulation of zone districts to regulate land use, and development of construction codes. Other techniques, such as nonconforming use procedures or tax incentives, can be incorporated into a zoning ordinance, or used separately to discourage improper use of hazardous areas.

There are a number of interrelated factors that need to be considered in planning to prevent seismic hazards.

1.   *Seismic history and fault location.* Assessment of potential fault displacement for purposes of land use planning requires earth science data on the history of surface rupture. The amount of geologic data required increases as decision making moves from general regional-area planning to actual development proposals. Consideration should be given to the frequency, location, and magnitude of all recorded seismic activity, the active or inactive nature of the existing faults, the extent of their surface traces, and the examination of features suggestive of faulting. Trenching and other field investigations should be conducted across suspicious features that may be related to fault activity.

2.   *Subsurface geologic structure.* Evaluation should be made of conditions existing

beneath the surface of the land, such as caves, sinkholes, or riverbeds, that may present future problems.

3. *Topography and surface instability.* Steep slopes are prone to failure, and landslides are common in many areas both adjacent to and some distance away from fault traces. Rock units within a fault zone are highly sheared and therefore are extremely susceptible to failure. Existing landslides and unstable slopes must be recognized and evaluated.

4. *Surficial foundation material.* The distribution of bedrock, consolidated and unconsolidated sediment and its grain size, and the existence of water-saturated material or artificial fill should all be delineated. Variations in bed thickness, grain size, and water content can lead to differential settlement or liquefaction during an earthquake.

5. *Structural design.* The final, but equally important considerations to be made are those of the structural design of buildings, their workmanship and materials, and existing and proposed building codes and standards. This is an area where information on seismic hazards has not been thoroughly integrated with engineering and economics.

Ground Displacement

Because few structures can withstand the effects of ground displacement, they should be located off fault traces. In considering land uses appropriate for areas subject to ground displacement, low density or open space uses should always be selected if fault rupture could result in unacceptable levels of damage or risk of life. One possible solution is to adopt an ordinance establishing a seismic zone district (see p. 67), which allows planned use of the potentially hazardous area but provides regulations over and above that provided by the existing land use ordinance. Where detailed site data is lacking, the ordinance certainly should require on-site geologic investigations prior to all development. If fault traces are identified, a setback of 35 to 70 meters should be designated for structures in accordance with their proposed use and the type of deformation expected. Setbacks should be larger if the fault has a history of shearing over a wide zone, or if it is poorly located from available studies (see figure 2.51; Nichols and Buchanan-Banks, 1974).

Ground Shaking

Although prediction of ground shaking is difficult, and methods for dealing with this class of hazard are more complicated than those for ground displacement, some planning precautions can be taken. Generalized "intensity" maps indicating degree of ground shaking for various areas can be compiled from information on local soil and geologic conditions and proximity to active faults to assess the gross effects of ground shaking for general planning purposes. These could be used to delineate areas appropriate for low density land uses and could also have some value in placing general constraints on specific proposals, such as an urban renewal program or a large-scale development plan. As an example, structural requirements might be more stringent, and high rise structures prohibited altogether in areas underlain by soft mud or thick, unconsolidated sedimentary materials. Local ordinances can require detailed geological, soil engineering, and structural engineering analyses for high occupancy buildings in areas subject to moderate or severe ground shaking. It is important, however, that rec-

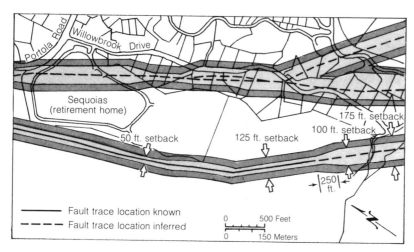

Figure 2.51. *Fault Zone Setbacks for Building. Example of minimum easements required for building setbacks from active fault traces by new ordinance in Portola Valley, California. All new building construction is prohibited within the 100 foot wide zone (50 feet on either side of the well-located portion of the San Andreas fault); structures with occupancies greater than single family dwellings are required to be 125 feet from the fault trace. Where location of the fault trace is less well known, setbacks of 100 feet for single family residences and 175 feet for higher occupancy structures are required.* REDRAWN FROM: *D R. Nichols and J. M. Buchanan-Banks, "Seismic Hazards and Landuse Planning,"* Geological Survey Circular 690 (1974).

ommendations made in these reports be carried out. As discussed earlier, it is difficult at the present time to predict ground acceleration from earthquakes. It is desirable, therefore, to adopt ordinances and procedures that are flexible enough to accommodate increasingly sophisticated methods of prediction.

Liquefaction and Related Types of Ground Failure

Formulating land use policy for ground failure hazards is also handicapped by difficulties in predictive capability. However, a general evaluation of ground failure potential may be made using background data on landslide distribution, slope, presence of loosely consolidated alluvial sediments, and groundwater conditions. Using these data, geologists and planners can compile interpretive maps (at scales of 1:12,000 to 1:62,500) that will identify problem areas (see figures 2.52 and 2.53).

As with other earthquake hazards, these maps can be used to designate areas of known or suspected instability as geologic hazard zones on official zoning maps. In such areas, detailed geologic and soil engineering reports should be required prior to approval of any development to demonstrate either that hazardous ground failure conditions do not exist or that they can be over-

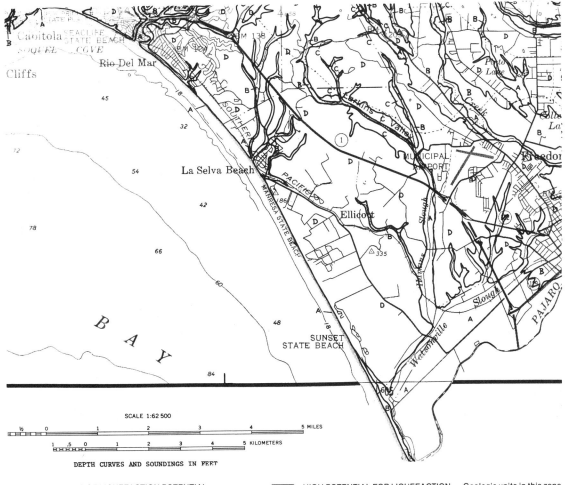

SCALE 1:62 500

½ 0 1 2 3 4 5 MILES

1 .5 0 1 2 3 4 5 KILOMETERS

DEPTH CURVES AND SOUNDINGS IN FEET

ZONES OF LIQUEFACTION POTENTIAL
The following zones express the general liquefaction potential of areas underlain by Quaternary deposits in Santa Cruz County. This information is suitable for general land-use planning but it is not authoritative in determining the relative hazard at any particular site. Presence of water in sandy layers near the surface of the ground could make a site highly susceptible to liquefaction during an earthquake even though the geologic unit generally has low potential. Similarly, local dewatering of a sandy deposit by pumping could make a site less susceptible to liquefaction. Site safety with respect to liquefaction should be determined after field investigations by qualified engineering geologists or soils engineers.

A HIGH POTENTIAL FOR LIQUEFACTION — Geologic units in this zone include younger flood-plain deposits (Qyf); some of the older flood-plain deposits (Qof) and alluvial deposits (Qal); basin deposits (Qb); beach sand (Qbs); and abandoned channel fill deposits (Qcf)

B MODERATELY HIGH POTENTIAL FOR LIQUEFACTION — Geologic units in this zone include some of the older flood-plain (Qof) and alluvial (Qal) deposits; dune sand (Qds); colluvium (Qc); and alluvial fan deposits (Qf)

C MODERATELY LOW POTENTIAL FOR LIQUEFACTION — Geologic units in this zone are alluvial fan deposits (Qf); colluvium (Qc); older flood-plain deposits (Qof); and alluvial deposits (Qal)

D LOW POTENTIAL FOR LIQUEFACTION — Geologic units in this zone include eolian deposits of Manresa Beach (Qem) and Sunset Beach (Qes); terrace deposits (Qwf, Qwa, Qcu, Qce, Qt, and Qcl); Aromas Sand (Qa, Qac, and Qaf); and continental deposits (QTc)

Figure 2.52. *Seismic Suitability Map for Zones of Liquefaction Potential, Santa Cruz County, California. The zones A-D express the general potential for liquefaction for areas underlain by Quaternary deposits.* PHOTO COURTESY: *Seismic Element (Santa Cruz, Calif.: Santa Cruz County Planning Dept., 1974).*

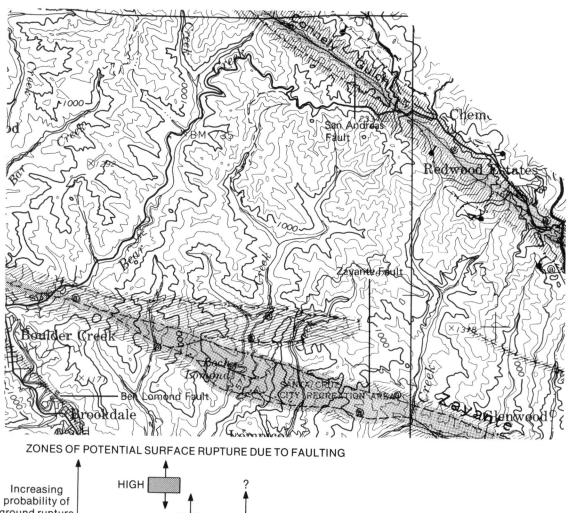

Figure 2.53. *Seismic Suitability Map for Zones of Potential Surface Rupture Due to Faulting, Santa Cruz County, California. Fault lines have been designated according to the potential for ground surface breakage during future faulting.* PHOTO COURTESY: *Seismic Element (Santa Cruz, Calif.: Santa Cruz County Planning Dept., 1974).*

come with site preparation or engineering design. If it is found that engineering techniques cannot mitigate hazards to within acceptable risk levels, then other locations for the development should be sought.

Other techniques can also be pursued. In areas where ground instability problems have been well documented, consideration should be given to adoption of a hazardous building abatement ordinance for existing developed areas. Tax assessment practices should be adjusted to allow deductions for owners with property in potential ground failure zones. Such a program can be designed to reduce the tax burden on existing structures to facilitate early removal through an abatement program. At the same time the tax incentive program on vacant land should be established to discourage new development unless the owner can demonstrate conclusively that all hazards have been eliminated (Nichols and Buchanan-Banks, 1974).

Earthquake Hazard Zoning

A logical planning program in seismic areas involves the use of risk zones, applied through a zoning ordinance after a thorough investigation of seismic hazards existing within a given jurisdiction. On a general basis, the designation of three zones might be appropriate.

Zone 1. Substantial risk or hazard zone. This zone would comprise areas having the following conditions:

a. an active fault trace and its "fault trace zone" (commonly called "fault zone"). It would encompass the maximum width of the exposed fault trace and could be modified to include the land affected by previous earthquakes, and the probable

damage zone expected to be involved in future shocks, if known.

b. compressible tidal muds or alluvial soil conditions where major settling or differential settlement is probable.

c. active landslides or areas of known slope instability problems, including areas subject to liquefaction, lurching, or other types of ground failure.

Zone 2. Moderate risk zone. Areas containing the following conditions:

a. buffer areas immediately adjacent to both sides of an active fault zone, or lying between its branches (*see* figure 2.51; this intermediate zone would be considered relatively safer than the fault zone but geographically near enough to be affected directly by earthquakes originating along it).

b. potentially active or inactive fault traces as identified by geologic investigations.

c. areas subject to intensified ground shaking. (It should be noted that these areas may be as hazardous as those identified in the "substantial risk" category; however because large geographical areas are usually involved, and the precise hazard is more difficult to evaluate, placing these areas in an intermediate risk zone may be more acceptable for land use planning purposes.)

Zone 3. Minimum risk zone. This designation would be applied to those areas outside known fault zones, and to areas that are not subject to landsliding, liquefaction, or accelerated ground shaking as indicated by accurate geologic mapping and background data. These areas would be the safest on the basis of existing data. Certain questionable seismically related hazards might be delineated within such areas where they are some distance from active faults.

The next step involves the assignment of certain structures or activities to specific areas or zones. The zones would provide for decreases in population density corresponding to increases in seismic risk. Certainly those structures that are destined to be used by large numbers of people or that need to be protected because of their postdisaster importance (hospitals and fire stations, for example) should be placed in the safest areas.

At this point, it must be said in retrospect that the present situation in the San Francisco Bay area, where numerous hospitals and schools are located along the Hayward and San Andreas faults, could scarcely be more poorly planned. The following guidelines suggest placement of specific structures and activities in the future according to the three zones of risk.

Zone 1. Determining the use of land in the fault trace zone demands extreme caution. Government purchase of this land is impractical in most areas, for the simple reason that the amount of land in question is extremely large and, therefore, would represent a significant expense. Although this technique should not be overlooked, serious consideration should be given to other solutions to the problem. In directing future growth these areas should only be utilized by those facilities that involve an absolute minimum of the population and required structures. This zone should be designated for the following uses:

1. cultivation and related agricultural pursuits

2. grazing and animal husbandry of cattle, sheep, fowl, and the like

3. land reserves, such as game or wildlife reserves and state parks

4. recreation for nondense sports, such as city parks and golf courses

5. automotive facilities, such as parking lots and sales lots

6. solid waste disposal

7. cemeteries

Zone 2. In this intermediate zone land use is somewhat more difficult to delineate. As a general guideline planning for this area should include those structures and activities that are primarily low density in terms of spacing and population. Included would be:

1. low density single family housing

2. light industry

3. light commercial development, such as stores, shops, office buildings

4. storehouses and warehouses

5. public buildings necessary to service the area where special design and construction techniques are utilized to minimize public risk

Zone 3. Outside of any risk or hazard zone would be the appropriate location for the bulk of the community's population and large structures. Structures that should be located here include:

1. high density housing units, such as apartments and hotels

2. educational and research facilities

3. emergency services, such as hospitals and police and fire stations

4. communication centers, such as broadcasting facilities, telephone systems, and post offices

5. water and power systems for generation, storage, and distribution

6. governmental buildings

7. transportational facilities and centers, such as airports, bus and train stations, and ports

8. high density recreational or entertainment facilities, such as theaters and stadiums

Implementation of these recommendations requires technical expertise from a variety of disciplines. A group of technically qualified personnel should be given the responsibility of establishing a seismic planning program along the lines stated above. This personnel should include a soils engineer, structural engineer, geologist, planner, architect, and possibly an economist to give suitable balance and provide for the rapidly changing state of environmental geology and planning. To keep the zoning maps, geologic report requirements, and design criteria current, this group should be a continuing body functioning within a duly constituted governmental agency (Steinbrugge, 1968).

Building Codes and Construction Methods

As seismic waves pass through the ground they are transferred to buildings through their foundations. As a result of vertical motion during an earthquake, the inertia of the building causes the walls to expand and compress. Horizontal motion, on the other hand, causes buildings to bend and sway. Most structures are designed to take considerable vertical loads, and vertical movement during an earthquake usually does not cause damage. Horizontal movement, however, can be far more damaging if not taken into account in building design. It was not until 1961 that the United States Uniform Building Code added lateral or horizontal design requirements that took into account the

dynamic characteristics of different types of buildings.

During past earthquakes certain types of buildings have consistently performed well, while others have either performed poorly or failed. Due to their inherent lightness and flexibility, wood and steel frame buildings, for the most part, have an excellent record. The inertial weight of such buildings is relatively small and they can accommodate motion during an earthquake without deformation. Some older wooden structures suffer from (1) lack of proper foundation or adequate connection to their foundation, or (2) lack of or insufficient lateral bracing. Structures must have adequate lateral support if they are to withstand the horizontal forces of an earthquake (see figures 2.54 and 2.55).

Masonry is heavy, inflexible, and has a poor seismic record. A mass of brick or stone creates an inertial load during shaking that cannot be resisted unless the masonry has been properly reinforced. Many old brick and stone buildings were constructed prior to the inception of building codes and suffer from the following deficiencies:

1. lack of steel reinforcing

2. mortar weakness due to age and poor quality

3. inadequate ties between horizontal and vertical elements of building

4. lack of lateral bracing or support

Much construction in seismically active areas of the world outside the United States and Japan, in regions such as Greece, Turkey, the Middle East, and Central and South America, consists of unreinforced brick, stone, or concrete. This is for the most part due to the lack of other building materials. Earthquakes in these areas have repeatedly

Figure 2.54. *Wood Frame House Collapse, 1971 San Fernando Earthquake. Tract houses collapsed from seismic shaking during the San Fernando earthquake due to lack of lateral bracing. Note that the intact structures had slight lateral reinforcement of wire mesh. Houses that had only uncovered framing collapsed completely.* PHOTO BY: *John Shadle, Los Angeles Dept. of Building and Safety.*

Figure 2.55. *Close-up View of Failure of Wood Frame House. Although wire mesh had been applied in preparation for stucco, this was insufficient lateral bracing in this instance.* PHOTO BY: *California Dept. of Highways.*

leveled entire towns, and even significant portions of large cities, such as Managua, Nicaragua, in 1973. Multistory concrete buildings have totally collapsed with large losses of life (*see* figures 2.56 and 2.57).

Reinforced concrete is potentially an ex-

cellent building material in that it combines the compressional strength of concrete with the tensional strength of steel. It can still, however, be inadequately designed and poorly fabricated.

One problem with building codes in the

Figure 2.56. *Agadir, Morocco, 29 February 1960. Collapse of the upper floors of the Sud building was caused by inadequate lateral bracing.* PHOTO COURTESY: *America Iron and Steel Institute.*

Figure 2.57. *Caracas, Venezuela, 1967. Total collapse of Mijagual high rise apartment in Caracas, Venezuela, during an earthquake in 1967. Each layer of concrete marks an additional floor of the building.* PHOTO COURTESY: *El Nacional and the California Div. of Mines and Geology.*

Figure 2.58. *The Embacadero Freeway and Civic Center Area of San Francisco. How many of these buildings will withstand the next major earthquake in the San Francisco region?* PHOTO COURTESY: *U.S. Army Corps of Engineers.*

United States is the national use of common building materials. Rules for structural design in the use of concrete, steel, wood, and the like are formulated nationally through the Department of Housing and Urban Development. The use of steel or concrete in places such as Chicago or Houston, which are essentially areas free of earthquakes, is quite different from their use in San Francisco or Los Angeles (Degenkolb, 1969). Although adoption of the Uniform Building Code is not mandatory, local jurisdictions, often without adequately trained staffs, normally adopt the code in its entirety without modifying it for local seismic conditions. Therefore these national standards are incorporated in, for example, California's building codes even though they may be completely unsuitable in earthquake prone areas.

A major limitation to all engineering and building codes is the previous lack of good geologic and geophysical data upon which to base them, in addition to the magnitudes of the forces that we are trying to accommodate or design for. A planning and design expert for the California Division of Highways, after an inspection of the 1971 San Fernando earthquake damage, said that "nothing highway engineers could have done would have saved the more than two dozen overpasses and roadways from bucking the force of the quake" (*San Francisco Chronicle*, 11 February 1971; see figure 2.43).

Prior to the San Fernando earthquake, as mentioned earlier in a discussion of ground shaking, the acceleration data with which engineers had to work was inadequate and, in close proximity to faults, values were significantly underestimated. It seems likely that many structures built in seismic areas prior to this earthquake, depending upon the safety factors employed, are underdesigned (see figure 2.58). Many of these will be subject to structural damage or failure during a great earthquake. Engineering is, therefore, only as good as the geologic data upon which it is based.

In Retrospect

To begin thinking seriously about earthquake or fault zone planning, all those involved must first recognize the seriousness of the threat. Competent geologists now have the tools and data at their disposal to study an area and to produce the kinds of information and maps that planners and engineers need to utilize. Portola Valley, located astride the San Andreas Fault in the San Francisco Bay area, is an example of a small community that has been working directly with this problem with encouraging results, such as producing a building moratorium on the fault trace itself. Unfortunately, shortsighted economics and politics commonly enter the planning picture at this point. But perhaps we are at the beginning of a period of change. It seems irrational, to say the least, not to take the most reasonable steps to minimize damage, injury, and death from future earthquakes in light of our past experience and present state of knowledge.

REFERENCES

References Cited in the Text

Albee, A. L., and Smith, J. L. "Earthquake Characteristics and Fault Activity in Southern California." In *Engineering Geology in Southern California*. Los Angeles: Special Publication of Association of Engineering Geologists, 1966.

Allen, C. R. "Geological Criteria for Evaluating Seismicity." *Geological Society of America Bulletin* 86(1975): 1041–57.

Alquist, A. E. "Seismic Hazards — A Question of Public Policy." In *Environmental Planning and Geology*. Edited by D. R. Nichols and C. C. Campbell. Washington, D.C.: U.S. Government Printing Office, USGS-HUD, 1969.

Atwater, T. "Implications of Plate Tectonics for the Cenozoic Tectonic Evolution of Western North America." *Geological Society of America Bulletin* 81(1970): 3513–36.

Aune, Q. A. "Quick Clays and California's Clays: No Quick Solutions." *Mineral Information Service* 19(1966): 119–23.

Bonilla, M. G. "Surface Faulting and Related Effects." In *Earthquake Engineering*. Edited by R. L. Wiegal. Englewood Cliffs, N.J.: Prentice-Hall, 1970, 147–74.

Clements, T. "St. Francis Dam Failure of 1928." In *Engineering Geology in Southern California*. Los Angeles: Special Publication of Association of Engineering Geologists, 1966.

Committee for Environmental Information. "Underground Nuclear Testing." *Environment* 11,6(1969): 3–13, 41–53.

Degenkolb, H. J. "An Engineer's Perspective on Geologic Hazards." In *Proceedings of the Conference on Geologic Hazards and Public Problems.* Edited by R. A. Olson and M. W. Wallace. Santa Rosa, Calif.: Office of Emergency Preparedness, 1969.

Ericksen, G. E.; Plafker, G.; and Concha, J. F. *Preliminary Report on the Geologic Events Associated with the May 31, 1970 Peru Earthquake.* U.S. Geological Survey Circular 639, 1970.

Gates, G. O. "Earthquake Hazards." In *Proceedings of the Conference on Geologic Hazards and Public Problems.* Edited by R. A. Olson and M. W. Wallace. Santa Rosa, Calif.: Office of Emergency Preparedness, 1969.

Greensfelder, R. "Seismological and Crustal Movement Investigations of the San Fernando Earthquake." *California Geology* 24(1971): 62–68.

Gutenberg, B., and Richter, C. F. "Magnitude and Energy of Earthquakes." *Annals of Geophysics* 9(1956): 1–15.

Hammond, A. L. "Earthquake Prediction: Breakthrough in Theoretical Insight." *Science* 180(1973): 851–53.

Healy, J. H.; Hamilton, R. M.; and Rayleigh, C. B. "Earthquakes Induced by Fluid Injection and Explosion." *Tectonophysics* 9(1970): 205–14.

Healy, J. H., and Marshall, P. A. "Nuclear Explosions and Distant Earthquakes: A Search for Correlations." *Science* 169(1970): 176–77.

Housner, G. W. "Intensity of Earthquake Ground Shaking Near the Causative Fault." *Proceedings of the Third World Conference on Earthquake Engineering,* Aukland, N.Z.: Earthquake Research Institute, 1965.

Iacopi, R. *Earthquake Country.* Palo Alto, Calif.: Lane Books, 1973.

Kerr, P. F. "Quick Clay." *Scientific American* 209,5(1963): 132–42.

Liebling, R. S., and Kerr, P. F. "Observations on Quick Clay." *Geological Society of America Bulletin* 76(1965): 853–77.

Morton, D. M. "Seismically Triggered Landslides Above San Fernando Valley." *California Geology* 29(1971): 81–82.

NAS-NRC. *The San Fernando Earthquake of February 9, 1971.* Washington, D.C.: National Academy of Sciences, 1971.

Nichols, D. R., and Buchanan-Banks, J. M. *Seismic Hazards and Land Use Planning.* U.S. Geological Survey Circular 690, 1974.

Ploessel, M. R., and Slosson, J. E. "Repeatable High Ground Accelerations from Earthquakes." *California Geology* 27(1974): 195–99.

Radbruch, D. H., Bonilla, M. G. et al. *Tectonic Creep in the Hayward Fault Zone, California.* U.S. Geological Survey Circular 525, 1966.

Savage, J. C., and Burford, R. O. "Geodetic Determination of Relative Plate Motion in Central California." *Journal of Geophysical Research* 78(1973): 832–45.

Steinbrugge, K. V. *Earthquake Hazard in the San Francisco Bay Area: A Continuing Problem in Public Policy.* Berkeley: University of California, Institute of Governmental Studies, 1968.

Whitcomb, J. H.; Garmann, J. D.; and Anderson, D. L. "Earthquake Prediction: Variation of Seismic Velocities Before the San Fernando Earthquake." *Science* 180(1973): 632–35.

Other Useful References

Anderson, D. L. "The San Andreas Fault." *Scientific American* 225,5(1971): 52–67.

Eckel, E. B. *The Alaska Earthquake of March 27, 1964: Lessons and Conclusions*. U.S. Geological Survey Professional Paper 546, 1970.

Executive Office of the President, Office of Science and Technology. "In the Interest of Earthquake Safety: Findings and Conclusions of the Task Force on Earthquake Hazards Reduction." Institute of Governmental Studies, University of California at Berkeley, 1971.

Kisslinger, C. "Earthquake Prediction." *Physics Today* 27(1974): 36–42.

Lomnitz, C. "Casualties and Behavior of Populations During Earthquakes." *Bulletin of the Seismological Society of America* 60(1970): 1309–13.

Press, F. "Earthquake Prediction." *Scientific American* 232(1975): 14–23.

Rothé, J. P. "Fill a Lake, Start an Earthquake." *New Scientist* 11(1968): 75–78.

Wallace, R. E. *Goals, Strategy and Tasks of the Earthquake Hazard Reduction Program*. U.S. Geological Survey Circular 701, 1974.

Wiegel, R. L., ed. *Earthquake Engineering*. Englewood Cliffs, N.J.: Prentice-Hall, 1970.

Yanev, Peter. *Peace of Mind in Earthquake Country*. San Francisco: Chronicle Books, 1974.

CHAPTER 3

Volcanic Activity

Contents

INTRODUCTION

WITHIN the lower forty-eight states of the United States volcanic activity in the recent geologic past (the last several million years) has been confined primarily to the Cascade Range and other areas in the western states, such as the Sierra Nevada and the California desert areas (see figure 3.1). The lack of recent historical eruptions and activity in these areas makes it difficult, however, to predict with any certainty what may occur in the future. In Hawaii, and in numerous other areas throughout the world (see figure 3.2), volcanic eruptions are still occurring, and we are learning continually from these events and the kind of activity that accompanies them. As with faults it is difficult to say simply that a volcano is "inactive" because we have not witnessed significant historic activity. Arenal volcano in Costa Rica had been dormant for hundreds of years and was thought to be extinct. Then in July 1968,

with very little warning, it erupted violently and killed more than seventy people in nearby villages (Crandell and Waldron, 1969). Paricutín volcano rose from a cornfield in western Mexico in 1943, literally beneath the feet of Indians working in the field. Ashes, gas, and cinders were first emitted from a crack in the ground. After a week of eruption a cone 140 meters (460 feet) high had formed; after a year it had reached 330 meters (1100 feet). Lava then erupted for the next eight years, followed by quiescence until the present. During the eruption, the surrounding farmland, and the villages of Paricutín and San Juan de Parangaricutiro were buried beneath ash, cinders, and lava. This eruption was unexpected. Similarly we do not know which of the Cascade volcanoes will be the next to erupt. However, by determining what has happened during past eruptions, and by observing recent activity,

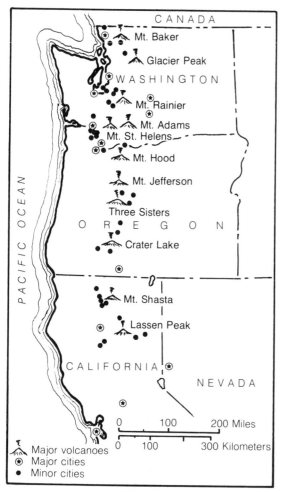

we can at least evaluate what may occur in the event of a future eruption.

TYPES OF VOLCANIC ACTIVITY

Lava flows, eruptions of volcanic ash, discharge of gases, formation of hot avalanches, mudflows, and floods are all phenomena or processes that characterize volcanic eruptions (Crandell and Waldron, 1969).

Lava flows are the streams of hot molten rock that are discharged from volcanoes and that usually flow down valleys or other topographic lows (see figure 3.3). These flows vary in their size, extent, and velocity, but are often slow enough so that people can get out of their way. In 1950, however, when the first flow came down the side of the

◀ **Figure 3.1.** *Volcanoes in the Western States. The map shows the location of major volcanoes in the Cascade Range of northern California, Oregon, and Washington and cities in their vicinity.* REDRAWN FROM: *D. R. Crandell and H. W. Waldron "Volcanic Hazards in the Cascade Range," Proceedings of the Conference on Geologic Hazards and Public Problems, R. A. Olson and M. W. Wallace, eds. (Santa Rosa, Calif.: Office of Emergency Preparedness, 1969).*

Figure 3.2. *Distribution of the Earth's Active Volcanoes. In addition to volcanoes on land, known submarine eruptions are shown.* DATA FROM: *The Center for Short-Lived Phenomena, 1968-1974.*

Figure 3.3. *Lava Flow, Kilauea Volcano, Hawaii, 1970. This flow partly covered Chain of Craters Road at Aloi Crater.* PHOTO BY: *D. A. Swanson, U.S. Geological Survey.*

Figure 3.4. *Lava Flow, Kilauea Volcano, Hawaii, 1959–1960 Eruption. A papaya orchard slowly being destroyed by a lava flow from Kilauea.* PHOTO BY: *J. P. Eaton, U.S. Geological Survey.*

Hawaiian volcano Mauna Loa at a speed of almost 10 kilometers per hour (6 mph), there was barely enough time for people to move out of the way. It is reported that the 1823 lava flow of Kilauea, also in Hawaii, advanced so rapidly on a coastal village that some old people and small children were caught and killed (MacDonald, 1972). The heat from the lava may start brush or forest fires, or in ice and snow, may cause rapid melting and consequent floods or mudflows. In Hawaii, the initial eruption of lava from Mauna Loa on the flank of Kilauea in 1969, quickly inundated about 5 square kilometers (2 square miles) of grassland, orchard (*see* figure 3.4), and lush hardwood forest; covered extensive stretches of two roads (*see* figure 3.5); and buried (literally) a volcano-observatory drilling program (Swanson et al., 1971).

The eruption of volcanic ash and other debris may affect vast areas, depending upon how much ash is ejected and the strength and direction of the prevailing wind. Mate-

rial ejected from the Indonesian volcano Krakatoa in its catastrophic eruption in 1883 is believed to have traveled completely around the earth many times in the atmosphere. The eruption of Mount Mazama (now occupied by Crater Lake, *see* figure 3.6) in southern Oregon about 6600 years ago, blanketed an area of about 500,000 square kilometers (200,000 square miles) of the Pacific Northwest (*see* figure 3.7).

Even 4 or 5 centimeters of ash may be enough to smother grass and other low plants and to cause temporary disastrous effects to agriculture. Grazing animals may die, partly of starvation and partly from eating ash-laden vegetation, which may clog their digestive systems. Water supplies can be contaminated and ash can also damage respiratory systems.

The accumulation of ash can overload roofs, possibly causing them to collapse (*see* figure 3.8). Rainfall will increase the load because the rain will soak in and saturate the ash. In the 1973 eruption of Kirkjufell on the

Figure 3.5. *Lava Flow Covering Highway. Asphalt burns as lava flow crosses Chain of Craters Road on the south flank of Kilauea Volcano, Hawaii, 1970. Flow also started a large fire in surrounding forest.* PHOTO BY: *J. D. Judd, U.S. Geological Survey.*

Figure 3.6. *Crater Lake Within the Caldera of Now Dormant Mount Mazama in the Oregon Cascades. The catastrophic explosion of Mount Mazama about 6600 years ago removed the entire top of the mountain and scattered ash over the Pacific Northwest.* PHOTO BY: *John Gilchrist.*

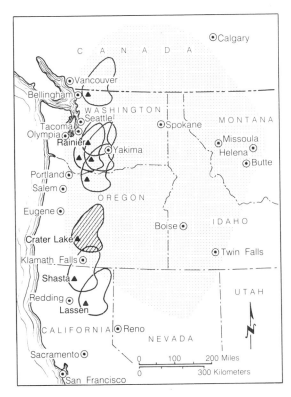

Figure 3.7. *Ash Distribution from Mount Mazama, Cascade Range. This area was covered by ash erupted from Mount Mazama (now occupied by Crater Lake) about 6600 years ago. Outer line delineates maximum limit of ash fall (about 500,000 square kilometers) and smaller area is that covered by 15 centimeters of volcanic ash. The same 15-centimeter pattern is shown superimposed on the other major Cascade peaks.* REDRAWN FROM: *D. R. Crandell and H. W. Waldron, "Volcanic Hazards in the Cascade Range," Proceedings of the Conference on Geologic Hazards and Public Problems, R. A. Olson and M. W. Wallace, eds. (Santa Rosa, Calif: Office of Emergency Preparedness, 1969).*

Figure 3.8. *Fallout of Volcanic Ash, 1959–1960 Kilauea Volcano, Hawaii. The trees shown here have been killed by the deposition of volcanic ash, which also can overload roofs.* PHOTO BY: *J. P. Eaton, U.S. Geological Survey.*

island of Heimaey, just south of Iceland, 1 to 3 meters of ash collapsed or buried over 100 houses in the adjacent village of Vestmannaeyjar (see figures 3.9 and 3.10). The density of wet **tephra** or compressed ash here

was 1 gram per centimeter; this means that 1 meter of ash imposes a load of 1000 kilograms per square meter. It is thus surprising that more houses did not collapse under the weight because most of them would have been built to take an imposed load of not more than 100 to 200 kilograms per square meter. It also became clear during the ash fall that houses with steeper roofs stood up to the weight better than those with flat roofs (Einarsson, 1974). Extensive ashfalls over

Figure 3.9. *Volcanic Ash Vestmannaeyjar, Iceland. Nearly 225 million cubic meters of volcanic ash and debris partially buried the important Icelandic fishing port of Vestmannaeyjar on the island of Heimaey, Iceland, in 1973.* PHOTO COURTESY: *Mal Og Menning, Reykjavik.*

Figure 3.10. *Accumulation of Volcanic Ash in Vestmannaeyjar, Iceland, 1973. The top of a 3 to 4 meter flag pole pokes through the volcanic ash in Vestmannaeyjar on the Icelandic island of Heimaey.* PHOTO BY: *James G. Moore, U.S. Geological Survey.*

populated and agricultural areas have been devastating throughout history.

The discharge of large volumes of gas by volcanoes is well recognized, although the composition and effect of each discharge may be variable. Carbon dioxide, carbon monoxide, water vapor, and sulfur and chlorine oxides are among their most common constituents. The oxides of sulfur and chlorine combine with water vapor to form sulfuric and hydrochloric acid, which can cause considerable damage. During the catastrophic eruption of Mount Katmai in Alaska in 1912, chlorine was liberated in such abundance that clothes on clotheslines as far away as Chicago were rotted by the gas drifting through the atmosphere and combining with water to form hydrochloric acid (MacDonald, 1972). Gas drifting westward from Masaya volcano in Nicaragua has done extensive damage to coffee plantations, and gas from Hawaiian eruptions has damaged fruit trees up to nearly 50 kilometers away.

Hot avalanches of ash and rock debris are

also fairly common and have now actually been recorded on film. These fiery avalanches, known as **"nuées ardentes,"** are composed of hot masses of rock fragments and gas, which erupts and flows rapidly as a fluid downhill. These avalanches may move at speeds of 60 to over 125 kilometers per hour (36 to over 75 mph). The temperatures in the avalanches and the cloud of smoke and gas that accompanies them may be hundreds of degrees centigrade so that everything in their path is completely incinerated. Hot avalanches have destroyed entire villages, as at Lamington volcano, New Guinea, in 1951, and even entire cities, as St. Pierre on Martinique, in 1902 (see page 86).

Mudflows and floods may also be generated by volcanic eruptions and can be another potentially serious hazard. The snow cover on the Cascades throughout most of the year, which could be rapidly melted by the heat accompanying an eruption, and the evidence of past mudflows around a number of Cascade peaks indicate the potential for mudflows is significant in this range. Mudflows are potentially very dangerous because they move very rapidly and can travel for many kilometers down valley floors if they have enough volume. Volcanic mudflows, especially large ones, can occur with little or no warning. They are most likely to start during an eruption when much of a volcano may well be hidden in smoke and steam.

The potential hazards of mudflows or floods generated as a result of volcanic eruptions from individual Cascade peaks have been discussed at considerable length by Crandell and Waldron (1969) of the U.S. Geological Survey. In the last 10,000 years Mount Rainier in the state of Washington has experienced at least fifty-five large mudflows, several hot avalanches of rock

debris, at least one period of lava flows, and at least twelve volcanic ash eruptions (see figure 3.11). The last major event occurred about 2000 years ago and involved an eruption of lava and ash accompanied by several large mudflows. One individual mudflow from Mount Ranier (the Osceola mudflow) occurred about 5000 years ago and contained 2 billion cubic meters of material. This is the equivalent of a square kilometer piled to a depth of about 2000 meters. The mud eventually covered an area of about 325 square kilometers (125 square miles) in the Puget Sound lowland, where at least 30,000 people now live (see figure 3.11). Another (the Electron mudflow) occurred only 500

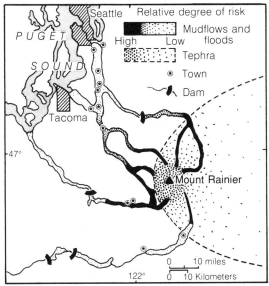

Figure 3.11. *Volcanic Hazards at Mount Rainier. Map of Mount Rainier, Washington, and vicinity indicating relative degrees of potential hazards from ash falls (tephra), mudflows, and floods that might occur during a future eruption.* REDRAWN FROM: *D. W. Crandell and D. R. Mullineaux, "Techniques and Rationale of Volcanic Hazards," in* Environmental Geology *(New York: Springer-Verlag, 1975).*

years ago and involved 150 million cubic meters of mud, which traveled out about 56 kilometers from the volcano. Between 2000 and 3000 people live on the surface of this flow today. If we assume that this peak will behave in the future as it has in the past, then the major hazard will be from mud flowing down valley floors (*see* figure 3.11).

Low-lying regions below other Cascade peaks contain populated areas in addition to dams and reservoirs that would be threatened by mudflows as well. A mudflow entering a reservoir may cause the water to overtop the dam and bring about a catastrophic flood downstream. This hazard exists in the case of two reservoirs at the base of Mount Baker in Washington. Mudflow-induced overflow of three reservoirs along the Lewis River, a tributary of the Columbia River, south of Mount St. Helens, might

Figure 3.12. *Lassen Peak in Northern California. The eruption of Mount Lassen in 1914 was the last major instance of any volcanic activity in the lower forty-eight states of the United States. The peak has been dormant since 1921, when steam was last observed.* PHOTO COURTESY: *University of California, Berkeley, Geography Dept. file.*

cause disastrous flooding along the heavily populated floodplain of the Columbia and in the large city of Portland, Oregon.

We can begin to estimate the kinds of damage that could occur in the event of an eruption today in the Cascades, and what areas may be affected, but the probability of such eruptions is nearly impossible to predict. The last eruption in the Cascades was that of Mount Lassen, which occurred from 1914 to 1917. Between 16 and 18 May 1915, lava flowed from the **crater** rim down the northeast and northwest slopes of the mountain for about 300 meters, melting snow and causing extensive mudflows. On 22 May of that year a horizontally directed blast of gaseous material and rock fragments exploded on the eastern side of the mountain, devastating the area (see figure 3.12). The last explosion on Mount Lassen occurred in 1917, and the last steam was observed in 1921. In 1974, however, after a

six-year USGS study of the area, the most heavily used portion of Lassen Volcanic National Park, the Manzanita Lake area, was closed. The threat of an eruption of hot **clastic** or fragmental material or of the triggering of large rockfall avalanches from the steep domes in the Chaos-Crags area by an eruption or earthquakes led to the decision. Recent mapping indicates that high velocity air cushioned avalanches of rock debris that occurred about 300 years ago had traveled over 4 kilometers. Either flows or rockfall avalanches could be hazardous to the most heavily used portion of the park, only 2 kilometers from the base of the crags (see figure 3.13; Crandell et al. 1974).

Most of the other major Cascade volcanoes, Mount Rainier, Mount Baker, and Mount St. Helens, have experienced at least minor volcanic activity as recently as the mid 1800s. In geologic terms this is very recent. In March 1974 Mount Baker, in north-

Figure 3.13. *Rockfall on the Chaos Crags, Lassen Volcanic National Park, California. Although major rockfalls occurred here about 300 years ago, the threat of future falls has led to the closing of portions of the park.* PHOTO BY: *D. R. Crandell, U.S. Geological Survey.*

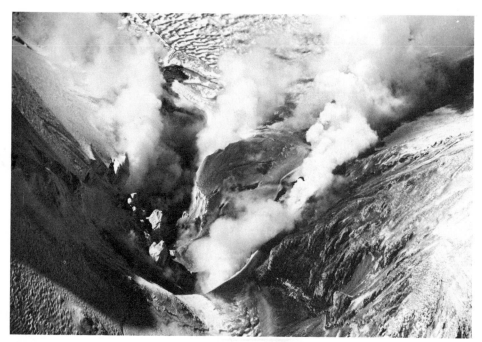

Figure 3.14. *Recent Activity at Mount Baker. Fumaroles in the eastern part of Sherman Crater, Mount Baker, Washington, 27 March 1975. Heat here has melted ice and snow and formed a lake.* PHOTO BY: *Austin Post, U.S. Geological Survey.*

ern Washington, began steaming and several months later was still discharging about 1300 kilograms (2800 pounds) of sulfurous gases along with steam each hour (see figure 3.14). Extensive aerial and ground monitoring began, but it is still uncertain whether this activity foreshadows the first major volcanic eruption in the lower United States in over half a century. The threat of debris-laden floods, rock avalanches, and mudflows descending into a valley below Sherman Crater, where a steam-melted lake has formed on the flank of the mountain led to the closing of the valley. Recent study of volcanic deposits around Mount St. Helens in southwestern Washington by USGS geologists has led to warnings about pos-sible eruptions of that peak before the end of the century.

EFFECTS OF VOLCANISM ON HUMAN ACTIVITY

As a result of the eruptions of lava or volcanic ash, the formation of fiery avalanches, or floods and mudslides, many widespread areas of the earth inhabited by people have been affected throughout history.

In some cases entire towns have been destroyed. Probably one of the best known, and also the best preserved example, is Pompeii, buried by the eruption of Vesuvius in AD 79.

Figure 3.15. *Pompeii, Italy. The well-preserved remains of the ancient city of Pompeii are seen in the foreground, with Mount Vesuvius towering in the background. Pompeii was buried for almost 1700 years before being discovered in 1595.* PHOTO BY: *Gary Griggs.*

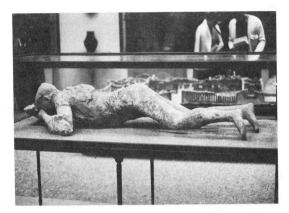

Figure 3.16. *A Victim of Mount Vesuvius. A cast made from a mold left by one of the victims buried in Pompeii from the eruption of Mount Vesuvius in* AD *79.* PHOTO BY: *Gary Griggs.*

The volcano Vesuvius, about 12 kilometers southeast of Naples, Italy, is about 1200 meters (4000 feet) in height (see figure 3.15). Prior to its eruption the ancient crater had been quiet so long that trees were growing within it. The area had been occupied first by the Greeks and then by the Romans for about 800 years. First signs of renewed ac-

tivity were intermittent earthquakes in the sixteen years preceding the eruption. In AD 79, however, an explosion blew off a large part of the cone and buried the cities of Pompeii, Herculaneum, and Stabiae. Very little if any lava was ejected, but a large volume of dust, ash, and steam was emitted, forming a pasty mud that flowed down the slopes and overwhelmed the surrounding cities. Over a period of two days, Pompeii, 8 kilometers from the eruption center, was completely buried to depths of over 6 meters. The entire city was forgotten until rediscovered in 1595. Initially, the eruption must have been quiet enough that people were able to leave; of those that stayed too long, however, many were poisoned or suffocated from gases emitted by the volcano. About 16,000 people perished and many of the bodies have been subsequently excavated, with perfect molds having been made, many clutching their mouths and others clutching bags of coins or jewels (see figure 3.16).

Essay

LOST ATLANTIS

The catastrophic eruption of a volcano in the eastern Mediterranean 3500 years ago has in recent years been convincingly tied to the ancient legend of the lost continent of Atlantis and the decline of Minoan civilization (Luce, 1968; Galanopoulos and Bacon, 1968). The remains of the volcano now appear as a picturesque Greek island named Santorini or Thera, about 125 kilometers (75 miles) north of Crete (see figure 3.17). Minoan culture was flourishing at the time of the eruption, principally on Crete, but also on a number of other Aegean islands, including Santorini. These people had a well-developed Bronze Age culture, characterized by a strong agricultural base, a navy, the use of metals, the spread of literacy, and a refined art.

The initial volcanic activity that gave rise to the island of Santorini can be traced back several million years and tied to the other well-known eastern Mediterranean volcanoes, Vesuvius, Etna, and Stromboli. Eruptions followed by long intervals of quiescence have been recorded up to the present day on the island. The most spectacular eruption, and one that may have been one of the greatest ever to occur on earth, took place about 1450 BC, at the peak of Minoan civilization. Evidence for the magnitude of the eruption comes from (1) the distribution of volcanic ash emitted, (2) the present configuration of the island, (3) the remains of the Minoan civilization that are still being unearthed, and (4) the dialogues of Plato. Although preliminary activity of the volcano about fifty years prior to the major explosion apparently led to an exodus from the island, the buried Minoan remains on Santorini are still being excavated and new information is being collected.

The closest comparisons that can be made to the magnitude and effects of the eruption are those of Krakatoa in 1883 in Indonesia, perhaps the largest historical volcanic eruption for which we have any detailed record. Krakatoa had been dormant for over 200 years prior to its eruption, which blasted away two-thirds of the island or 5 to 10 cubic kilometers (1.2 to 2.4 cubic miles) of material. The crater created had an area of 21 square kilometers (8 square miles) and was 180 to 275 meters deep. Windows were broken and walls cracked up to 160 kilometers (95 miles) away. The explosion was heard 4800 kilometers (3000 miles) away, and the dust and ash emitted rose to heights of 80 kilometers, falling in Japan, Africa, and Europe. Sunsets were red for months thereafter. Tsunamis generated by the eruption reached heights of 40 meters in the East Indies and killed 36,000 people along the coasts of Java and Sumatra.

By comparison, the crater on Santorini covers an area of 83 square kilometers and is 275 to 400 meters deep. The **caldera** volume or amount of material removed is estimated to have been about five times greater than Krakatoa! The thickness of ash deposited from Krakatoa does not exceed 40 centimeters (15 inches), whereas on Santorini it reaches thicknesses of 40 meters (130 feet). The ash has also been traced through the eastern Mediterranean and recovered in deep sea cores up to 700 kilometers to the southeast. Tsunamis generated by the explosion carried volcanic ash 250 meters up the side of one island 24 kilometers away.

From the projected magnitude of this explosion, relative to the well-documented Krakatoa eruption, the damage brought about to a confined eastern Mediterranean world from the earthquake accompanying the blast, the fallout of volcanic ash, and the tsunami would have been immediate and catastrophic (Luce, 1968; and Galanopoulos and Bacon, 1968).

Santorini itself, one of the centers of Minoan culture, was decimated with only a deep crater and the flanks left from the original volcano (see figure 3.18). The tsunamis would have no doubt destroyed many of the harbor areas and boats, as well as the low-lying coastal cities of Crete. The earthquake was probably responsible for shaking down the palaces on Crete, which were never rebuilt. The volcanic ash would have ruined crops, decimated the soil for years, and thereby led to agricultural decline and perhaps famine. Thus, nearly overnight, a mighty civilization was brought to its knees and essentially disappeared, as did much of the island of Santorini.

Figure 3.17. *The Eastern Mediterranean Island of Santorini. This island has been recently related to the fabled lost continent of Atlantis.*

GREECE

CRETE

SANTORINI (THERA)

ANAFI

Figure 3.18. *Santorini, Greece. The volcanic island of Santorini in the Aegean Sea, with a young, still active volcano in the center of the older caldera. The opposite rim of the crater can be seen in the distance.* PHOTO BY: Gary Griggs.

▼

Along the crest of the Mid-Atlantic Ridge, the eruption of Skapter Jokul in 1783 was a national disaster for Iceland. One-fifth of the entire island's people (10,000) died from its effects. One-half the cattle, three-fourths of the horse, and four-fifths of the sheep population were killed as well. A thousand kilometers away, volcanic ash and dust destroyed crops in Scotland.

In Indonesia, the eruption of Tambora in 1815 is believed to have killed 12,000 people by direct effects. In addition, the fallout of volcanic ash up to 500 kilometers (300 miles) away ruined vast areas of land and crops, which led to a famine resulting in the deaths of thirty to forty thousand more people. One of the most far-reaching eruptions of recorded history occurred when another Indonesian volcano, Krakatoa, blew up in 1883. The climax of the eruption was reached when an ash cloud rose to a height of 80 kilometers (50 miles) and the noise of the explosion was heard nearly 5000 kilometers (3000 miles) away. About a half-hour after the explosion, a tsunami, which reached a height of 40 meters in some bays, swept the neighboring coasts of Java and Sumatra, wholly or partially destroying 295 towns and killing 36,000 people, mostly by drowning. The tremendous amount of ash blown into the air plunged the surrounding regions into darkness, which affected areas as much as 450 kilometers (280 miles) away. The fine dust in the upper atmosphere traveled around the earth many times and remained in the atmosphere for months, causing sky glows, which were widely observed all over Europe and the United States (Bullard, 1962). Krakatoa, however, was not noted to be active from the time it first appeared in European records (1680) until it catastrophically exploded in 1883.

Another well-known, more recent eruption that destroyed an entire city was that of Mount Pelee on the island of Martinique in the West Indies in 1902. Prior to this time only two minor eruptions had occurred, neither of which was very serious or resulted in loss of life. On this occasion the volcano had been emitting ash and gases for over a month; birds and horses had died as a result. However, the government in power, due to instability and an impending election, prevented most of the people from leaving St. Pierre, the capital city. Ultimately, after days of warning, a massive eruption occurred and a dark cloud of dust, ash, and hot gases rolled down the mountainside like a hurricane at a velocity of over 150 kilometers per hour (90 mph) and swept through the entire city in minutes. All the houses were unroofed and otherwise demolished either in part or totally. The force of the blast tore walls of stone and concrete 1 meter thick to pieces. The cloud was so dense that it seemed to act as a liquid, staying close to the ground, which increased its capacity for destruction. Most of the deaths apparently resulted from the burning action of the hot gases and dust. Two men, one a prisoner in an underground dungeon, were the sole survivors in the city of 30,000 people (Bullard, 1962).

Volcanoes continue to erupt and are continuing hazards to people and their activities, indicating that the earth's surface (**lithosphere**) is still very active. In 1951 the eruption of Mount Lamington in Papua, New Guinea, killed 6000 people. A volcano in Bali took the lives of 1500 people, and more recently, the eruption of Taal volcano in the Philippines took almost 200 lives in 1965 (Moore et al., 1966). Active volcanoes are widely distributed around the world, and a volcano is nearly always erupting somewhere (see figure 3.2). As far back as history has been recorded, Stromboli, a volcano in the Tyrrhenian Sea between Sicily and the Italian mainland, has been in nearly a constant state of eruption. The name given to the volcano, "the lighthouse of the Mediterranean," has been due to the constant discharge of fiery molten material.

The short period of historical record in many places and the long time spans between eruptions have made it difficult to be certain whether an individual volcano is "active" or "inactive." There is no way of being absolutely certain that a volcano will not erupt again. Previous disasters are sad testimony to this point. In contrast to the active or inactive state of an individual volcano, however, is the series of events that precedes a major eruption of an active cone. Today in Hawaii, apparently as a result of a hot plume or mass of material rising from the **mantle**, continued eruptions of Mauna Loa and Kilauea are being closely studied by the Hawaiian Volcano Observatory and have been studied nearly continuously since 1912.

ERUPTION PREDICTION

Damage to life and property can be eliminated or reduced if volcanic eruptions can be successfully predicted. People and animals can be evacuated and certain types of property can be protected or moved. Ideally, the goal in such predictions would be the warning of where and when an eruption would occur and the effect it would have on the surrounding area. Although we are nowhere near this point, a good deal of progress has been made. Eruption predictions could be either general or specific. The knowledge or belief that any volcanoes of the Cascades, for

example, are likely to erupt in the future is a general prediction, but one that still has great value in future planning.

Years of volcano observations have also led to the discovery of various phenomena that may be taken as warning signs of impending eruptions. These are seldom certain and frequently there is little time between warning and eruption. Some of the more general signs observed in the past include the melting of snow caps, the disappearance of crater lakes, the drying up of wells and springs, the death of surrounding vegetation, and the migration of animals and birds. In addition to these observations that depend upon special circumstances, more routine methods have been used and are being perfected to the point where the eruptions of certain well-monitored volcanoes in Japan and Hawaii are being accurately predicted.

Seismic Methods

Earthquakes themselves are now the most common method of eruption prediction. The ascent of **magma** or hot molten material in the cone of a volcano causes tremors, and there is usually an increase in shallow focus earthquakes prior to eruption. When an elaborate array of seismographs is utilized, remarkable forecastings can be achieved, as the Hawaiian Volcano Observatory has shown. The eruption of Kilauea in December 1959 was forecast six months in advance by recording tremors from depths as great as 50 kilometers (30 miles). As time went by, the focal depth of the earthquakes became less and less and by measuring the speed of the rise of the focal depth, the date of the eruption could be forecast. The exact point and time of eruption have been predicted with unprecedented accuracy. At Asama volcano in Japan, a network of ten seismographs distributed over the entire summit area of the mountain is used to record the frequency and location of microearthquakes. A clear relationship has been found between number, depth, and magnitude of shallow earthquakes and the state of the volcano (McBirney, 1970). Unfortunately, earthquakes themselves occur as a result of a quite different set of forces than volcanic eruptions and our predictive abilities for them are not nearly so advanced as for volcanoes.

Tilt Measurements

As the lava pushes up inside a volcano prior to an eruption, the volcano surface itself is liable to swell (see figure 3.19). After erup-

tion the volcano normally sinks back to its original position. This swelling can be measured by leveling devices known as **tiltmeters**. These instruments consist of brass water pots (mounted with micrometer screws and glass lenses), which are joined by water and air tubes. They operate like a water trough, so that when one end is elevated, the water level is raised in one pot and lowered in the other. Although simple, they are remarkably sensitive. Such an instrument could detect the tilting produced by lowering one side of Hawaii only 2.5 centimeters with respect to the other side, 150 kilometers away (Waesche and Peck, 1966).

Horizontal Distance Measurement

At the Hawaiian Volcano Observatory inflation and deflation of Kilauea can also be de-

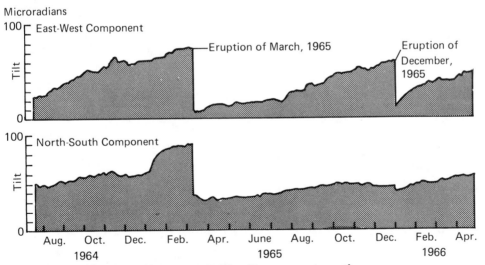

Figure 3.19. *Volcanic Tilt in Hawaii. The diagram portrays the east-west and north-south component of ground tilt recorded from 1964 to 1966, on Kilauea, Hawaii. Notice change in tilt prior to eruption.* REDRAWN FROM: *R. S. Fiske and R. Y. Koyanagi, U.S. Geological Survey Professional Paper 607 (1968).*

tected by precise measurement of horizontal distances across the summit. Utilizing a **geodimeter** (a device employing a carefully adjusted beam of light for ultra-precise distance measurements), scientists detected a shortening of about 28 centimeters in a 3050 meter line across the summit caldera during the deflation accompanying the March 1965 eruption (Waesche and Peck, 1966).

Temperature Measurements

The temperatures of crater lakes, and hot springs and **fumaroles** may, show a sharp increase before an eruption. With regular monitoring of these temperatures, a possible eruption may be predicted. There may be some temperature anomalies, however, so that this method is not completely reliable. The use of remote-sensing techniques, such as infrared or temperature sensitive photography, enables vulcanologists to detect hot spots on the flanks or at the crest of volcanoes.

Volcanic Geochemistry

The study of various exhalation products, such as gases and magmatic solutions, is another means of investigating volcanic activity. Studies of the relative abundance, occurrence, composition, and relationships of these volcanic emissions may provide some indications of impending eruptions. In some cases there is likely to be an increase in the amount of hydrochloric acid gas emitted from fumaroles or craters. Japanese scientists have been able to predict several eruptions with reasonable accuracy, through detailed

study of the amounts of chlorine, fluorine, and sulfur compounds discharged from several volcanoes (Chesterman, 1971).

Summary of Prediction Efforts

There is no single method or technique in the forecasting of volcanic eruptions that can be utilized worldwide. In Japan, geochemical techniques have been effective in predicting volcanic eruptions, and these warning systems have greatly reduced loss of life and property. In Hawaii, the utilization of elaborate seismic networks and tilt measurements has enabled scientists to predict eruption time and place with considerable accuracy.

Professor A. R. McBirney (1970) of the Center of Volcanology at the University of Oregon states:

> In order to evaluate various techniques and select those that can provide an effective warning system, it is necessary to observe a volcano through several cycles of activity. This is obviously impossible for volcanoes that erupt with violence after long periods of dormancy. Much more must be learned about the basic causes and controls of volcanism before these problems can be dealt with intelligently.

With continuous progress in this field the development of warning systems may provide a degree of safety for the communities in close proximity to active or potentially active volcanoes.

PROTECTIVE MEASURES

Regardless of the sophistication of a warning system, however, and the completeness of evacuation, a community itself cannot easily be removed. Throughout history, people threatened by volcanic eruptions have attempted to prevent the destruction of their homes and villages with varying degrees of success.

Lava Flow Diversion

The blocking, containment, or diversion of lava flows by various types of walls has been a common method utilized to protect towns and villages in many diverse areas of the world. A wall around a cemetery on the flank of Vesuvius, a similar structure in Samoa, a concrete building at O Shima, Japan, and loosely laid stone walls in Hawaii have all partially blocked or diverted advancing tongues of lava (Mason and Foster, 1953). The observation of these events and the subsequent recognition of the ability to engineer structures to withstand and divert lava flows indicate a very direct way of partially dealing with the hazard of lava flows advancing into populated areas or toward structures or farmland.

Just as with containing flood waters, the higher and more substantial these structures are, the greater is the degree of protection. In most instances, walls should be designed to divert and not to stop a flow, as lava would soon overtop a wall acting as a dam. Both sides of a wall may be sloped to obtain strength, but the upslope side should be steep enough to prevent overriding of the wall by the momentum of the flow. Both of these goals can be accomplished by locating a wall in a strategic topographic position and setting it diagonally to the slope to direct the flow into a selected channel where little damage will be done (Mason and Foster, 1953).

Several new methods of lava diversion have been somewhat successfully employed with recent eruptions on the volcanically active island of Heimaey, just south of Iceland (Williams and Moore, 1973). In early 1973 the eruption of Kirkjufell volcano began threatening Vestmannaeyjar, Iceland's most important fishing village (see figure 3.20). The eruption, with the production of large amounts of tephra and lava, forced the evacuation of nearly all the town's 5300 residents. Many houses, public buildings, and businesses were buried by heavy falls of tephra, set afire by lava bombs, or overridden by the advancing lava flows (see figures 3.21 and 3.22). Although many structures collapsed from the weight of the tephra, the shoveling of accumulated material from roofs saved dozens of others. Perhaps because they were not strangers to volcanism, the islanders evacuated their city without a single human injury; not a trace of panic was seen.

In addition to increasing the size of the island, the lava flows began to fill the harbor (see figure 3.23), a disaster for a fishing village. This harbor is the best along the southern coast of Iceland, lies in the midst of a very important fishing grounds, and processes approximately 20 percent of Iceland's fish catch. The lava flows advancing into the harbor severed a major fresh water pipe from the mainland and a 30,000-volt submarine power cable, and destroyed a large fish-

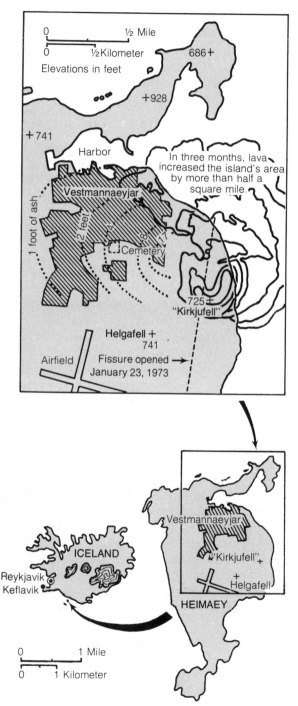

freezing plant and the local power-generating station.

Because of the importance of this area to Iceland, a determined effort was made to protect it by restricting the flow of lava, through the bulldozing of barriers and the pumping of large amounts of seawater to cool the flows. A massive pumping operation both from within the harbor and on land was undertaken (see figure 3.24). Spraying cold seawater lowered the temperature of the lava, causing the front of the flow to thicken and solidify. Following the cooling of the surface and margins of the flow, bulldozers made tracks and then pulled plastic hoses up onto the slowly moving flow so that water could be pumped on the lava behind the advancing front (see figure 3.25). This pumping was used in conjunction with bulldozed diversion barriers of **scoria** adjacent to the flow margin. This effort in Iceland at restricting lava flows was without question the most ambitious ever undertaken anywhere and did have an important effect in saving the harbor (Williams and Moore, 1973).

Other Protective Measures

Lava flows themselves are, however, one of the less violent volcanic phenomena and perhaps the easiest to avoid or counteract. The fallout of volcanic ash, seismic sea waves, or the fiery avalanches, mudflows, or floods that can accompany other more

◀ **Figure 3.20.** *Kirkjufell Eruption, Iceland, 1973. The map shows the deposition of volcanic ash erupted from Kirkjufell over the town of Vestmannaeyjar on the Icelandic island of Heimaey.* REDRAWN FROM: *N. Grove "A Village Fights for Its Life,"* National Geographic 144, 1(1973).

Figure 3.21. *Lava Flow on the Island of Heimaey, Iceland. Lava from the eruption of Kirkjufell invaded the fishing port of Vestmannaeyjar stopping here against and between two fish canneries.* PHOTO BY: *Richard S. Williams, Jr., U.S. Geological Survey.*

Figure 3.22. *Lava Flows on Heimaey Inundating and Destroying Houses. Before being officially declared "dead" on 3 July 1973, the volcano had destroyed or severely damaged over 800 buildings and forced the 5300 islanders to temporarily abandon their homes and the island itself.* PHOTO COURTESY: *Maĺ Og Menning, Reykjavick.*

Figure 3.23. *Lava Flow Control on Island of Heimaey, Iceland. Steam rises on Heimaey as rescue workers pump nearly a million gallons of seawater per hour over lava advancing into the harbor. Through these successful efforts the lava was prevented from blocking off the fishing port.* PHOTO COURTESY: *Mal Og Menning, Reykjavik.*

Figure 3.24. *Lava Flow Control in Vestmannaeyjar, Iceland. Spraying seawater in an attempt to stop the lava flows advancing on the homes of Vestmannaeyjar.* PHOTO COURTESY: *Mal Og Menning, Reykjavik.*

Figure 3.25. *Efforts to Control a Major Volcanic Eruption, Iceland. Rescue workers on the Icelandic island of Heimaey lay water pipe and bulldoze a road across piles of volcanic debris in their remarkably successful efforts to chill and slow the flows of lava endangering the major fishing port of Vestmannaeyjar.* PHOTO BY: *James Moore, U.S. Geological Survey.*

violent eruptions are another matter. Areas downwind from major eruptions of ash are relatively helpless; low-lying coastal areas will, if in the wrong locations, be inundated by seismic sea waves. Both can affect areas far away from the actual eruption site and are difficult to counteract. Mudflows and floods or nuées ardentes are restricted to the slopes and valleys or low areas around volcanoes and these sites can at least be identified (Crandell and Waldron, 1969).

As with any geologic hazard, however, identification or recognition is only the first step in dealing with the problem. Many of the dormant volcanoes within the United States are within park or forest land and are not, fortunately, sites of intensive develop-

ment. Nevertheless, areas that are set aside for recreation and that seasonally are visited by millions of persons, and those that provide locations for public utilities and water storage and supply systems, are all expanding in the Cascades, which stretch for almost 1000 kilometers (600 miles) from northern California through Oregon and Washington.

The recent surge in geologic hazards research has led to more active study of the Cascades, both by federal and state geologists. Deposits of early eruptions are being mapped and interpreted and changes in volcanic behavior are being monitored. These programs should all be encouraged, continued, and expanded.

Since volcanic eruptions cannot yet be

prevented, one safe approach is to fence them off from encroachment by development, but to leave them open for observation, study, and enjoyment. Any intensive use of hazardous areas should be carefully considered in light of past volcanic activity. The extensive mudflow deposits that occupy vast lowland areas around some of the northern Cascade peaks are examples of such areas that are difficult to deal with now that some habitation already exists. Perhaps close monitoring of volcano behavior and a workable warning system are the best solutions to these existing developments. Any expansion in use of potentially hazardous areas, especially involving permanent or critical structures or habitation, should be discouraged, or at least sited in the safest locations. Fiery avalanches, mudflows, and other volcanic discharges normally travel down valleys, which should obviously be avoided. Building codes can ensure that structures in volcanic areas can withstand the seismic shaking associated with eruptions and also will not collapse under heavy ashfall. These precautions are the relatively simple and straightforward ones that are necessary if damage and loss of life in future eruptions are to be minimized. Planning done on a regional or statewide scale should take cognizance of volcanic hazards and encourage recreational or other open space use of these areas.

REFERENCES

References Cited in the Text

Bullard, F. M. *Volcanoes in History, in Theory, in Eruption*. Austin: University of Texas Press, 1962.

Chesterman, C. W. "Volcanism in California." *California Geology* 24(1971): 139–47.

Crandell, D. R.; Mullineaux, D. R.; Sigafoos, R. S.; and Rubin, M. "Chaos Crags Eruptions and Rockfall Avalanches, Lassen Volcanic National Park, Calif." *USGS Journal of Research* 2(1974): 55–57.

Crandell, D. R., and Waldron, H. W. "Volcanic Hazards in the Cascade Range." *Proceedings of the Conference on Geologic Hazards and Public Problems*. Edited by R. A. Olson and M. W. Wallace. Santa Rosa, Calif.: Office of Emergency Preparedness, 1969.

Einarsson, P. *The Heimaey Eruption*. Reykjavik: Heimskringla, 1974.

Galanopoulos, A. G., and Bacon, E. *Atlantis: The Truth Behind the Legend*. London: Thomas Nelson and Sons, 1969.

Luce, J. V. *Lost Atlantis: New Light on an Old Legend*. San Francisco: McGraw Hill Book Company, 1969.

MacDonald, G. A. *Volcanoes*. Englewood Cliffs, N.J.: Prentice-Hall, 1972.

Mason, A. C., and Foster, H. L. "Diversion of Lava Flows at O Shima, Japan." *American Journal of Science* 251(1953): 249–58.

McBirney, A. R. "Some Current Aspects of Volcanology." *Earth Sciences Reviews* 6(1970): 337–52.

Moore, J. G.; Nakamura, K.; and Alcarez, A. "The 1965 Eruption of Taal Volcano." *Science* 151(1966): 955–60.

Swanson, D. A. et al. "Mauna Ulu Eruption, Kilauea Volcano." *Geotimes* 16,5(1971): 12–16.

Waesche, H. K., and Peck, D. L. "Volcanoes Tell Secrets in Hawaii." *Natural History* (March 1966): 20–29.

Williams, R. S., Jr., and Moore, J. G. "Iceland Chills a Lava Flow." *Geotimes* 18,8(1973): 14–17.

Other Useful References

Crandell, D. R., and Mullineaux, D. R. "Volcanic Hazards at Mount Rainier." *U.S. Geological Survey Bulletin 1238,* 1967.

Crandell, D. R., and Mullineaux, D. R. "Technique and Rationale of Volcanic Hazards." *Environmental Geology* 6,1(1975): 23–32.

Keller, G. V.; Jackson, D. B.; and Rapolla, A. "Magnetic Noise Preceding the August 1971 Summit Eruption of Kilauea Volcano." *Science* 175(1972): 1457–58.

Maiuri, A.; Bianchi, P. V.; and Battaglia, L. E. "Last Movements of the Pompeians." *National Geographic* 120(1961): 651–59.

Swanson, D. A. et al. *The February 1969 East Rift Eruptions of Kilauea Volcano, Hawaii.* U.S. Geological Survey Professional Paper 891, 1976.

Wilcox, R. E. "Some Effects of Recent Volcanic Ash Falls with Special Reference to Alaska." *U.S. Geological Survey Bulletin 1028-N* (1959): 409–76.

CHAPTER 4

Landslides and Mass Movement

Contents

INTRODUCTION

WITH increasing frequency, the sounds of snapping two-by-fours, shattering glass panels, and the cries of the home owner have been heard as an accompaniment to a very old and common geologic process — the landslide. Not long ago a naturally steep hillside would slide into a canyon with a dirgelike groan and no one would much care. Now as the level areas have become urbanized and developed, the hillsides have been invaded and landslides have become frightening and expensive events (Cleveland, 1967). The flattened surfaces of some older slides, which may add character to an otherwise featureless slope, often seem to provide excellent building sites to the prospective home owner or contractor. Although these ancient slides may now appear stable, people have found themselves capable of re-triggering their movement through many of their own activities.

Mass downslope movements may occur in a wide variety of geologic materials. Soil, rock, or combinations of the two may fail or move downhill under many different environmental conditions. Some slides may involve only a small amount of loose surface material, while others may result from deep failure of large masses of solid rock. Movement can occur on very gradual slopes as well as in steep terrain and under diverse climatic conditions. The more common smaller slides, although not a serious threat to life, cause hundreds of millions of dollars damage to property, both public and private, annually in the United States. Large landslides, on the other hand, have completely buried towns with heavy loss of life.

Although landslides are generally considered a detriment to the works of man, this is not always the case. Dams have in the past been built where river courses have been constricted by slides. The Bonneville Dam site on the Columbia River was formed by the deflection of the river by an old slide. The Cheakamus Dam in British Columbia was founded on a landslide, and the dam, an earth and rockfill structure, was constructed of landslide-derived material. Building in such instances requires intensified geologic and engineering investigations, however (Morton and Streitz, 1967).

In California, particularly, concern with landslides has grown with increasing population. More people and their housing "needs" have added pressure to develop land in progressively steeper and more unstable terrain surrounding existing urban areas. Extensive hillside developments have been damaged by movement on old slides and by the triggering or initiation of new slides due to poor construction and development practices (see figure 4.1).

Landslides are part of a more general erosional or surficial process known as **mass wasting,** which is simply the downslope movement of earth or surface materials due to gravity. The velocity spectrum of mass

Figure 4.1. *Landslide Damage to Residential Development in San Fernando Valley Area of Southern California. The underlying shale here was weak and failed along with the overlying fill.* PHOTO COURTESY: *Los Angeles Dept. of Building and Safety.*

wasting ranges from soil or rock creep, where rates of movement are measured in centimeters per year, to debris avalanches, where velocities are believed to reach over 400 kilometers per hour (250 mph). Landslides, in the strict sense, lie between these two velocity extremes and are further distinguished by the presence of a surface of rupture or a slip surface.

Various kinds of downslope movement occurring under the force of gravity are extremely important processes in the formation of the present California landscape (summarized by Morton and Streitz, 1967). In the Coast Ranges, landslides are common because of the instability of the geologically young rocks that make up these mountains (see figure 4.2). In northern California, especially among the rocks of the unstable Franciscan Formation, landslides are a pervasive feature. Andrew Lawson, almost seventy years ago (1908), noted that landslides were important "in the evolution of the geomorphology of the Coast Ranges of California to an extent equaled in few other regions." The

coastal areas, including the sea cliffs of California and Oregon, are also highly susceptible to various kinds of slope failures. In "solid" crystalline rocks, such as granite, material is actively being shed downslope, commonly in the form of rockfalls. Even desert areas have their characteristic process of downslope mass movement, the mudflows, which are common after heavy rainfalls.

The approach used in this chapter will be initially to describe and illustrate the various kinds of mass downslope movements. The recognition of landslides and related features in the field and the difficulties in mapping them will then be discussed. To understand why slope failure occurs, it is necessary to have a clear understanding of the factors, both natural and artificial, that affect slope stability. Specific examples and illustrations of landslides and slope failures should then serve to make the discussions meaningful. The impact of human activity is of principal importance here. Ultimately, we should learn from past observations and di-

Figure 4.2. *Typical California Coast Range Hummocky Landslide Topography Near Cambria, California.* PHOTO BY: *Venetia Bradfield.*

sasters and make every effort to avoid, prevent, or stabilize slope instability where
lives and property are involved.

PRINCIPAL TYPES OF MASS MOVEMENTS

Depending upon the type of movement —
creeping, falling, sliding, or flowing — and
the kind of material involved — rock, earth,
or mud — mass movements can vary considerably in their shape, rate, extent, and effect
on surrounding areas.

Soil or Bedrock Creep

Creep involves the slow downslope movement or the gradual plastic deformation of
the **soil mantle** and/or fracturing of the bedrock at imperceptible rates. There is no
single surface along which slippage occurs.
Rather the motion involves minute displacements of individual particles that are
moving at different rates. Creep is commonly
due to the expansion of the surface layer
followed by its contraction. This alteration
can be the result of warming and cooling,
freezing and thawing, or swelling of certain
clays or **humus**-rich soils after seasonal rainfalls. The outward movement in expansion
will be normal, or at right angles, to the
slope. However, during compaction the particles do not fall back to their original positions, but under the influence of gravity shift
slightly downhill (see figure 4.3). Creep occurs in materials as diverse as **adobe soils**
and **talus** slopes and can be observed to
some degree on almost any moderately
steep, soil-covered slope. Usually this
phenomena is observable in the tilting of
trees, fences, and utility poles (see figure

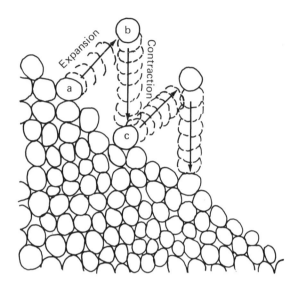

Figure 4.3. *Soil Creep. Individual soil or rock
particles are raised at right angles to the slope
by swelling or expansion, and then settle vertically downward during compaction or contraction. Net result is slow downslope creep.*

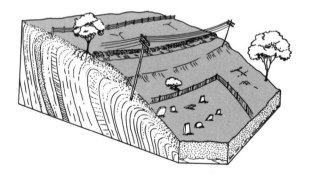

Figure 4.4. *Surface Evidence of Soil Creep.
Leaning utility poles and gravestones may be
evidence of soil creep.*

4.4). Damage from soil or bedrock creep is usually minimal.

Rockfalls

Rockfalls consist of abrupt free fall or downslope movement, such as rolling or sliding, of loosened blocks or boulders of solid rock (see figure 4.5). A rockfall differs from a slide in that free fall is the main mode of movement and no marked slide surface develops. This type of slope failure occurs in caverns and along steep canyons, sea cliffs (see figure 4.6), and steep road cuts through

Figure 4.5. *Cross-Section of a Rockfall.* RE-DRAWN FROM: *"Studies for Seismic Zonation of the San Francisco Bay Region," R. D. Borcherdt, ed.,* Geological Survey Professional Paper 941-A, *(1975).*

Figure 4.6. *Rockfall in Mudstone. Due to weathering and undercutting by wave action, this well-fractured mudstone along the central California coast periodically fails by rockfall.* PHOTO BY: *Gary Griggs.*

unstable bedrock. The bedding, jointing, and fracturing of the bedrock are the important basic factors affecting slope stability. The effects of weathering, such as freezing of water in joints, the pressure of water in fissures, and root pressure may initiate failure in weak rocks.

The combination of jointing patterns, percolation of surface water from street runoff and yard watering, wedging of tree roots, and impact and undercutting by waves has led to massive rockfalls along the sea cliffs flanking northern Monterey Bay (see figure 4.7). These rockfalls commonly occur after intense storms when rainfall and runoff have been high and the surf heavy.

Figure 4.7. *Rockfall Near Capitola, California. This rockfall is due to joint patterns in siltstones and sandstones. Note joint patterns at right angles to each other in the cliff.* Photo by: *Gary Griggs.*

Sections of sea cliff 3 to 5 meters wide have collapsed and fallen instantaneously to the beach below. Continuation of the rockfalls has led to undercutting of houses and abandonment of roads (*see* chapter 6).

The scale or magnitude of rockfalls may vary from the breaking off of isolated small rocks to the fall of enormous masses. Large scale failures in Europe, particularly in the Alps, have completely dammed river valleys creating lakes and destroying parts of towns (Zaruba and Mencl, 1969). On a smaller scale, the talus commonly found at the base of slopes in mountainous and desert areas is the end result of numerous small rockfalls over many years.

The continued expansion of the highway system through youthful rugged topography such as the Rocky Mountains, Sierra Nevadas, Cascades, and Coast Ranges of the western United States and Canada has been accomplished by blasting and grading through rocky areas with considerable relief.

The creation of road cuts with steep slopes in regions of extreme weather conditions, such as ice and snow and freezing and thawing, has led to numerous rockfalls that can be extremely hazardous (see figure 4.8). An excellent example is the rockfall combined with a rock slide that occurred in January 1965 along a new mountain highway connecting Hope and Princeton in British Columbia. The rocks that failed contained **schistosity**, or planes of mineral orientation or weakness, which was nearly parallel to the 30 degree slope of the mountainside. Two small earthquakes associated with heavy snow avalanches and some landslides appear to have been the triggering mechanisms. About 45 million cubic meters of rock together with some earth and snow crashed down a 1950 meter (6400 feet) high mountain range. A road length of 3.0 kilometers (1.9 miles) was buried up to a depth of 80 meters along with automobiles containing four people.

Figure 4.8. *Rockfall, Jefferson County, Colorado. Failure along joints and foliation planes in Precambrian gneiss and heavy rainfall led to this rockfall which obstructed all of eastbound traffic and part of westbound lane of Interstate 70 before cleanup began. Estimated weight of largest boulders exceeded 200 tons.* Photo by: *W. R. Hansen, U.S. Geological Survey.*

True Landslides

Landslides, in a strict sense, are characterized by failure of material at depth and then movement along a rupture zone or slip surface. The mass of rock or material involved either moves as a block or a series of blocks bounded by several slip surfaces. Several different types of movement may occur. If the motion occurs along a nearly planar surface, such as a **bedding plane**, joint, or fault, the failure is called a **block glide** (*see* figure 4.9). In the case of a slump, there is rotational motion and either a single or a series of curved slip surfaces that will produce individual **slump** blocks (*see* figure 4.10). Failure or rupture zones can occur within the bedrock, between the bedrock and the **overburden** or soil (when all the surface material moves), or within the overburden, which in some cases may consist of artificial fill. This type of failure is probably the most common and overall the most destructive to the hillside developments of California and other West Coast areas.

Wherever steep mountain or hillside slopes occur or are altered, the possibility of large landslides and consequent disaster exists. The Turtle Mountain landslide, which partly buried the town of Frank, Alberta, is an example of the possible effects of a large slide. Turtle Mountain is comprised of massive steeply sloping limestone underlain by somewhat weaker shale, sandstone, and coal beds (*see* figure 4.11). In April 1903 a sudden slide developed along joint planes in the limestone, sending approximately 21 million cubic meters (27 million cubic yards) of material down 900 meters (3000 feet) to the valley below. The slide killed seventy people, buried 1.6 kilometers of railroad, and did severe damage to the town.

Figure 4.9. *Cross-Section of a Block-Glide Landslide. Motion in this case occurs along a planar surface.*

Flows

The fourth type of downslide movement involves the deformation of an entire mass that then flows downslope as a viscous or sticky fluid. A high water content or perhaps seismic shaking may lead to liquefaction or generate such a fluid flow. The slopes necessary for this type of movement need not be very steep. Downslope motion may be relatively slow, in the case of an earth flow, to very rapid, in the case of a debris flow or avalanche.

Earth Flows

Earth flows are very common where moderate or steep slopes are overlain by soil or overburden that seasonally may become saturated by heavy rains. The material slumps away from the top or upper part of a slope, leaving a scarp, and flows down to form a bulging toe (see figure 4.12). Earth flows may be very small, or may involve hundreds of tons of material that can ooze onto highways or into homes (see figure 4.13).

Figure 4.10. *Cross-Section of a Slump. Material moves downslope by rotational motion along individual slip surfaces.* REDRAWN FROM: *"Studies for Seismic Zonation of the San Francisco Bay Region," R. D. Borcherdt, ed., Geological Survey Professional Paper 941-A (1975).*

Debris Flows or Mudflows

Debris flows or mudflows, terms that are used interchangeably, involve the relatively rapid but viscous flow of mud and other surficial material. These usually advance downslope in channels for considerable distances until the slope decreases or the chan-

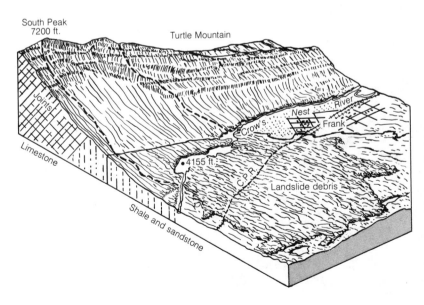

Figure 4.11. *Turtle Mountain Landslide, Alberta, Canada, 1903. Approximately 21 million cubic meters of limestone broke loose along joints and descended to the valley below.* REDRAWN FROM: *An original drawing by A. N. Strahler in Physical Geography, 3rd ed. Copyright © 1969 by John Wiley & Sons, New York.*

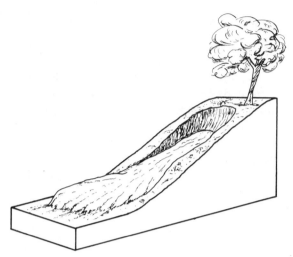

Figure 4.12. *Cross-Section of an Earthflow. Soil and other surficial material commonly move downslope as a viscous flow.* REDRAWN FROM: *"Studies for Seismic Zonation of the San Francisco Bay Region," R. D. Borcherdt, ed., Geological Survey Professional Paper 941-A (1975).*

nel widens, at which point they fan out. In 1941 a mudflow, at Wrightwood in the San Gabriel Mountains near Los Angeles, traveled about 24 kilometers (Sharp and Nobles, 1953). Mud or debris flows commonly begin in steep terrain where the vegetation and organic litter that usually stabilize the soil and retain the rainfall and runoff have been removed by fires, logging, grading, or other processes. Intense precipitation then may trigger mobilization of the surface material. Such flows have caused great damage in the Alps, where they have disrupted railway lines and roads and destroyed valuable farmland. It has been recognized for over 100 years that the flows there are frequently the result of deforestation of the mountain slopes (Zaruba and Mencl, 1969). These flows are potentially more dangerous than other types of mass movements because they

Figure 4.13. *Large Earthflows in Hills East of San Francisco Bay. No damage has been caused by these particular flows. Note cattle and fence in foreground for scale.* PHOTO BY: *Gary Griggs.*

can form very quickly and move at velocities up to 80 kilometers per hour (50 mph). Their greater density provides mudflows with a relatively higher destructive capacity than floodwaters, and unlike some floodwaters, the mud does not recede after the storm.

Mudflows can also be generated on the flanks of volcanoes, when great volumes of water from heavy rain, melting snow or glaciers, or steam from volcanic vents are rapidly introduced into poorly consolidated deposits of ash or other volcanic debris. The extensive mudflows associated with the eruptions of various Cascade volcanoes in Oregon and Washington have been treated in detail by Crandell and Waldron (1969) and discussed in chapter 3.

The loss of vegetation from large brush fires in recent years in the steep Santa Monica Mountains, in the Big Sur area, and behind Santa Barbara, California has led to disastrous mudflows during the following winters. In the Big Sur area on the central California coast, a brush fire in the late summer of 1972 burned over the vegetation on about 17 square kilometers (4300 acres) of steep slopes. The mountains normally receive 100 to 150 centimeters (40 to 60 inches) of rainfall annually. Short periods of high intensity rainfall following longer periods of steady saturating rainfall in November of that year led to the initiation of debris flows within the burned-over basins. The individual debris flows, estimated at up to 7650 cubic meters in volume, partially covered the small village of Big Sur, including houses, businesses, automobiles, house trailers, and the highway (see figure 4.14). Blocks of rock up to 2.5 meters across, and giant redwood trees over 1.0 meter in diameter, were carried along by the flows (see figure 4.15; and Cleveland, 1973, for summary). There is very little that can be done to

Figure 4.14. *Mudflows at Big Sur, California. Several stores, a service station, garage, and a post office were all inundated by these flows.* PHOTO BY: *Gary Griggs.*

Figure 4.15. *Large Boulder Carried by Big Sur Mudflows. In addition to this car which was demolished, a number of other automobiles and mobile homes were destroyed.* PHOTO BY: *Gary Griggs.*

prevent or predict such an event after a major fire has burned off the vegetation in a steep mountainous area subject to high intensity rainfall. Southern California has repeatedly experienced these disastrous mudflows.

Essay

SLOPE FAILURES IN MINING AREAS

At 9 AM on the morning of 21 October 1966 a spoil bank from a coal mine failed on a hillside above the town of Aberfan in Wales. The material flowed downslope at a speed of nearly 32 kilometers per hour (20 mph), engulfing a farm house, destroying the village school, and ruining some homes in the village. Of the 144 persons who died in the disaster, 116 were schoolchildren and five were teachers who were directly in line with the flow of material (Bolt et al., 1974).

The dumping of debris from coal mining operations in banks or piles, known as **tips**, is a common practice in Wales as well as in the United States and other countries. The material commonly consists of **clay, shale**, and overburden material that is often wet or muddy. A number of tips had built up over time on a 14 degree slope above Aberfan. The hillside here consists of downslope-dipping sandstone, which contains some impermeable beds and is overlain by a thin layer of boulder clay. Groundwater springs and seeps emerge on the slope because of this impermeability and the tips were built on top of these springs and also over some small stream courses. The interruption of the natural drainage and 150 centimeters (59 inches) of annual precipitation led to saturation and instability of the lower portions of these banks. Failure of this material had occurred in the past with subsequent downslope flow toward the village. The failure that led to the disaster of 21 October occurred in a tip about 156 meters above and 600 meters away from the village. Water from the underlying sandstone appears to have led to liquefaction of the material, followed by flowage.

The investigation following the 1966 disaster discovered that no soil mechanics investigations had ever been carried out prior to tip construction in Wales; in fact, no stability investigations had even been made on existing tips despite the occurrence of a number of slides. Subsequent to the failure at Aberfan, it was recommended that a National Tip Safety Committee be established to assess the hazards associated with existing disposal of such industrial waste.

Failures of structures composed of mine waste or tailings have also occurred in Chile and the United States among other places. Following the failure in Wales, sixty waste banks in Virginia, West Virginia, and Kentucky were studied (Davies, 1967). Some of these structures are up to 250 meters high and as long as 1600 meters (1 mile). A number of them are used as dams to contain settling ponds for wash water from coal-processing plants. Many of the banks are unstable and subject to failure due to (1) slippage along shear planes within the banks; (2) saturation of the debris by heavy rainfall; (3) explosion of burning banks; (4) overloading of the foundations beneath banks; (5) overtopping and washout of the dams that close off valleys; (6) rock and soil slides on valley walls breaking through banks; (7) deep gullying; and (8) excavation of toes of banks. In the forty years preceding the study, there had been nine refuse banks failures in the three states, claiming at least twenty-five lives. Thirty-eight of the sixty banks examined showed signs of instability. In February 1972, a refuse dam in Buffalo Creek, West Virginia, failed from heavy rainfall (see chapter 6). The resultant flood took the lives of 118 people living downstream.

The hazards created by unplanned spoil banks can no longer be neglected. Wash basins should not be constructed or allowed above or upstream from populated areas. Existing structures of this type need to be drained or removed and attempts need to be made to stabilize existing unstable banks in hazardous locations.

A **debris avalanche** is a related phenomena that may occur in very steep mountain ranges where large quantities of ice, snow, rock, and loose rubble may be perched at high altitudes in unstable positions. Avalanches of this type, often initiated by earthquakes such as the one that followed the 1970 Peru earthquake (*see* chapter 2), can move at speeds of several hundred kilometers per hour or faster and destroy or bury everything in their path with no warning. Relatively few events of this type have ever been observed; where they have occurred, destruction of life and property has been extensive.

Snow Avalanches

A snow avalanche is the rapid downslope movement of snow, ice, and associated debris such as rock and vegetation. In the Rocky Mountain region, particularly Colorado, in the Sierras, and in many parts of Europe, such as the Alps, snow avalanches are common events that take many lives and cause considerable damage to highways, railroads, power lines, and other structures. The occurrence of avalanches is directly related to topography, steepness and orientation of slope, vegetation, and climate — including amount of snowfall, temperature variations, and, therefore, the condition of the snow. Extensive reviews and evaluations of avalanche hazards have recently been published and form the basis of this discussion (Martinelli, 1974; and Rogers et al., 1974).

The complex route of an avalanche as it moves downslope consists of the following:

1.   a starting zone — the point on a slope where the unstable snow breaks away from the more stable snowpack;

2.   a **track** — the path of the material downslope; and

3.   a **runout zone** — the base of the slope where the snow and associated avalanche material comes to a halt.

Avalanches can vary in size and significance from small events in uninhabited areas to large flows that can bury entire villages. They start most frequently on slopes with average gradients of 30 to 45 degrees. Along the track of an avalanche the slope may be considerably less, while in the runout zone the terrain may be flat.

The most reliable method of locating avalanche areas is by studying past events, their locations, frequency, and severity. Although long historic records exist for many avalanche areas in Europe, much less is known about similar areas in the United States. The placement of highways, ski or winter resorts, home sites, and mining operations often is determined without adequate avalanche information. Collecting all existing data on past snow avalanches, utilizing aerial photographs for mapping avalanche paths, and estimating the frequency and severity of this hazard are logical beginnings to averting future disasters.

Field Evidence for Avalanches

Areas of past avalanches can be delineated either with or without snow cover, using some characteristic features (Rogers et al., 1974). Under summer conditions avalanche paths in forested areas may appear as downslope strips characterized by a different age or type of vegetation; patches of trees that are broken or scarred, that lean downhill, or that are down are good indicators of winter avalanches, as are accumulations of logs and trees at the base of a slope.

In winter, field evidence for avalanches may consist of the following:

1. a fracture line where unstable snow has broken away;

2. a change in snow depth and in the texture and features of the snow surface (snow surface of avalanche area may be cleaner or dirtier than surrounding snow);

3. isolated mounds or blocks of snow; and/or

4. deep grooves in the snow-oriented downslope.

As with other mass downslope movements, snow avalanches are not hazards until land use and human activity interfere and are adversely affected by them. Recreation, housing, transportation, and mining are different types of development that may be exposed to avalanches with consequent property damage, maintenance costs, injury, and death. In the state of Colorado, for instance, there have been forty-three recorded deaths from avalanches since 1950. Roads, highways, and railroads may become blocked by avalanche snow and debris (see figure 4.16). In addition to delaying highway and rail travelers, avalanches are costly to clear from transportation routes.

Mitigation, Avoidance, and Control of Avalanches

It is difficult to predict the location, time, and extent of snow avalanches. Because potentially destructive avalanches are common in mountainous areas like Colorado, any construction of new facilities or land use change should avoid avalanche-prone sites, or, if possible, provide for adequate protection.

A prerequisite for avoiding areas subject to avalanches, however, is the recognition of these locations and their delineation on maps. Land use that is not in conflict with the hazards associated with snow avalanches is another alternative. Land used for agriculture or seasonal recreation that does not occur during those periods when avalanches are common are two examples, of avoidance.

Attempts to minimize or control snow avalanches have involved the following:

1. explosive devices. Dynamite and small artillery have been used for years in an effort to trigger small controlled avalanches and thereby avoid the larger destructive events. Considerable study of snow cover and avalanche conditions should precede any explosive detonation so that an uncontrolled disaster does not occur.

2. structures. These vary somewhat in their placement and purpose (see figure 4.17). Supporting structures are often built in the rupture or failure zone with the objective of preventing avalanches from starting or retarding movement before it gains momentum. Along the path or track of the avalanche, sheds may be constructed over highways and railroads to allow the snow, ice, and debris to pass without damage to, or disruption of, traffic. In the runout zone, deflecting walls, dams, and earth barriers serve to retard or redirect a snow avalanche away from buildings. A combination of these structures is often used to provide

Figure 4.16. *Avalanche Path Near Denver, Colorado. Interstate 70, a major transportation artery, crosses the run-out zone of this avalanche about 80 kilometers west of Denver on the north side of Clear Creek Valley.* PHOTO BY: *Arthur I. Mears.*

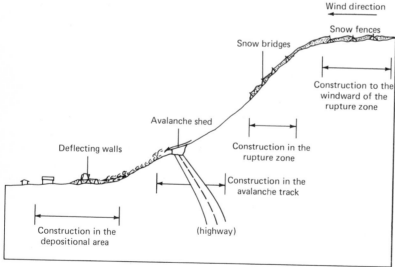

Wind direction

Snow fences

Snow bridges

Construction to the windward of the rupture zone

Avalanche shed

Construction in the rupture zone

Deflecting walls

Construction in the avalanche track

Construction in the depositional area

(highway)

Figure 4.17. *Avalanche Control Structures. A number of different types of structures can be built in the rupture zone, path, and depositional site of an avalanche.*

maximum protection. With more information, advanced planning, strict enforcement of land use regulations, and appropriate structural control, risks from avalanches can be reduced.

FACTORS PRODUCING LANDSLIDES AND MASS MOVEMENTS

To avoid, minimize, or possibly control landslides and downslope movements, it is important that we recognize why an area is susceptible to sliding. We can then try and understand what factors may trigger hillside failure or downslope movement and can begin to initiate appropriate planning solutions. When it is stated that a certain landslide came as a surprise, it is probably more accurate to say that unfavorable geologic and soil conditions in the area were either undetected or poorly understood.

Mass downslope movement occurs when the component of weight along a surface exceeds the frictional resistance and/or cohesion of the material; in other words, hillside failure occurs when the strength of the material making up a slope is overcome by a downslope stress. The **shear strength** of a material is defined as the maximum resistance to failure or **shear stress** and can be broken down into two components:

1. internal friction due to the interlocking of granular particles. Any granular material has the ability to stand at some angle or maintain some slope due to grain-to-grain friction.

2. cohesion due to the forces that tend to hold particles together in a solid mass

such as clay. Commonly these may consist of electrostatic attractions or chemical bonds.

Shear stress is defined as the component of gravity that lies parallel to a potential or actual surface of slippage or rupture (see figure 4.18). As slopes become steeper, the shear stress exerted on the material increases because the downslope component of gravity increases.

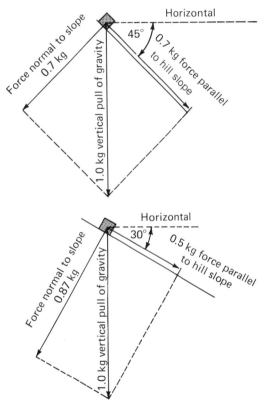

Figure 4.18. *Components of Gravity on a Slope. The resolution of the vertical force of gravity into a force parallel to the slope and a force normal to the slope on which the object or material rests. Notice increase in component directed downslope with increase in slope angle from 30 degrees to 45 degrees.*

The problem of stability of slopes, both natural and disturbed, has to be considered in any human activity involving areas with significant relief. When slope stability is disturbed, any of the diverse types of downslope movements previously described may occur. There are a number of processes, some operating at the surface, some active at depth, that affect slope stability. The surface or external factors affect the stress acting on a slope, whereas the internal factors alter the strength of the material.

External Factors

Change of Slope Gradient

Changes in slope can be caused either by a natural process or by human activity. A stream may erode or undermine the foot of a slope, or in the excavation for a road, a building site, or any of a number of other construction activities, the basal portion of a slope may be removed. More often, however, a slope is simply graded and steepened. This increase in slope gradient produces a consequent change in stress in the rock or soil mass, and the existing equilibrium may be disturbed. With the removal of this support and the penetration of water, the slope material may simply not be strong enough to stand at the steeper angle. In terms of human activity, grading slopes too steeply is probably the single most important factor producing landslides. Oversteepened and therefore unstable excavations for either roads or building sites are unfortunately very common occurrences in many areas (see figure 4.19).

Excess Loading

Loading can generally be thought of as an artificial or human-induced factor rather than a natural one. Whether from dumping, filling, or piling up of material, or as a result

Figure 4.19. *Large Landslide Along a New Section of California Highway 1 Near Santa Cruz. This failure occurred due to oversteepening a cut in poorly consolidated sand.* PHOTO BY: *Gary Griggs.*

of massive construction, the overloading of a slope may lead to an increase in stress. The added weight or load may also increase the water pressure in the pores of clay-rich rocks, which in turn produces a decrease in shear strength. In a normal situation the load on a rock is carried by the grain-to-grain contacts; however, by overloading and compressing this material, the load can be transferred to the water between the grains or in the voids. The result is an unstable material with much less shear strength than the original material. The more rapid the loading, the more dangerous is the instability.

Change in Vegetative Cover

The removal of vegetation from a slope, whether from grading, logging, overgrazing, or fire often leads to a change in slope stability. This may be caused by the loss of soil binding by roots and surface organic matter or by changes in percolation, runoff, and drainage of surface water.

Shocks and Vibrations

Earthquakes, explosions or blasts, and large machines produce vibrations or disturbances within surface and subsurface materials. These may lead to temporary changes in stress that can disturb the equilibrium of a slope. In fine sands, **silts**, and clays that are water saturated, shocks can result in a displacement or rotation of grains that may lead to liquefaction and then failure.

Vast landslide-covered slopes adjacent to major faults, such as the San Andreas, are clear evidence of the ability of earthquakes to induce slope failure and subsequent sliding (see figure 4.20). The overall danger from these types of landslides may be of greater significance in some areas than the actual surface disruption from a fault itself. The 1906 San Francisco earthquake triggered

Figure 4.20. *Hillside Landslide Topography Along the San Andreas Fault, Near Watsonville, California. Dashed lines delineate main fault zone.* PHOTO BY: *Richard Farrington, U.S. Forest Service.*

numerous landslides over an area of approximately 34,000 square kilometers (13,000
square miles). The 1952 Kern County
(California) earthquake generated hundreds
of landslides, most in the vicinity of the
White Wolf Fault; however, 80 to 95
kilometers away, rockfalls partially blocked
the Angeles Crest Highway in the San Gabriel Mountains. The massive landslides in
Anchorage, 125 kilometers from the epicenter of the 1964 Alaskan earthquake, probably were some of the most damaging that
have ever occurred (see figure 4.21).

Internal Factors

Change in Water Content

The importance of water to the process of
landsliding and other mass downslope

Figure 4.21. *Landslide Damage During 1964 Alaska Earthquake. This dwelling in Anchorage was destroyed when a landslide scarp formed beneath it during the 1964 earthquake.* PHOTO COURTESY: *U.S. Geological Survey.*

movements cannot be overemphasized. Precipitation, runoff, or artificial application and infiltration of water can penetrate into joints and cracks of rocks and soils. This added water increases the stress on a slope by raising the water pressure within the rock. The extra weight that water may add to surficial deposits or loose bedrock on a slope may be sufficient in itself to initiate downhill movement.

Water can also change the consistency of the material on a slope, resulting in a loss of cohesion and internal friction, and leading to a wet, unstable mixture or slurry. Certain clay minerals, such as **montmorillonite**, expand when water is present and this swelling may serve to lubricate layers in a sequence of rocks. Many southern California hillside failures have been attributed to the presence of these **expansive clays.** Recurrent sliding commonly occurs in years of unusually high rainfall and most commonly during winter months, which attests to the importance of excess water in downslope movement and slope failure. A short period of high intensity rainfall occurring after the ground has been previously saturated can trigger many landslides. In addition, an abrupt change in water level, along the bank of a reservoir or stream, for example, may induce displacement of grains (especially in silt or sand) and possible liquefaction.

Effects of Groundwater

Groundwater flow may exert pressure on soil particles, which can impair the stability of the slope. In addition, it can wash out or dissolve soluble cement such as calcium carbonate and thus weaken internal bonds or cohesion.

Inherent Properties or Weathering of Materials

The presence of certain materials with perfect cleavage such as **gypsum, serpentine**, or shist may lead to unstable slopes and eventual failure, depending upon the orientation of **cleavage** or **foliation** relative to the slope. A number of other kinds of materials are susceptible to mechanical and/or chemical weathering, which can gradually disturb or decrease internal cohesion. In some landslides, chemical changes (such as the expansion of certain clays like montmorillonite) induced by percolating water are deleterious factors. An ideal setting for eventual slope failure occurs where a permeable rock or soil unit overlies a weak or **expansive clay** layer that dips downslope. Water will penetrate to the clay, weakening, lubricating, or expanding it, with resultant downslope movement of the overlying material (see figure 4.22).

RECOGNITION AND MAPPING

Many geologists in the past were concerned primarily with unraveling earth history and spent the major part of their field time looking for fresh outcrops, mapping formations and their contacts, looking for folds and faults, and trying to determine the deformation and geologic history of a particular area. Surface materials and recent history were usually given little consideration or study, except by Pleistocene or glacial geologists and geomorphologists. More recently with growing population, the influx of people into such areas as southern California, the San Francisco Bay area, and the Puget Sound lowland, more field work and experience has

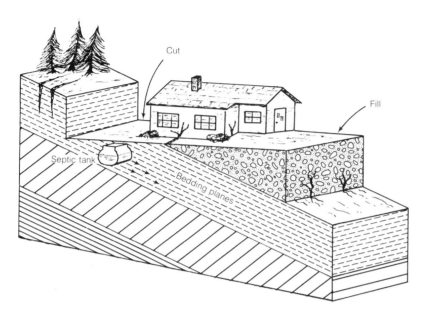

Cut

Fill

Septic tank

Bedding planes

Figure 4.22. *Dipslope Failure. Many landslides in southern California hillside subdivisions are due to this general situation. The formation of extensive cracks is one of the first warnings of sliding. Cutting away the support of the dipslope or weakening or lubricating a subsurface layer with the penetration of water can produce instability.*

begun to reveal that ongoing surface processes are of considerable importance to human endeavors. Landslides and other mass downslope movements certainly are of major importance. Beginning in 1970 the United States Geological Survey, with funding from the Federal Housing and Urban Development office (HUD), began a program of environmental geologic mapping in the nine-County San Francisco Bay area that has focused on these kinds of geologic hazards. This program has been so successful that it has now been extended into a number of other urban complexes throughout the nation (Greater Pittsburg, Greater Denver, Puget Sound, the Connecticut River Valley, the Washington-Baltimore, and the Phoenix-Tucson metropolitan areas). One important aspect of the work has involved the mapping and recognition of landslides and their extent in the hills surrounding the bay area. Initially this landslide mapping had been of a reconnaissance nature. Much of the work

was accomplished utilizing aerial photographs and topographic maps and was followed by field checking of selected areas (see figure 4.23; Nilsen, 1973). This type of analysis places some limitations on the interpretations, and for accurate documentation of individual areas, field work is essential. Among the factors that affect the reliability of photo interpretation maps of landslide distribution are the following:

1. The date of the photography. Older photographs will obviously not include recent slope failures or sliding.

2. The scale of photographs and resultant maps. Where photographs or maps cover large regions or are on a small scale, their resolution and reliability in individual areas may be very limited.

3. The forest or brush cover present. Wherever heavy brush or tree cover exists, it

Figure 4.23. *U.S. Geological Survey Landslide Reconnaissance Map. This map is an example of those produced by the Geological Survey for the San Francisco Bay area. Large areas can be mapped in a reconnaissance or preliminary fashion utilizing stereo aerial photographs and topographic maps.* PHOTO COURTESY: *T. H. Nilsen, "Preliminary Photointerpretation Map of Landslide and Other Surficial Deposits of the Livermore and Part of the Hayward 15-Minute Quadrangles; Alameda and Contra Costa Counties, California,"* U.S. Geological Survey Miscellaneous Field Studies Map MF-519, (1973).

is very difficult to delineate individual landslides unless they are very large.

4.   The degree of urbanization and farming. Where cultivation or development occurs, the land surface may have been so altered that any surface evidence of slides is now totally obliterated.

5.   The limitations of photo interpretation. The quality of the photographs, the age of the slides, the experience of the observer, all

affect the interpretation of the ground surface topography.

Some of the features that have proven to be useful in the recognition of recent or young landslides are shown in figure 4.24 and include:

1.   Scarps. These are commonly arcuate or circular in shape and very steep to vertical in some cases. There is usually a main scarp at

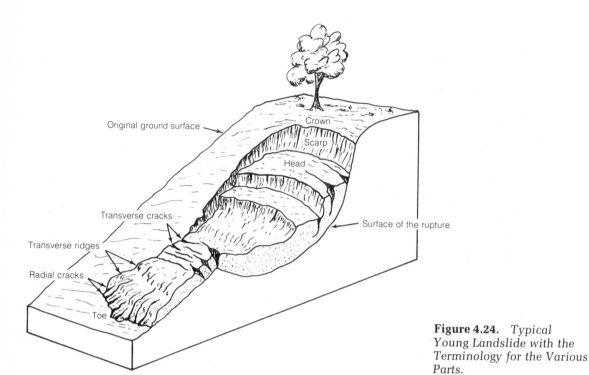

Original ground surface

Crown

Scarp

Head

Transverse cracks

Surface of the rupture

Transverse ridges

Radial cracks

Toe

Figure 4.24. *Typical Young Landslide with the Terminology for the Various Parts.*

the head of the slide and often a number of minor scarps along the slide surface.

2. Hummocky surfaces. The toe, or lowermost portion of a slide, is commonly an area of crumpled topography (*see* figure 4.25); this may show up on detailed topographic maps as a hummocky irregular area of anomalous contours.

3. Isolated swamps or ponds. Because of the disruption of the ground surface as well as material at depth, and the development of closed depressions, surface water or groundwater may be impounded and collect at or near the surface of the landslide; springs may also occur.

4. Vegetation or moisture differences. For the same reasons (i.e., disruption of either surface or subsurface water flow), the vegetation may vary considerably; even subtle variations in grass color or moisture may show up in aerial photographs, particularly those taken with infrared film.

5. Anomalous flat areas. Within unstable regions, or on hillsides, the presence of abundant flat areas that might appear suitable for construction sites, and sometimes have spectacular views, are commonly surfaces of landslides (*see* figure 4.26).

6. Leaning trees or utility poles. The disruption of trees or poles is easy to recognize

Figure 4.25. *Hummocky Topography Common at the Toe of an Earth Flow or Slide. Notice the anomalous bumpy topography at the base of flow in contrast to the smooth hillsides.* PHOTO BY: *Richard Farrington, U.S. Forest Service.*

Figure 4.26. *Anomalous Flat Area at the Head of a Large Landslide. Notice the anomalous flat areas in the center of the photograph with scarps on the right (marked by shadows), and on the left (covered with trees).* PHOTO BY: *Richard Farrington, U.S. Forest Service.*

on the ground and may be good criteria for identifying landslides or downslope movement (*see* figure 4.27); these features probably are somewhat more difficult to detect on aerial photographs, however. The tilting of trees downslope may also be due to surficial creep so this line of evidence needs substantiation. In those cases where individual slump blocks have been rotated, trees may actually be tilted upslope.

7. Disturbed cultural features. Within developed or urbanized areas the cracking of roads, the breaking of water pipes, and the downslope movement of entire houses may be more noticeable and more readily detected than any natural features of a landslide (*see* figure 4.28).

Among the factors that produce and shape landslides, time is important as well.

Figure 4.27. *Trees Leaning at Severe Angles on the Displaced Blocks of a Large Slide. The tilting and growth patterns of trees if used with care can be good slope stability indicators.* PHOTO BY: *Gary Griggs.*

Figure 4.28. *Offset of Road by Landslide Scarps. Smaller cracks of this type are common along many mountain and coastal highways and are continually being repaired.* Photo by: *Gary Griggs.*

A slide usually passes through several phases of development. Initially, conditions exist for the origin of the slide and signs of disequilibrium or instability develop, such as cracks in the upper part of the slope. As the loosened mass moves downslope, all the features of recent slides just discussed are very apparent and are unaffected as yet by erosion and other surficial processes. If the mass is still unstable or conditions change, additional movement may occur, but eventually a stable position will be reached. Inactive or dormant landslides will have had their surfaces revegetated or modified by erosion so that traces or evidence of movement are less obvious (see figure 4.29). Young trees that were tilted by the landscape may add new growth in a vertical direction.

The recognition of a landslide whose form has been greatly altered by surface processes is a definite problem to the earth scientist. Such a landscape may go completely undetected except to an experienced field geologist who has actually worked with landslides. The lack of time in which to study and delineate individual slides throughout a large unstable or potentially unstable area may lead a geologist to label the entire area as unstable or as a potential trouble site. The U.S. Geological Survey is beginning to follow up their regional landslide reconnaissance program with slope stability or landslide susceptibility maps, which serve to direct attention to individual areas.

Figure 4.29. *Subdued Landslide Topography in the California Coast Ranges. Notice the anomalous hummocky or bumpy topography.* Photo by: *Venetia Bradfield.*

Although many of the factors that influence slope stability occur on most geologic maps, the preparation of the maps is not carried out with this in mind. It is impossible for the nongeologist to make such interpretations. There is a clear need for **derivative maps**, or maps that can take the pertinent data and portray them in a form indicating slope stability or susceptibility to landsliding. Such maps are again, however, limited by the reliability and scale of the maps from which they are derived and also by the experience and interpretive skill of the scientist involved. A map of landslide susceptibility (see figure 4.30) is designed for use in regional planning or for solving problems of slope stability in large blocks of undeveloped land. A map of this type can be used in the preliminary selections of sites or corridors for roads and other transportation systems, utilities, sewer and water lines, reservoirs, large industries, schools, and other large scale development. The map would indicate how extensive and expensive the landslide problem may be in an area and would provide a guide to alternative, and less costly and problematic, areas.

A landslide susceptibility map was prepared of all San Mateo County, California (Brabb et al., 1972) following the completion of a landslide distribution map (Brabb and Pompeyan, 1972). The principal factors controlling landslide occurrence here and probably in many regions are (1) degree of slope, and (2) nature of bedrock. The relationship between existing slides and these two factors can be determined from either a grid system of accounting, or a transparent overlay utilizing the existing maps of bedrock geology, slope, and landslide distribution. In this study, individual formations or distinctive members of formations were first analyzed

for the area of outcrop within the county, and then the amount of this outcrop area that had failed was determined. From this preliminary analysis, relative landslide susceptibility numbers were given to each formation or unit (see table 4.1). Each geologic unit was then analyzed for landslides within a series of slope intervals (0–5 percent, 5–15 percent, 15–30 percent, and so forth) and relative susceptibility numbers (I through VI in this study) were assigned to each unit in each slope category (see table 4.2). From these determinations a derivative map can be compiled and the relative landslide susceptibilities explained. This type of map has distinct advantages over a descriptive map, which only delineates landslide distribution. The relative stability of every area in the entire county is now plotted on the basis of slope and rock type regardless of the actual presence of slides. Thus a derivative map becomes a composite of all the factors that may influence sliding and slope failure. Although bedrock geology and the degree of slope are the principal factors in slope failure in general, the distribution of landslides in San Mateo County is also related to the following phenomena (Brabb et al., 1972):

1.   the relationship between orientation of bedding, foliation, or cleavage, and slope direction;

2.   the amount, spacing, and type of jointing and faulting in the rocks;

3.   the extent of undermining of surficial deposits and bedrock by streams;

4.   the type and amount of vegetation present;

5.   the amount and distribution of groundwater and precipitation;

Figure 4.30. *Landslide Susceptibility Map. This map is for a portion of San Mateo County in northern California and is based primarily on slopes and rock types (see tables 4.1 and 4.2 for explanation). Such a map is useful for general planning purposes where large areas of land are being considered.* PHOTO COURTESY: *E. E. Brabb, E. H. Pompeyan, and M. G. Bonilla, "Landslide Susceptibility in San Mateo County, California."* U.S. Geological Survey Miscellaneous Field Investigation Map MF-360, (1972).

Table 4.1. *Landslide Failure Record for Rock Units in San Mateo County*

Surface extent of the rock unit that has failed by landsliding		Rock unit on geologic map by Brabb and Pampeyan (1972a), in order of increasing proportion of surface having failed by landsliding	Map symbol	Approx. area in County (sq mi)	Approx. area that has failed (sq mi)	Relative susceptibility numbers	Susceptibility numbers in each slope interval					
							0–5	5–15	15–30	30–50	50–70	>70
Little of the rock unit has failed	Percent 0–1	(No data for surficial deposits, undivided, Qu; alluvium, Qal; San Francisco Bay mud, Qm; windblown sand, Qd; beach deposits, Qb; artificial fill, Qaf; terrace deposits, Qt; Page Mill Basalt, Tpm; unnamed volcanic rocks, KJv; marble, m; shale near Palo Alto, Ksh; conglomerate, fcg; or metamorphic rocks, fm, but extent of landsliding probably small)				I						
		Limestone	fl	.30	.00		I	I	I	I	I	I
		Colma Formation	Qc	11.11	.01		I	I	I	I	I	I
	2–8	Sandstone at San Bruno Mountain	KJs	4.77	.10		I	I	I	I	I	II
		Butano(?) Sandstone	Tb?	10.56	.19		I	I	II	II	II	II
		Unnamed sandstone	Tus	1.81	.04		I	I	I	I	I	II
		Granitic rocks	Kgr	24.61	.90		I	I	I	I	I	II
		Serpentine	sp	4.76	.09	II	I	I	I	II	II	II
		Sandstone of Franciscan assemblage	fs	22.19	.74		I	I	I	II	II	II
		Slope wash and ravine fill	Qsr	4.51	.18		I	I	I	II	II	II
		Greenstone of Franciscan assemblage	fg	11.70	.61		I	I	I	II	II	II
		Chert of Franciscan assemblage	fc	1.43	.10		I	I	I	II	II	II
		Lompico Sandstone of Clark (1966)	Tlo	.40	.03		I	I	I	I	I	II
	9–25	Sheared rocks of Franciscan assemblage	fsr	9.99	.83		I	I	II	III	III	III
		Pigeon Point Formation	Kpp	7.77	.84		I	I	II	II	III	III
		San Lorenzo Formation, undivided	Tsl	1.00	.11		I	I	I	II	II	III
		Merced Formation	QTm	7.91	1.01	III	I	I	II	II	III	III
		Sandstone, shale and conglomerate	Tss	3.34	.51		I	I	I	I	II	III
		Butano Sandstone, Skylonda area	Tb	22.55	4.33		I	I	II	III	III	III
	26–42	Santa Clara Formation	QTs	6.97	1.85		I	I	III	IV	IV	IV
		Rices Mudstone Member of San Lorenzo Formation of Brabb (1964)	Tsr	1.37	.43		I	I	II	III	IV	IV
		Vaqueros Sandstone	Tvq	7.60	2.41		I	I	II	III	III	IV
		Monterey Shale	Tm	5.11	1.76		I	I	II	IV	IV	IV
		Purisima Formation, undivided	Tp	23.06	7.81	IV	I	II	III	IV	IV	IV
		Lambert Shale	Tla	19.95	7.25		I	I	II	III	III	IV
		Mindego Basalt and other volcanic rocks	Tmb	10.80	4.01		I	I	II	III	III	IV
		Butano Sandstone along Butano Ridge	Tb	20.18	7.66		I	I	II	III	IV	IV
		Santa Cruz Mudstone of Clark (1966)	Tsc	19.25	7.98		I	I	I	III	IV	IV
	43–53	San Gregorio Sandstone Member of Purisima Formation of Cummings and others (1962)	Tpsg	2.41	1.06		I	I	I	IV	V	V
		Tunitas Sandstone Member of Purisima Formation of Cummings and others (1962)	Tptu	2.76	1.24		I	I	III	V	V	V
		Tahana Member of Purisima Formation of Cummings and others (1962)	Tpt	33.46	16.08	V	I	II	III	V	V	V
		Pomponio Member of Purisima Formation of Cummings and others (1962)	Tpp	11.97	5.76		I	I	II	V	V	V
		Twobar Shale Member of San Lorenzo Formation of Brabb (1964)	Tst	.80	.42		I	I	II	III	IV	V
Most of the rock unit has failed	54–70	Santa Margarita Sandstone	Tsm	.65	.41		I	I	I	III	III	VI
		San Lorenzo Formation and Lambert Shale, undivided	Tls	6.83	4.56	VI	I	I	III	V	VI	VI
		Lobitos Mudstone Member of Purisima Formation of Cummings and others (1962)	Tpl	3.71	2.57		I	II	II	VI	VI	VI
	100	Landslide deposits	Qls	83.88	83.88	L	L	L	L	L	L	L

SOURCE: E. E. Brabb; E. H. Pompeyan; and M. G. Bonilla, "Landslide Susceptibility in San Mateo County, California," U.S. Geological Survey Miscellaneous Field Studies Map, MF-360, 1972.

Table 4.2. *Explanation of Susceptibility Map Units*

Least	I	Areas least susceptible to landsliding. Very few small landslides have formed in these areas. Formation of large landslides is possible but unlikely, except during earthquakes. Slopes generally less than 15%, but may include small areas of steep slopes that could have higher susceptibility. Includes some areas with 30% to more than 70% slopes that seem to be underlain by stable rock units. Additional slope stability problems; some of the areas may be more susceptible to landsliding if they are overlain by thick deposits of soil, slopewash, or ravine fill. Rockfalls may also occur on steep slopes. Also includes areas along creeks, rivers, sloughs, and lakes that may fail by landsliding during earthquakes. If area is adjacent to area with higher susceptibility, a landslide may encroach into the area, or the area may fail if a landslide undercuts it, such as the flat area adjacent to sea cliffs.
	II	Low susceptibility to landsliding. Several small landslides have formed in these areas and some of these have caused extensive damage to homes and roads. A few large landslides may occur. Slopes vary from 5–15% for unstable rock units to more than 70% for rock units that seem to be stable. The statements about additional slope stability problems mentioned in I above also apply in this category.
	III	Moderate susceptibility to landsliding. Many small landslides have formed in these areas and several of these have caused extensive damage to homes and roads. Some large landslides likely. Slopes generally greater than 30% but includes some slopes 15–30% in areas underlain by unstable rock units. See I for additional slope stability problems.
	IV	Moderately high susceptibility to landsliding. Slopes all greater than 30%. These areas are mostly in undeveloped parts of the County. Several large landslides likely. See I for additional slope stability problems.
	V	High susceptibility to landsliding. Slopes all greater than 30%. Many large and small landslides may form. These areas are mostly in undeveloped parts of the County. See I for additional slope stability problems.
	VI	Very high susceptibility to landsliding. Slopes all greater than 30%. Development of many large and small landslides is likely. Slopes all greater than 30%. The areas are mainly in undeveloped parts of the County. See I for additional slope stability problems.
Most	L	Highest susceptibility to landsliding. Consists of landslide and possible landslide deposits. No small landslide deposits are shown. Some of these areas may be relatively stable and suitable for development, whereas others are active and causing damage to roads, houses and other cultural features.

Definitions: Large landslide — more than 500 feet in maximum dimension
  Small landslide — 50 to 500 feet in maximum dimension

SOURCE: E. E. Brabb; E. H. Pompeyan; and M. G. Bonilla, "Landslide Susceptibility in San Mateo County, California," U.S. Geological Survey Miscellaneous Field Studies Map, MF–360, 1972.

6. the frequency, location, and magnitude of earthquakes; and

7. those human activities that alter slope stability, including landscape irrigation, septic tank drainage fields, artificial drainage devices, road building, placement of fill and structures on marginally stable slopes.

Recently the U.S. Geological Survey developed a method for relating all the variable factors contributing to slope instability through the use of computer analyses. Although more costly, the resultant map gives a more accurate approximation of landslide potential.

An example of a more detailed landslide susceptibility analysis was an investigation

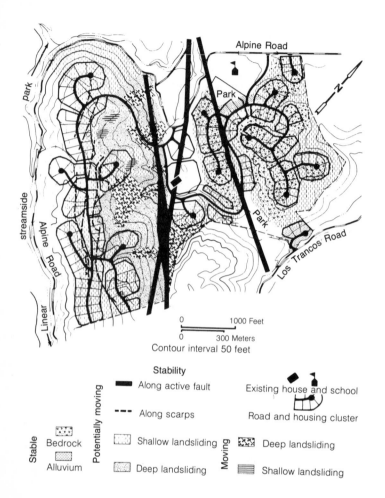

Alpine Road

Park

Park

Park

Los Trancos Road

Alpine Road

Park

streamside

Linear

0   1000 Feet
0   300 Meters
Contour interval 50 feet

Stability

Potentially moving

Along active fault

Existing house and school

Along scarps

Road and housing cluster

Stable

Bedrock

Shallow landsliding

Moving

Deep landsliding

Alluvium

Deep landsliding

Shallow landsliding

Figure 4.31a. *Geologic Map of Bovet Properties, Portola Valley, California, 1968. A relatively inexpensive reconnaissance survey delineated faults and landslides from an examination of outcrops, typography, and from study of lineations and vegetation differences on aerial photographs. This superimposed subdivision plan shows somewhat fewer lots than the original plan (Figure 4.31b). But is somewhat more sensitive to geologic hazards on the property. The new design, however, still shows building sites on potentially unstable ground and uncomfortably close to active fault traces.* ADAPTED FROM: *G. G. Mader and D. F. Crowder, "An Experiment in Using Geology for City Planning — the Experience of the Small Community of Portola Valley, California,"* Environmental Planning and Geology, *D. R. Nichols and C. C. Campbell, eds. (Washington, D.C.: U.S. Geological Survey, HUD, 1969), pp. 176–89.*

done by Arvid Johnson (1968) for a property in Portola Valley, California (reported in Patri et al., 1970). Geologic formations and stability conditions were studied both from aerial photographs and on the site and then mapped at a scale of 1″ = 200′ (*see* figure 4.31A). The investigation indicated almost two-thirds of the 1.8 square kilometer (450 acres) property either had some potential for ground failure or was failing at the time of the study. A residential development was proposed for the property in 1969. After re-

view by the town's planning staff, the plan was later amended to avoid areas designated as unstable. An earlier subdivision plan proposed for the same property in 1956 showed the entire area covered with roads and lots, with no attention given to geological problems (see figure 4.31B). Had the earlier plan been followed and the subdivision built in 1956, fifteen homes and a major road would have been built on actively moving ground, and a substantially greater number of homes built on potentially unstable ground.

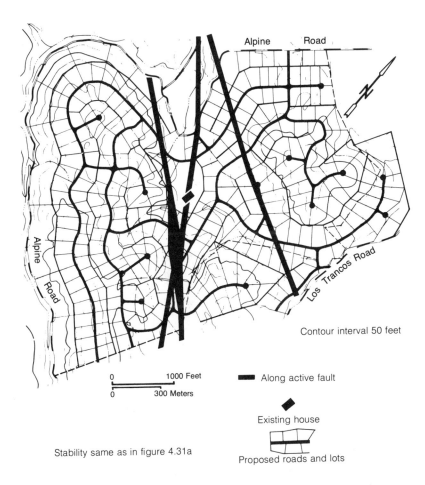

Figure 4.31b. *Subdivision Map of Bovet Properties, Portola Valley, California, 1956. These lots were proposed prior to any geological studies. About fifteen homes and several roads were proposed on land likely to move nearly every year.* ADAPTED FROM: *G. G. Mader and D. F. Crowder, An Experiment in Using Geology for City Planning — the Experience of the Small Community of Portola Valley, California,"* Environmental Planning and Geology, *D. R. Nichols and C. C. Campbell, eds. (Washington, D.C.: U.S. Geological Survey, HUD, 1969), pp. 176–89.*

LANDSLIDES AND URBAN DEVELOPMENT

Along the Pacific Coast of the United States, landsliding in clastic sediments in the coastal mountain ranges is becoming an increasingly grave problem. There are a number of reasons for this increase, some geological, some human. To begin with, much of this area has considerable relief and is seismically very active. The mountains consist of young, clastic rocks, which may be exposed to intense, short-term precipitation. These factors alone, in the absence of any people, would lead to more than a normal distribution of landslides. Human activity has, however, significantly accelerated the rate of downslope movement and slope failure. When people initially settled in places like the Los Angeles Basin and the Santa Clara Valley, they built on the flat alluvial plains and valley floors, where water was available and soil was fertile. However, as more people moved to the West Coast, construction began to move into surrounding hillsides. With time, the stable and flatter sites were used, and excavation and construction became bolder, as steeper hillsides were subdivided (see figure 4.32).

We have therefore seen in recent years an increase in the frequency of landsliding. Grading, logging, roadbuilding, utility corridor routing, selection of less favorable building sites, and the improvement in our ability to recognize slides have all contrib-

uted to this increase. A sobering aspect of California's landslide problem is the real possibility that it cannot be completely controlled within a realistic economic framework.

Slope Stability Analysis

Laboratory investigations by soils engineers have improved our ability to assess problems

Figure 4.32. *Hillside Terracing for a Residential Subdivision. This type of grading and construction is typical of numerous southern California hillside subdivisions.* PHOTO COURTESY: *California Div. of Mines and Geology.*

of landsliding and slope failure, but there are still a number of problems that make this extremely difficult. Many slopes are underlain by materials that may have vastly different physical properties within a distance of a few meters. It is relatively easy to collect a soil sample, run it through a series of laboratory tests, and arrive at some very precise values for various engineering parameters. Although these numbers are very precise, how representative or useful are they? The accuracy of the results depends upon the ability of the investigator to determine the mechanical properties of the rock and overburden in relation to the actual geologic setting of the site. Soil samples may, for example, be collected down to a depth of 5 meters, whereas failure may occur in very different material at a depth of 15 meters.

In many cases bedrock materials never seem to behave in the same manner in the field as they do in laboratory tests. The continued failure of "engineered" slopes or road cuts is clear evidence of this problem. A number of factors are usually operating simultaneously in the natural environment that do not operate in a laboratory sample. The percolation of groundwater, for instance, which may affect the very properties being measured in the laboratory, is removed from the system. Rainwater, frost, and other climatic agents attack slopes and significantly change their characteristics throughout the year. The removal of vegetation, the denudation and erosion of natural slopes, and the excavations for highways, railroads, and other construction will result in a reduction of weight on the slope, and a consequent unloading and decompression of material. As a result, the strength and deformation parameters in the natural environment will differ significantly from those values determined in the laboratory.

Although advances in soils engineering and testing continue to be made, the larger geologic setting may be of much greater overall importance. Geologists have acquired considerable knowledge about the mechanisms of landsliding, but this is only one aspect of the problem. No matter how much we learn about landsliding, a detailed knowledge of the geologic unit or province in question is also essential. These provinces owe their character largely to the rocks and secondarily to the numerous other geological features and processes that may occur or be operating in a given region. Geologic formations have a strong influence on determining slope stability characteristics. Both the Monterey and Franciscan formations, for example, which stretch for hundreds of kilometers through California, constitute distinct provinces, and are troublesome in terms of slope stability.

Hillside Construction and Associated Problems

Mariners never embarked upon more uncharted seas than the hillside developer in southern California who meticulously blueprints everything down to the last stud and nail and overlooks the ground he or she is building on. Concentrating on the building as the major element in hillside areas to the exclusion of the ground stability is like concentrating on the waves at sea in the midst of a mine field (Leighton, 1966). In one storm period in 1962 alone, landslide disasters in the greater Los Angeles area forced evacuation of over 100 homes, took two lives, and resulted in millions of dollars in property damage to hillside homes and their associated pools, patios, garages, driveways, and utilities.

It took many thousands of years to produce many of the hillside areas in southern California but only within this last half-century has civilization acquired the tools to grade them. With modern engineering and grading tools and practices, and appropriate financial incentive, few hillsides appear too rugged for future development. No hillside can withstand the assault by grading and blasting activities that almost overnight can create a series of roads and flattened, terraced building sites (Leighton, 1966).

Grading a hillside development will significantly alter the natural landscape and its slope properties. This generally involves the cutting or excavation of the high areas and the filling of the lows, or alternating cutting and filling to create a number of flattened house pads. Prime requirements for stability are that the slopes be graded at a safe angle and that the fill be engineered properly. By grading at a 1:1 slope angle (horizontal:vertical) rather than a 3:1 angle, the lot depth can be increased considerably; this gain in flat lot space acquired by steepening the slope angle can be used to increase the number of lots in the subdivision and therefore the financial returns to its developer (see figure 4.33). This financial incentive is no doubt the major reason behind the cutting of oversteepened and, therefore, unstable slopes in typical large scale housing tracts.

Hundreds of slope failures in southern California and other areas are traceable to this general situation. A number of different types of failure have occurred in developments where grading left unstable slopes:

1. Failure due to settlement of fill. If not properly compacted and engineered, any or all of the following can contribute to the

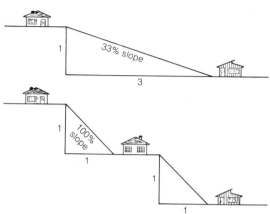

Figure 4.33. *Variations in Hillside Grading Producing Different Slopes. A safe cut slope angle is the most essential requirement for slope stability. There is obviously a tendency to cut steeper slopes because this will produce more level pad space. This also decreases slope stability and increases risk of subsequent failure.*

settlement or total failure of a fill (see figure 4.34):

a. Failing to remove vegetation, compressible soils, or trash material

b. Placing the fill in layers that are too thick and cannot be adequately compacted

c. Placing the fill without adequate subdrainage of groundwater, which may result in the saturation of the fill

2. Failure along slippage planes. This may occur either within the fill itself due to excessive moisture from yard watering, swimming pools, septic tanks, or from loading or shaking. It may occur within the original soil above the bedrock if this material was never removed, or it may occur within the bedrock itself if it represents a dip slope (see figure

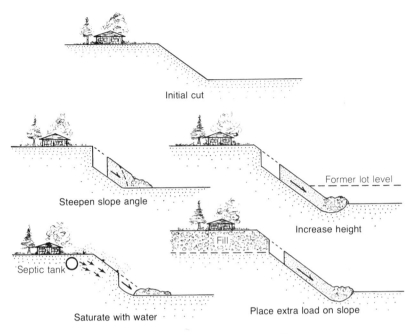

Initial cut

Steepen slope angle

Former lot level

Increase height

Septic tank

Fill

Saturate with water

Place extra load on slope

Four ways to make a stable slope unstable

Figure 4.34. *Making a Stable Cut Unstable. Any of these practices can lead to failure of an initially stable cut slope.*

4.22). This same kind of failure may occur on the cut slope forming the rear of an individual construction site where inclined beds "daylight," or have been exposed.

3. Failure due to structural weaknesses. Even if the dip of the underlying rocks is favorable for grading, the presence of faults, intrusions, jointing, or some structural weakness may lead to failure when the slope has been cut to a steep angle.

A striking example of southern California hillside instability is the Pacific Palisades area (see figure 4.35). This area consists of moderately steep terrain underlain by clastic Tertiary sediments cut by sea cliffs and deeply incised canyons, containing numerous zones of inherent instability. Some of the underlying rock in the area of one major slide and home construction consists of broken and brecciated debris from prehistoric landslides. In addition, much of the underlying sediment is not well drained and acts as a water trap. The grading of roads and construction of expensive houses occurred in this area partially on top of an ancient landslide that was locally saturated.

Although the slope of the area (with a gradient average of about 50 percent) was within existing municipal codes, there were apparently no stability analyses done on the underlying rock and soil conditions prior to construction. It would seem inevitable that a landslide would eventually occur here. With the (1) overloading of the slope by homes and apartments, (2) saturation and, perhaps,

Figure 4.35. *Pacific Palisades Landslide of 5 April 1958. Note the blocked highway in the lower left.* Photo courtesy: *California Div. of Mines and Geology.*

(3) lubrication of underlying sediments by increased drainage and runoff, (4) undercutting for roads and building sites, and, perhaps, (5) triggering by traffic, surf, or seismic vibrations, a slide finally did occur over about a three-week period in May and June 1965. Four expensive homes and two buildings of an apartment complex were either partially or totally destroyed. Efforts to salvage buildings, to stabilize the slope, and to lower the water table were all futile. An accurate and thorough geologic investigation prior to construction would have illuminated these potential problems. Today information on landslide potential or susceptibility would be collected before build-

ing could take place, and it is unlikely that construction in such an area would be allowed at all.

Recent work in the southern Puget Sound lowland area of Washington has identified a number of large ancient slumps in unstable Pleistocene glacial sediments (Smith, 1974). The presence of impermeable silts and clays beneath coarser material, and the melting of the ice that buttressed the slopes led to the failure of the saturated sediments as massive slumps. Recent movement on one large slump and the destruction of houses, roads, and utilities indicate these old slumps could be reactivated by natural processes or improper land development practices.

COSTS OF LANDSLIDE DAMAGE

A recent court ruling found Los Angeles County to be partly responsible for the reactivation of an old landslide in the Palos Verdes Hills, not far from Pacific Palisades (see figure 4.36). Part of the Portugese Bend slide, covering about 1.5 square kilometers (400 acres), began moving in 1956 and continued to move. In the following three years, over $10 million in property damage occurred (see figure 4.37). Possible causes for the reactivation of the slide include:

1. increased groundwater due to surface watering and sewage effluent,

2. increased natural precipitation, and

3. loading at the landslide head by highway fill.

A court judgment against Los Angeles County for its role in the landslide reactivation amounted to $5,360,000 (Morton and Streitz, 1967). Landslides can be very expensive problems.

The San Francisco Bay area is similar to the Los Angeles area in the incidence of landsliding that has accompanied continued development on unstable hillsides (see figure 4.38) The U.S. Geological Survey as part of the HUD environmental geologic mapping program (see p. 124) investigated the economic cost of structurally damaging landslides during a single winter (1968–1969) in the nine bay-area counties (Taylor and Brabb, 1972). The landslide costs were at least $25 million during the winter, of which about $9 million was direct loss or damage to private property, principally by lower market value (see table 4.3); in addition $10 million damage occurred to public property, chiefly for repair or relocation of roads and utilities; and about $6 million of miscellaneous costs were incurred that could not easily be classified in either the public or private sector. The rainfall that winter was somewhat above average, but was not excessive, and the total amount could be expected every third or fourth year. Shorter period intensities (hourly, daily, or weekly) may be of much greater significance than yearly totals; however, these data were not analyzed so that precise recurrence intervals for this landslide activity are not well established. The economic costs are quite clear, nevertheless.

The point is simply that landslides are both common and widespread and that landslide damage to both public and private property has been both extensive and expensive. Until we begin to recognize the reasons for slope instability and failure, and act with these in mind, this trend will continue. A detailed investigation of slope stability, including a delineation of existing landslides, conducted by a qualified geologist, should be mandatory and precede any major construction planned in areas of significant relief and unstable bedrock. This investigation should describe areas that should not be built on and measures that should be taken to overcome less hazardous geologic liabilities that may occur on other parts of a site.

Table 4.3. *Costs of Structurally Damaging Landslides in the San Francisco Bay Region, California — Winter of 1968–1969*

Public costs		$10,184,948
state highways	$4,995,800	
county costs	5,177,148	
tax revenue lost	12,000	
Private costs		9,088,808
depreciated property	7,105,546	
other	1,983,262	
Miscellaneous		6,120,200
Total		$25,393,956

SOURCE: F. A. Taylor and E. E. Brabb, "Map Showing Distribution and Cost by Counties of Structurally Damaging Landslides in the San Francisco Bay Region, California — Winter of 1968–1969," U.S. Geological Survey Miscellaneous Field Studies Map MF-327.

Figure 4.36. Portuguese Bend Landslide. View northward of a portion of the Palos Verdes Hills, Southern California, showing approximate known limits of large ancient landslide. The outline and direction of movement of the Portuguese Bend landslide are shown on the eastern part of the ancient landslide. PHOTO COURTESY: Los Angeles Dept. of County Engineer.

Figure 4.37. Home Damaged by Portuguese Bend Landslide. Ten million dollars in property damage occurred during the 3-year period, 1965–1968. PHOTO COURTESY: Los Angeles Dept. of County Engineer.

Figure 4.38. *Destruction of Homes by Landsliding in Oakland Hills, California. These houses are resting on the upper surface of a large slide. The headward scarp is just below the white railing.* PHOTO BY: *John Gilchrist.*

LANDSLIDE PREVENTION OR STABILIZATION

In addition to our ability to generate or accelerate landsliding or downslope movement, we also have the ability to decrease the potential for slope failure and stabilize some existing slides. The methods used for both slide prevention and slide stabilization are the same, but their utilization depends upon the individual area involved.

The drainage of hillsides is one of the most often used methods to either lower the potential for slope failure or to stabilize an existing slide. All surface or subsurface water-courses in such areas should ideally be diverted or channeled into drain pipes, lined ditches, or culverts (see figure 4.39). Water from gardening and septic tank leach fields should be directed away from potentially unstable hillsides. Water within the slope or slide, if possible, should be pumped

out as well. Because groundwater is one of the major causes of slope instability, subsurface drainage is a very effective remedial measure. It complements or substitutes for the adjustment of slopes, because a drained slope may be stable at a steeper angle than an undrained one (Zaruba and Mencl, 1969).

Sealing the ground surface will prevent infiltration and may aid in stabilizing slopes as well. The Ventura Avenue oil field, in Ventura County, California, is developed in a hilly area with considerable downslope movement, which has caused extensive damage to producing oil wells. To increase slope stability the vegetation was removed and the area was covered with asphalt. This, combined with an extensive system of horizontal subdrains to the interiors of the slides, has been effective in decreasing landslide activity (Morton and Streitz, 1967). The cost and aesthetics of such a program make its application somewhat limited, however.

Figure 4.39. *Stabilization of a Cut Slope by Drainage. The insertion of perforated drain pipes into a hillside and the removal of water can aid in slope stabilization.* PHOTO BY: *Gary Griggs.*

The revegetation of slopes by the planting of grasses, shrubs, and trees is also a commonly used method to prevent slope failure. Where grading or land clearing is proposed, trees and other vegetation should be retained on slopes greater than 30 percent. Landscaping on hilly terrain should utilize native plants, which do not require supplemental water for establishment and growth. The vegetation serves two functions: drying out the surface layers and binding the surface by root structure. Slides with deep-lying rupture surfaces will most likely not be stabilized by vegetation, although this can lower the infiltration of surface water into a slope and contribute indirectly to stability.

In the case of small earthflows in some road cuts, the construction of concrete retaining walls at the foot of a slope may aid in stabilization (see figure 4.40). The installation of concrete cribbing, which consists of individual precast concrete units set up in cells that are backfilled with gravel or coarse rock, is commonly utilized to prevent slope failure (see figure 4.41). These are advantageous in that they can be erected rapidly and allow for complete drainage.

Where cuts have to be made in unstable material, the benching of the slope, or the cutting of lower angle slopes, will increase the stability of the cut. Excavations should not be made into slopes where bedding is inclined or dipping downslope. In extreme cases, the complete removal of hills or slopes has totally eliminated the potential for downslope movement, along with the removal of the natural topography.

In certain instances, the temporary stabilization of a slope or a slide may be achieved by freezing, by cement grout, or by other chemical means. These are expensive techniques and should be considered adaptable to emergency conditions where temporary stability, as in certain construction jobs,

Figure 4.40. *Stabilization of a Slope with Concrete Cribbing. The installation of cribbing and then backfilling behind it with gravel for drainage can help stabilize cut slopes.* PHOTO BY: *Gary Griggs.*

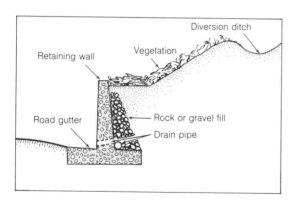

Figure 4.41. *Cross-Section of Retaining Wall. A cut slope, if properly planted, supported, and drained, can be completely stabilized. The proper precautions, however, are not always taken.*

is required. During the building of Grand Coulee Dam on the Columbia River, between the states of Oregon and Washington, numerous slides began to impede construction. The silt that formed the floor of the river repeatedly failed during excavations for the base of the dam. Ultimately, one large section of silt, which acted as an arch to hold back additional material, was completely frozen with the aid of a portable refrigeration system and hundreds of freezing points. Construction was successfully completed at considerable savings in time and additional excavation.

An understanding or awareness of the stability of the surface and subsurface material prior to any activity is of critical importance in eliminating future slope failures. The distribution of landslides within an area can be considered as a rough indication of the importance of slope failure as an erosional process, and, therefore, as a measure of overall slope stability. A map delineating landslide deposits should not be used indiscriminately to determine the probability of future landsliding. Many older slides may have been in part more a result of different climatic and geologic conditions than those presently in existence. However, these kinds of determinations are often difficult to make (Brabb et al., 1972). In an area characterized by landsliding, even with a very thorough investigation, it is nearly impossible to absolutely determine future slide potential. Slopes in such areas that have not failed may in fact be more unstable than slopes on which slides have occurred. Certainly human activity can overcome prevailing stable conditions, so that both previously undisturbed areas and old slides, if care is not taken, may eventually fail.

REFERENCES

References Cited in the Text

Bolt, B. A. et al. *Geological Hazards.* New York: Springer-Verlag, 1975.

Brabb, E. E., and Pompeyan, E. H. "Preliminary Map of Landslide Deposits in San Mateo County, California." U.S. Geological Survey Miscellaneous Field Studies Map MF-344, USGS-HUD, 1972.

————, and Bonilla, M. G. "Landslide Susceptibility in San Mateo County, California." U.S. Geological Survey Miscellaneous Field Studies Map MF-360, USGS-HUD, 1972.

Cleveland, G. B. "Why Landslides?" *Mineral Information Service, California Div. Mines and Geology* 20(1967): 115–20.

————. "Fire + Rain = Mudflows, Big Sur 1972." *California Geology* 26(1973): 127–35.

Crandell, D. R., and Waldron, H. W. "Volcanic Hazards in the Cascade Range." *Proceeding of the Conference on Geologic Hazards and Public Problems.* Edited by R. A. Olson and M. W. Wallace. Santa Rosa, Calif.: Office of Emergency Preparedness, 1969.

Davies, W. E. *Geologic Hazards of Coal Refuse Banks.* New Orleans: Abstract Volume— Geological Society of America Annual Meeting, 1967.

Lawson, A. C. et al. *The California Earthquake of April 18, 1906: Report of the State Earthquake Investigation Comm.* Washington, D.C.: Carnegie Institute Publication 87, 1908.

Leighton, F. B. "Landslides and Hillside Development." In *Engineering Geology in Southern California.* Edited by R. Lung and R. Proctor. Los Angeles: Association of Engineering Geologists Special Publication, 1969.

Martinelli, M., Jr. *Snow Avalanche Sites: Their Identification and Evaluation.* U.S. Forest Service Agriculture Information Bulletin 360, 1974.

Morton, D. M., and Streitz, R. "Mass Movement." *Mineral Information Service, Calif. Div. Mines and Geology* 20(1967); 123–29.

Nilsen, T. H. "Preliminary Photo-interpretation Map of Landslide and Other Surficial Deposits of the Livermore and Part of the Hayward 15-Minute Quadrangles, Alameda and Contra Costa Counties, California." U.S. Geological Survey Miscellaneous Field Studies Map MF-519. USGS-HUD, 1973.

Rogers, W. P. et al. *Guidelines and Criteria for Identification and Land Use Controls of Geologic Hazard and Mineral Resource*

Areas. Denver: Colorado Geological Survey Special Publication no. 6, 1974.

Sharp, R. P., and Nobels, L. H. "Mudflow of 1941 at Wrightwood, Southern California." *Geological Society of America Bulletin* 64(1953): 547–60.

Smith, M. "Landsliding and Slope Stability in the Southern Puget Lowland, Washington." *Geological Society of America Abstracts with Programs* 6,3(1974): 255.

Spangle, W. E. et al. "Application of Earth Science to Land-Use Planning in the United States: A State-of-the-Art Study." In *Geology, Seismicity, and Environmental Impact.* Los Angeles: Association Engineering Geology Special Publication, 1973.

Taylor, F. A., and Brabb, E. E. "Map Showing Distribution and Cost by Counties of Structurally Damaging Landslides in the San Francisco Bay Region, California — Winter of 1968–69" U.S. Geological Survey Miscellaneous Field Studies Map MF-327, USGS-HUD, 1972.

Varnes, D. J. "Landslide Types and Processes." In *Landslides and Engineering Practice.* Edited by E. B. Eckel. National Research Council, Highway Research Board, Special Report 29, 1958.

Zaruba, Q., and Mencl, V. *Landslides and Their Control.* Amsterdam: Elsevier, 1969.

Other Useful References

Arora, H. S., and Scott, J. B. "Chemical Stabilization of Landslides by Ion Exchange." *California Geology* 27(1974): 99–107.

Eckel, E. B., ed. "Landslides and Engineering Practice: Highway Research Board Special Report 29." *National Academy of Science-National Research Council Publication 544,* 1958.

Kiersch, G. A. "Vaiont Reservoir Disaster." *Civil Engineering* 34(1964): 32–39.

CHAPTER 5

Subsidence and Collapse

Contents

INTRODUCTION

IN numerous places throughout the world, land levels are dropping, sometimes gradually, sometimes suddenly. Yet, to many people, the subsidence, collapse, or settlement of the ground surface is almost an unknown phenomenon. It usually is not considered a major geologic hazard or problem, even by those aware of it. Yet the problem is both serious and solvable. In the United States the quickly growing region of Texas between Houston and Galveston Bay, which includes some 2 million inhabitants and 10,000 square kilometers (2.5 million acres) of land, is sinking. The seawater from the bay is claiming the land. In Baytown's showplace subdivision of Brownwood, many waterfront homes are inundated or have been abandoned, and the land under other homes is often so saturated that it quivers like gelatin. Given the 4-meter tides that sometimes accompany hurricanes, and continued subsidence, parts of the Houston Space Center could be awash in ten years. Subsidence is causing growing concern and damage in New Orleans, Louisiana, in Las Vegas, Nevada, and in sections of Arizona and California. Nearly half of the nation's states have reported at least some subsidence (Schiller, 1975).

Although the occurrence is gradual in

most places, collapse can be sudden and catastrophic, as it has been in the southeastern states. In Bartow, Florida, for instance, two houses dropped into a **sinkhole** during one night in 1967. Near Weeki Wachee, in Florida, an 11-meter high drilling rig, two trucks, and a load of steel pipe worth more than $100,000 were gobbled up in less than ten minutes. In Birmingham, Alabama, a warehouse three stories high was destroyed when a sinkhole collapsed abruptly beneath it.

Subsidence elsewhere in the world is equally serious. Venice has been slipping into the Adriatic for hundreds of years. Parts of London, Shanghai, New Zealand, Venezuela, and Taiwan are threatened with collapse. Mexico City has sunk so far into its valley — 9.5 meters — that sewage must be pumped uphill to the mains and you must walk downstairs to reach the ground floors of older buildings. In Tokyo, 2 million people live below sea level and are protected by dikes.

Surface collapse and subsidence are extremely widespread phenomena that can occur as a result of many natural processes and now, to a greater extent than ever before, as a result of human activities. Wherever the supporting subsurface material is altered or removed, the ground surface may subside or collapse. In uninhabited areas this is of no serious consequence. Wherever people or structures are involved the results may be destructive, either immediately or progressively over time.

The major causes of land surface subsidence or collapse are varied and include the following.

1. the withdrawal of large volumes of fluids (water, natural gas, or oil) from weakly consolidated sediments

2. the application of water to moisture-deficient deposits above the water table (**hydrocompaction**)

3. tectonic activity

4. the solution or leaching of soluble subsurface material (**limestone,** for example) by circulating groundwater

5. the removal of subsurface mineral deposits (coal, for example) without leaving sufficient support for the surface

6. the melting or disturbance of **permafrost**

SUBSIDENCE DUE TO FLUID WITHDRAWAL

Fluid withdrawal is by far the most common and widespread cause of human-induced regional subsidence. The removal of subsurface deposits such as oil, natural gas, or water, in large volumes or at high rates results in an overall depression of the land surface. With modern technology, such as **artificial recharge** to replace withdrawn water or oil, subsidence can be slowed and even stopped. There is no known method yet for raising the land surface back to its former elevation once it has subsided. As a result the damage produced by subsidence is permanent. In coastal areas this can require the construction of extensive systems of dikes, floodwalls, and pumping stations to protect industrial and urban areas from flooding. Subsidence problems in California and Texas have been well documented, but similar problems in varying degrees have been developing in parts of Louisiana, Arizona, and Nevada. There are probably many more areas of subtle land subsidence in the United

States that we have not yet detected. We can also expect an increase in subsidence problems in the future if we withdraw greater and greater quantities of water, oil, gas, and mineral resources from the subsurface.

Subsidence Due to Water Withdrawal

Considerable subsidence has occurred in recent years in the western United States from intensive groundwater withdrawal for both agricultural and domestic needs. Much of this has been concentrated in central and southern California where continually increasing numbers of people have placed greater and greater demands on the groundwater reservoirs (see figure 5.1).

As groundwater is pumped from an aquifer, the fluid pressure within that layer decreases, and more of the overburden weight must be carried by the granular framework of the sediment. The additional weight compacts the unconsolidated sediments and squeezes water out of pore spaces in the fine-grained clays and silts into the coarse-grained aquifer. The reduction of **porosity** of the fine-grained sediments leads to a reduction in volume, which causes subsidence of the land surface. In coarse-grained sediments such as sand or gravel, depending upon the degree of consolidation, compaction is usually small and it occurs nearly instantaneously (see figure 5.2). The withdrawal of water from silts and clays, which may either be aquifer interbeds or confining layers, creates different results. A clay or clay layer is much more likely to undergo significant compaction with fluid withdrawal than coarser-grained sediments (see figure 5.2). The escape of water and the adjustment of pressure within the pore

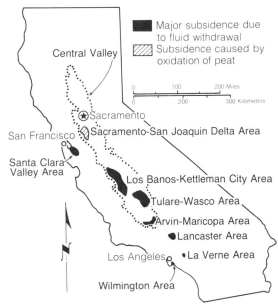

Figure 5.1. *Areas of Land Subsidence in California. Major areas of subsidence due to fluid withdrawal are outlined. Subsidence in Sacramento-San Joaquin Delta area is caused by oxidation of peat.* REDRAWN FROM: *J. F. Poland, "Land Subsidence in the Western United States," Proceedings of the Conference on Geologic Hazards and Public Problems, R. A. Olson and M. W. Wallace, eds. (Santa Rosa, Calif.: Office of Emergency Preparedness, 1969).*

spaces of the clay are both slow and time dependent. This time dependence of the pore pressure decline makes subsidence or compaction prediction very complicated (Poland, 1969).

Subsidence usually occurs quite slowly and is closely related to the rate at which fluid is being removed. Two methods are commonly used to measure subsidence: (1) repeated leveling of bench or survey marks at the land surface (this is the most common way and the only method for determining total areal subsidence); and (2) at a single site, equipment can be set up to measure

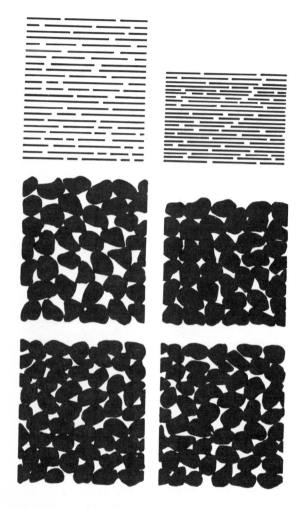

Figure 5.2. *Differential Consolidation in Fine- and Coarse-Grained Materials. Clay-rich layers within the porous sands, which yield oil or water, can compress considerably when the fluids are removed. Left: before fluid withdrawal with fluid separating particles. Right: consolidation following removal of fluid.*

Poorly consolidated rocks will undergo some compaction when fluids are removed because grains are only weakly held together and they can therefore be compressed with additional pressure.

The grains in well-consolidated materials are strongly bound together and fluid withdrawal has little effect on their volume. REDRAWN FROM: *H. W. Menard, Geology, Resources, Society. W. H. Freeman and Company. Copyright* © *1974.*

compaction directly. A compaction recorder measures the change in distance between some deep anchor weight and a recorder connected to it at the ground surface by a tight cable.

Continued population growth and increased groundwater withdrawal for domestic and industrial uses in the Santa Clara Valley area at the south end of San Francisco Bay have led to considerable subsidence. This valley, as many others in California,

traditionally had been a very fertile and productive agricultural area. However, in the 1950s and 1960s urbanization proceeded at a rapid rate and the character of the land and the demands on it changed rapidly. Pumping of groundwater for a variety of uses increased from about 50 billion liters in 1916 to about 225 billion liters per year (163 million gallons per day) during the early 1960s. The result was a drop of as much as 75 meters in the artesian water level. The loss of

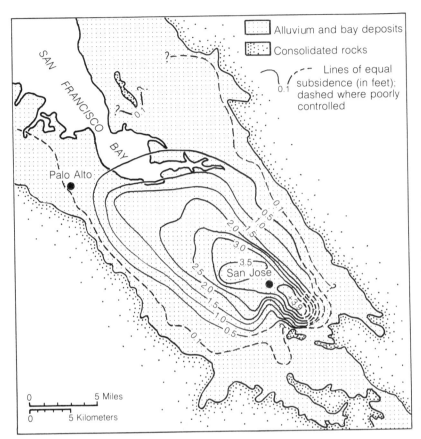

San Francisco Bay

Palo Alto

San Jose

☐ Alluvium and bay deposits
▨ Consolidated rocks
⌒ Lines of equal subsidence (in feet); dashed where poorly controlled

0   5 Miles

0   5 Kilometers

Figure 5.3. *Land Subsidence from 1960 to 1967, Santa Clara Valley, California. This was the first area in the United States where land subsidence due to excessive groundwater withdrawal was recognized.* REDRAWN FROM: *J. F. Poland, "Land Subsidence in the Western United States," Proceedings of the Conference on Geologic Hazards and Public Problems, R. A. Olson and M. W. Wallace, eds. (Santa Rosa, Calif.: Office of Emergency Preparedness, 1969).*

artesian support led to aquifer compaction and settlement of the land surface. This was the first area in the United States where land subsidence was recognized as having been caused solely by excessive groundwater removal. Approximately 650 square kilometers (160,000 acres) of land in the Santa Clara Valley have been affected by subsidence with a maximum decline of about 4 meters (13 feet) reached in downtown San Jose by 1967 (see figure 5.3). In the seven years from 1960 to 1967, the time of peak growth rates in the valley, maximum subsidence reached nearly 1.2 meters (Poland, 1969). This can be

an expensive problem. About $9 million has been spent on levee construction and other remedial work on stream channels to prevent flooding as a result of land subsidence, especially along the southern shore of San Francisco Bay. Estimates on the cost of redrilling and repairing several hundred damaged wells amount to at least $4 million.

A direct relationship has been observed between groundwater levels and subsidence in this area (see figure 5.4). Plots of bench mark elevations and water level in an adjacent well show nearly parallel patterns for a fifty-year period. As the water table recov-

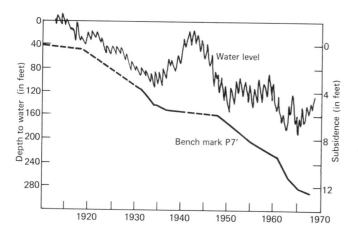

Figure 5.4. *Corresponding Subsidence of Bench Mark and Decline in Water Table. Progressive change in level of bench mark P7 and drop in level of water in nearby well between 1920 and 1970. Note stabilization of bench mark during period of water table rise from 1940 to 1950.* REDRAWN FROM: *J. F. Poland, "Land Subsidence in the Western United States," Proceedings of the Conference on Geologic Hazards and Public Problems, R. A. Olson and M. W. Wallace, eds. (Santa Rosa, Calif.: Office of Emergency Preparedness, 1969).*

ered and rose in the period from 1936 to 1943, the subsidence leveled off. A subsequent decline in the water table due to the demands of increasing numbers of people has continued until very recently, followed by increased subsidence. In recent years, the construction of groundwater recharge basins has been initiated, and large quantities of surface water have been imported. To pay for these efforts and to encourage the use of imported water, Santa Clara County levies a countywide tax and a charge for water extracted from the groundwater basin. The import of surface water has permitted a decrease in groundwater withdrawal. As a result the artesian water level has risen about 18 meters (60 feet). This has reversed the trend of increasing grain-to-grain stress in the sediments and thus has slowed the rate of subsidence.

The world's largest area of intense land subsidence has occurred in the San Joaquin Valley in central California, where 1100 square kilometers (270,000 acres) have settled more than 30 centimeters (1 foot, see figure 5.1).

An area of the valley 110 kilometers long has subsided over 3 meters, and maximum values exceed 8 meters. The intensive withdrawal of groundwater for agricultural usage had drawn the water table down as much as 135 meters. In the Los Banos-Kettleman Hills and Tulare-Wasco areas, each 10- to 25-meter decline in the water table has resulted in 1 meter of ground surface subsidence. Rates of subsidence on the west side of the San Joaquin Valley increased until the mid-1950s when the maximum observed rate was 55 centimeters per year. This decreased to about 34 centimeters per year from 1963 to 1966 with a gradual reduction in the rate of groundwater decline. More recent data indicate the subsidence rate in 1973 was almost zero in certain areas. This decline has been a result of the reduction of groundwater withdrawals due to the arrival of irrigation water from the California aqueduct. This importation of northern California water, which began in 1968, has replaced much of the groundwater formerly pumped, with a consequent rise in groundwater levels by as much as 60 meters.

From 1850 to 1950 as much as 4 meters of subsidence occurred in the Sacramento-San Joaquin Delta region of central California as the land was drained for cultivation. Approximately 1150 square kilometers (284,000 acres) of this area was underlain by peat up to 12 meters thick. As the area was drained and cultivated, drying, **oxidation**, and compaction as well as burning and wind erosion all contributed to general subsidence (Poland, 1969). The land surface of many of the islands in the delta was initially very close to sea level, but is now, in some cases, 3 to 4.5 meters below sea level. As the heights of the levees have been raised to protect the subsiding islands, the stress on the levees that retain the delta channels has been increasing. In the event of levee failure during periods of high stream flow, extensive areas will be subjected to flooding (see figures 5.5 and 5.6).

These differential changes in elevation of the ground surface in subsiding areas create severe problems in the construction and maintenance of water transport structures, such as canals, and in irrigation and drainage systems, and they also affect stream channels with low gradients. Although subsidence was recognized in the San Joaquin Valley in 1935 and correlated with excessive pumping, the United States Bureau of Reclamation, unaware of the problem, built the large Delta-Mendota Canal through this area in 1951. The consequence of this lack of information and, therefore, inadequate planning, was the sinking of 55 to 65 kilometers (34 to 40 miles) of canal more than 2 meters by 1966 with extensive damage (Prokopovich, 1969).

When the San Luis Canal was constructed in the San Joaquin Valley from 1963 to 1967, the areas of subsidence were

Figure 5.5. *Flooding of Low-Lying Lands in the Sacramento River Delta Area, California, from Levee Failure, 1972. Although excessive precipitation was the primary reason for the high river stage, the flooding itself occurred because the level of the surrounding land has slowly subsided below the river level over the years.* PHOTO COURTESY: *U.S. Army Corps of Engineers.*

Figure 5.6. *Inundation of Houses and Mobile Homes During 1972 Levee Failure, Sacramento River Delta, California.* PHOTO COURTESY: *U.S. Army Corps of Engineers.*

mapped and settlement was predicted. Although over 5 meters of subsidence has already occurred in places, as much as 4.5 meters of additional settlement were predicted. Over 30 kilometers (18 miles) of the route was flooded in an attempt to reach equilibrium conditions, which led to soil structure collapse and additional sinking (*see* pp. 165–166). The flooding was carried out over twelve to eighteen months at a cost of $4 million (Prokopovich, 1969). Subsidence will no doubt continue to occur, however, and will continue to affect both existing and future water transport and distribution systems in this area. The seriousness of the problem is easily understood when one realizes that highly populated southern California has unfortunately become dependent on imported water that must pass through the valley.

About 2.3 billion liters per day (600 million gallons per day), on the average, are withdrawn from the aquifer known as the Beaumont clays in the area around Baytown, Texas. About half of this water is used domestically and the remainder is used by industry (chiefly paper and petrochemical plants) along the Houston ship canal. As water is withdrawn, the pore pressure in the sandy aquifer has decreased, causing water to be forced in from the surrounding clays and clay interbeds. As the clay has gradually been compressed, the ground has slowly subsided. The area around Baytown has subsided more than 2.5 meters since 1920 and could subside another meter by 1980. Subsi-

Figure 5.7. *Flooding of Low-Lying Area in the Brownwood Subdivision, Baytown, Texas, as a Result of Ground Surface Subsidence. Subsidence causes waters of Galveston Bay to inundate nearshore property. Many waterfront homes have been abandoned.* PHOTO BY: *C. W. Kreitler, Bureau of Economic Geology, University of Texas at Austin.*

dence in 1971 alone was more than 15 centimeters. Tidal flooding has already occurred (see figure 5.7). If surface water use is not substituted for groundwater in the Baytown area, or withdrawals do not decrease by 1980, groundwater levels will continue to decline about 1.8 meters per year until 1995. Total subsidence would then reach about 4.5 meters below the original land surface. Some of the subsidence is believed related to active faults in the area (Van Siclen, 1967). A vertical displacement of 4 centimeters (about 1.5 inches) per year and a maximum of 105 centimeters (41 inches) since 1928 has occurred along these faults. It has been concluded that the movement has been triggered by the withdrawal of groundwater. Among the local problems that have resulted from this movement have been the fracture

of a natural gas line, the distortion of highways, and the need to give special treatment to other pipelines in the vicinity of the faults (Leggett, 1973).

The predicted subsidence would be extremely serious because unusual but expectable high tides, such as those produced by Hurricane Carla in 1961, can flood everything at an elevation of less than 4 meters. About 500 square kilometers (123,000 acres) of an area near Galveston are less than 6 meters in elevation. Because the land is very flat and also close to sea level, subsidence to date has led to the abandonment of homes and the construction of floodwalls. Sections of some coastal communities are almost continually under water. One small town is elevating a roadway 2 meters to enable people to drive to and from their homes.

Although not many afflicted cities have yet taken remedial action, Houston has (Schiller, 1975). The city plans to complete in 1976 a 96-kilometer-long (60 miles), $200 million system of canals and aqueducts to import water from the Trinity River that will supply much of the needs of the area. The Texas legislature in early 1975 established the Harris-Galveston Coastal Subsidence District, whose board will control groundwater withdrawals in the area. Although these measures will help to stabilize the region, it is still not completely out of danger. Because of the excessive cost ($70 million) of the dike proposed by the U.S. Army Corps of Engineers to protect Brownwood, a subdivision between Houston and Galveston, this plan has been abandoned. Instead the corps has advocated that the federal government spend $16 million to buy up the 435 homes in Brownwood and relocate the families. In the meantime, residents study tide tables, keep emergency pumps primed, and work on evacuation plans. A storm today of similar magnitude to Hurricane Carla, with the land 1.5 meters lower than in 1961, would wash Brownwood away.

It seems clear that as the withdrawal of groundwater continues to increase, subsidence will continue to expose more land to seawater. A program of groundwater recharge or injection along with recycling of industrial and/or domestic water supplies will be demanded to produce a stable situation. The most critically subsiding areas need to be identified, the costs of recycling waste water or obtaining water of adequate quality for recharge need to be determined, and the responsibility needs to be fixed so that a recharge and subsidence-control program can get underway. In areas undergoing development, groundwater recharge areas must be protected, and communities must develop policies to limit water extraction to a "safe yield" so that groundwater reserves are not depleted (see chapter 7). In some regions the availability of water may place an upper limit on overall growth.

Subsidence Due to Petroleum Withdrawal

Subsidence due to petroleum withdrawal has occurred in such diverse locations as Lake Maracaibo, Venezuela, Niigata, Japan, and in the Wilmington oil fields at Long Beach, California. The subsidence at Long Beach has been extensive and damaging because it is centered in the midst of a highly industrialized area that includes a major international port and a large naval shipyard.

The subsiding area is located over the crest of an underground oil reservoir known as the Wilmington oil field. Small amounts of regional subsidence related to groundwater withdrawals or other causes were noted in the area as early as 1928. Significant subsidence, however, did not occur until after the oil field development began in the late 1930s. Initially a number of causes were proposed for the settlement:

1. lowering of the water table due to withdrawals

2. oil reservoir compaction due to fluid withdrawal

3. surface loading by buildings

4. vibrations due to land usage

5. tectonic movements or motion along faults

Most authorities, however, agree that the withdrawal of fluids from the oil zones and

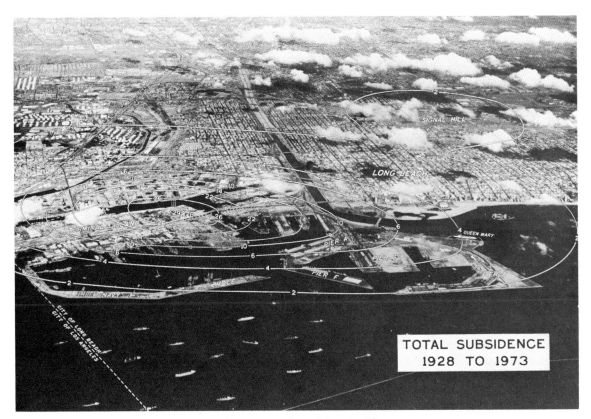

Figure 5.8. *Subsidence in the Area of Long Beach, California. The withdrawal of petroleum from the Wilmington Oil Fields has led to continual subsidence here since the 1930s.* PHOTO COURTESY: *Dept. of Oil Properties, City of Long Beach, California.*

the consequent lowering of pressure caused compaction in the oil sands and the interbedded silts and shales.

A large circular depression eventually formed with maximum vertical movement of almost 9 meters (30 feet) and horizontal movement of up to 3 meters (10 feet; *see* figure 5.8). The bowl of subsidence is centered under a portion of the naval shipyard where maximum rates of subsidence reached 66 centimeters (26 inches) per year in 1951. Extensive damage totaling perhaps $100 million has occurred. Wharves are flooded at

high tide; pipelines, buildings, railroad tracks, streets, and bridges have been cracked or displaced due to the horizontal and vertical stresses from subsidence (*see* figures 5.9 and 5.10). Although remedial surface work such as diking, raising of land areas, and repairs to wharves, oil wells, and other facilities have kept most of the area in operation, it became obvious to most observers by 1960 that the ultimate answer here, and in other areas with similar problems, had to be the arrest of the subsidence. The solution recommended by geologists was the

Figure 5.9. *Damage to Railroad Tracks Due to Horizontal Movement Associated with Subsidence over the Wilmington Oil Fields, California.* Photo courtesy: *Dept. of Oil Properties, City of Long Beach, California.*

Figure 5.10. *Flooding of Wharf at High Tide Due to Oil Withdrawal, Long Beach Harbor Area, California.* Photo courtesy: *Dept. of Oil Properties, City of Long Beach, California.*

repressuring of the oil field by water injection. The initial problems of land ownership, responsibility, and costs were eventually resolved. Most of the Long Beach Harbor area is now being continuously flooded. Approximately 700,000 barrels of seawater per day are being injected into the oil reservoirs; and there are plans to raise this to 1,000,000 barrels per day. Subsidence has been stopped over a large portion of the field. The area affected by subsidence has been reduced from 52 square kilometers to 10 square kilometers near the center of the bowl by the water injection and repressuring. A small amount of surface rebound has occurred in the area of heaviest injection, although the center is still subsiding at a rate of 6 centimeters per year (Mayuga and Allen, 1966). The estimated cost of the repressuring installations is about $30 million.

The Baldwin Hills Reservoir, southwest of downtown Los Angeles, was constructed in 1951 on the northeast flank of the Inglewood oil field, nearly adjacent to the Long Beach-Wilmington area. The reservoir was cut out of a hilltop consisting of poorly consolidated late Cenozoic silts and sands, which are extremely loose and erodible. The active Newport-Inglewood Fault zone passes within 150 meters of the reservoir site, and evidence in the form of fresh **slickensides** (polished rock surfaces) and fault gouge from the reservoir site itself indicated fault traces were present before construction began (see figure 5.11). In addition, detailed observations and measurements in this area had defined a bowl of subsidence centered about 1 kilometer west of the reservoir. Maximum subsidence of almost 3 meters had occurred over the forty-five years prior

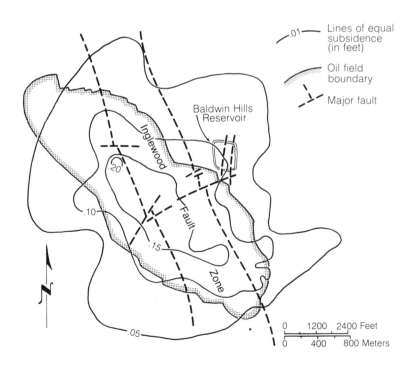

Lines of equal subsidence (in feet)

Oil field boundary

Major fault

Baldwin Hills Reservoir

Inglewood

Fault

Zone

N

.20

.10

.15

.05

0   1200   2400 Feet
0   400   800 Meters

Figure 5.11. *Faults and Subsidence Rates Around the Baldwin Hills Reservoir, California. The reservoir was constructed over several faults associated with the active Newport-Inglewood Fault zone, and within an area of known and measured subsidence.* REDRAWN FROM: *L. B. James, "Failure of Baldwin Hills Reservoir, Los Angeles, California," in Engineering Geology Case Histories, no. 6, G. A. Kiersch, ed. (Boulder, Colo.: Geological Society of America, 1968), pp. 1–12.*

to the reservoir's dam failure in 1963 (James, 1968). An average subsidence rate of approximately 1.25 centimeters per year had been occurring at the site itself. This deformation was probably influenced both by tectonic activity and production from the Inglewood oil field. A plot of epicenters made in 1965 near Los Angeles shows a cluster of points near Baldwin Hills, indicating that the area is still tectonically active.

The dam and reservoir were constructed even though the known engineering geology before and during the construction indicated that the site was in an area of known active subsidence, close to a major fault, and that the site itself was cut by parallel active faults. The reservoir lining and dam incorporated some of the most modern design innovations of their day, including an elaborate subsurface drainage system. In De-

Figure 5.12. *Baldwin Hills Reservoir (California) after Failure. Subsidence along one of the faults crossing the reservoir cracked the waterproof lining of the bottom of the reservoir and led to failure.* PHOTO COURTESY: *State of California, Dept. of Water Resources.*

cember 1963, with less than four hours of warning, the dam failed and the resulting flood claimed five lives and caused $15 million in damage (see figure 5.12). Immediately after failure, a continuous crack, with a vertical displacement of 5 centimeters, was visible on the floor of the reservoir. The crack was later found to correspond to one of the faults. The state engineering board that investigated the disaster concluded that "the earth movement which triggered the reservoir failure was caused primarily by subsidence" (Jansen et al., 1967).

These often poorly understood and indirect effects of fluid extraction, including the costs of property damage and remedial repairs, must be given equal consideration in future planning along with the direct benefits or economic returns of fluid withdrawal.

SUBSIDENCE DUE TO WATER APPLICATION: HYDROCOMPACTION

It is remarkable that the application of water to the ground, in addition to its withdrawal, can lead to subsidence or settlement. In arid or semiarid regions, where rainfall has seldom been sufficient to penetrate below the root zone of the soil, the introduction of water into dry, low density sediments can cause reorientation of the sedimentary particles and collapse of the internal sedimentary structure. Along the western and southern edges of the San Joaquin Valley in California, moisture-deficient clastic sediments, consisting chiefly of mudflow and alluvial fan deposits, have subsided 1.5 to 4.5 meters after the application of irrigation water (Poland, 1969). The clay bonds supporting the

voids in the sediments were weakened by the wetting and collapsed.

This near-surface compaction and subsidence have been serious problems resulting in sunken irrigation ditches and undulating fields. Transmission line towers, buildings, gas and oil pipelines, and canals and roads have been extensively damaged (see figures 5.13 and 5.14). To minimize future subsidence along the aqueduct in this area, about 115 kilometers (72 miles) of the alluvial deposits were preconsolidated by prolonged wetting (hydrocompaction) prior to canal construction. Protection of the freeway in this same area has required that adequate

Figure 5.13. *Hydrocompaction Damage, San Joaquin Valley, California. These concrete slabs were a portion of a test canal that buckled due to subsidence. Subsidence was caused by the collapse of the underlying soil structure due to the addition of water.* PHOTO COURTESY: *N. P. Prokopovich, U.S. Bureau of Reclamation.*

Figure 5.14. *Hydrocompaction Damage, San Joaquin Valley, California. These concrete block walls were cracked as a result of ground subsidence from hydrocompaction.* PHOTO COURTESY: N. P. *Prokopovich, U.S. Bureau of Reclamation.*

drainage facilities be designed to prevent irrigation water and storm runoff from reaching soils near the freeway that are susceptible to this type of subsidence.

Moisture-deficient alluvial deposits that compact on wetting have also been reported in Wyoming, Washington, Montana, and Arizona where up to 2 meters of subsidence after the application of water has created problems with engineering structures (Lofgren, 1969). Loess up to 30 meters thick covers extensive areas in the Missouri River Basin and has caused problems in the construction of dams, canals, and irrigation structures. Precompaction by water application has been the most common solution to this problem after it has been recognized. As pressure continues to mount on the more

marginal land surrounding presently developed and cultivated areas, or as networks of highways, electrical transmission lines, and water transport facilities expand, additional problems of this sort will be encountered. Regions covered by dry soils or sediments that are susceptible to compaction with water application need to be delineated and shown on constraint maps. Site-specific geologic reports based upon thorough field analysis of subsurface geologic conditions should be required as a precondition of any major development. Depending upon the structures and land usage being considered, the sensitive areas should either be totally avoided or appropriate preconsolidation practices should be carried out.

EXPANSIVE CLAYS

The study of clays by high power microscopes and x-ray diffraction has shown them to be composed of individual clay minerals. Although many types of clays exist, two end members of distinct or extreme character can be identified (*see* figure 5.15):

1. **Kaolinite** has a definite or rigid crystal structure.
2. **Montmorillonite** can expand by adding water within its crystal structure.

Bentonitic clays, which are commonly derived from deposits composed of volcanic ash, consist principally of montmorillonite; this explains the ability of bentonite zones to expand or swell, which can lead to slope failure along such a plane or layer.

In some warm or arid areas, such as Texas, southern California, and other parts of the Southwest, problems are encountered over clay soils that have desiccated or dried to a considerable depth (up to 6

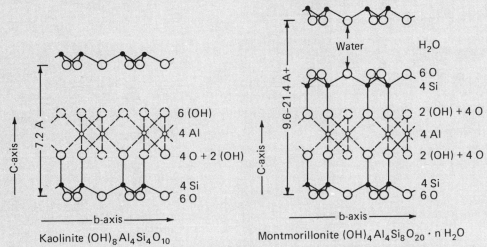

Figure 5.15. *Comparative Clay Mineral Structure. Note that the C-axis spacing for kaolinite is fixed at 7.2 angstroms 1 Å = 10^{-7} millimeters), whereas montmorillonite can expand considerably with the uptake of water.* REDRAWN FROM: Sedimentary Rocks, 2nd Edition by F. J. Pettijohn: After Fig. 46, p. 132 (After Grunner, from Grim, 1942) and after Fig. 47, p. 133 (After Hofmann, Endell and Wilm, Grim, 1942). Harper & Row, 1957.

meters) and then have come into contact with water. This may be a seasonal process due to the percolation of rain water or a change in the water table elevation below the dry or desiccated zone. As water is absorbed, the clay begins to swell; the increase in volume is followed by the rise of the ground surface. During drying, contraction occurs, and the ground usually splits, leaving gaping cracks.

Moisture has a tendency to migrate from warmer to colder soil zones. In hot or warm climates a building or road shades the soil surface area that it overlies and thereby cools that soil. Due to the impermeability of the overlying structure, evaporation of the moisture is prevented, so that the water content of the clay will increase very slowly. Eventually heaving of the surface beneath the structure occurs, resulting in cracked slabs and walls in buildings and deformation of roads (Tschebotarioff, 1973). This happens irrespective of the weight of the slab or building because of the tremendous swelling pressures that can develop in some soils.

The types of swelling soils that are troublesome can be readily recognized by laboratory testing. Local evidence of expansion or heaving may also be present in developed areas. Depending upon the range of estimated surface heave, soils have been graded from very good (0.0 to 6.3 millimeter heave), through good (6.3 to 12.7 millimeter heave), fair (12.7 to 51.0 millimeter heave), bad (5.1 to 10.2 centimeter heave) to very bad (> 10.2 centimeter heave).

Depending upon the structure planned, the following precautions in design can be taken when the existence of heave has been recognized:

1. use of heavily reinforced concrete, especially at critical points in the structure
2. provision for evaporation through ventilation channels from beneath buildings
3. use of pilings or casings for foundation support of heavy buildings that extend through the swelling layers of clay and are anchored below
4. drainage of water away from structures.

SUBSIDENCE DUE TO TECTONIC ACTIVITY

Tectonic activity is probably one of the earliest recognized causes of land surface subsidence. In general, changes due to tectonic activity are much slower than those due to other causes, although considerable vertical movement has taken place almost instantaneously during some large earthquakes along major faults. During the 1872 Owens Valley, California, earthquake (magnitude 7.75) a vertical offset of 7 meters occurred. Although the area was essentially unpopulated at the time, a similar event today would certainly cause considerable damage.

More recently, during the 1964 Alaska earthquake, nearly 200,000 square kilome-ters (77,000 square miles) underwent crustal deformation, the largest such area known to be associated with a single earthquake in historic time. A zone of subsidence totaling 110,000 square kilometers (42,500 square miles), which included most of the Kenai Peninsula and almost all of Kodiak Island, has been recognized (see figure 5.16). This downwarp, about 800 kilometers (500 miles) long and 150 kilometers (90 miles) wide, subsided 1 meter on the average. The village of Portage, at the head of Turnagain Arm of Cook Inlet, is now flooded at high tide as a result of 2 meters of tectonic subsidence (see figure 5.17). A number of roads along the coast are also covered at high tide and subject to wave erosion. Fringes of dead trees now border the coastline in many areas

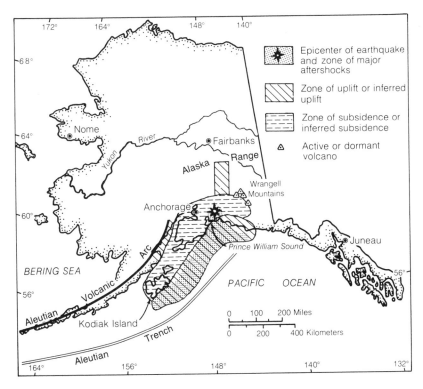

Figure 5.16. *Areas Affected by Tectonic Subsidence During the 1964 Alaska Earthquake. Approximately 110,000 square kilometers underwent subsidence averaging 1 meter.* Redrawn from: *G. Pflaker, "Tectonic Deformation Associated with the 1964 Alaska Earthquake," Science 148 (1965): 1675–87. Copyright 1965 by the American Association for the Advancement of Science.*

Figure 5.17. *Flooding at Portage, Alaska, Due to Tectonic Subsidence. Two meters of subsidence during the 1964 earthquake now leave the village of Portage flooded at high tide.* PHOTO COURTESY: *U.S. Geological Survey.*

where roots were dropped below high tide and exposed to repeated seawater inundation (Pflaker, 1965); *see also* chapter 2).

Much slower and more gradual uplift and subsidence has also been documented in southern California. Repeated levelings for thousands of stations in the Los Angeles area over a twenty-five-year period show the Santa Monica and San Gabriel mountains are rising at rates of up to 0.6 centimeters per year, while the center of the San Fernando Valley is subsiding at 1.4 centimeters per year. Although these rates may be thought of as very slow they are of considerable significance in the design of flood-control channels, sewers, aqueducts, and other types of engineering structures where maintaining consistent gradients is important (Miller, 1966).

SOLUTION AND COLLAPSE IN LIMESTONE TERRAIN

Regions underlain by limestone, **dolomite**, or other **calcareous** rocks, because of the solubility of carbonates in acidic groundwater, commonly contain extensive solution cavities. With time, characteristic **karst** topography results, marked by closed depressions (sinkholes) of varying dimensions, streams that abruptly disappear underground, and caves or caverns. This type of terrain is very common in Indiana, Kentucky, Tennessee, Alabama, and Florida. **Carbonate rocks**, which can undergo solution, collapse and subsidence are common in many other areas of North America and the world as well.

Formation of sinkholes often results from

collapse of cavities in residual clay as the clay migrates downward through openings in the underlying carbonate rocks (see figure 5.18). This process, known as **spalling**, may be caused or accelerated by:

1. a lowering of the water table resulting in a loss of support to clay overlying openings in the bedrock

2. fluctuation of the water table against the base of residual clay

3. downward movement of surface water through openings in the clay

4. increase in the velocity of water moving through the subsurface

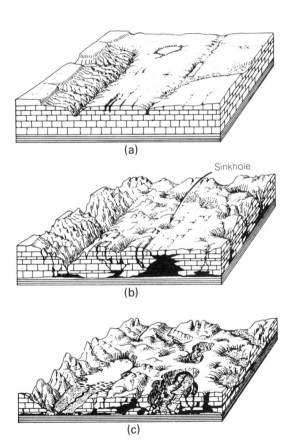

Figure 5.18. *Sinkhole Formation, Sinkholes often form when surface clay and soil collapse into subsurface cavities or caves.* Reproduced from: *A. N. Strahler, Physical Geography, 3rd ed. Copyright © 1969 by John Wiley and Sons, New York. Original drawing by E. Raisz.*

Collapse has occurred where spalling and resulting cavity enlargement has progressed upward until the overlying clay could not support itself, and where sufficient vibration, shock, or loading over cavities caused the clay to be jarred loose or forced down (Lamoreaux and Warren, 1973). Thus natural phenomena such as heavy rainfall, seasonal fluctuations in the water table, in addition to various human-imposed effects such as water withdrawal, artificial drainage, breaks in water or sewage pipes, and vibrations or shocks may result in eventual collapse.

Construction in this kind of terrain needs to be preceded by test holes or borings to adequately determine the nature of the subsurface geology, including the distribution and extent of caves or caverns beneath a site. If not, subsidence and collapse of the material underlying the imposed load may lead to eventual building failure (see figure 5.19). The building in California in which much of this book has been written was constructed on limestone, in part over an old sinkhole and adjacent to another one. Unfortunately, the building was located according to a master plan and on the basis of tree distribution, and little regard was taken of the geologic conditions of the site despite warnings by a geologist. Although differential settlement was apparently taken into account during construction of the structure, a small amount of subsidence of one segment of the building

Figure 5.19. *House Destruction from Sinkhole Collapse. This area is underlain by limestone, which commonly dissolves in groundwater leading to collapse of the overlying material, including any structures.* PHOTO COURTESY: *U.S. Dept. of Housing and Urban Development.*

relative to another has led to cracked floors and repeated repairs.

There are numerous examples in the southeastern United States of failure due to subsurface collapse. Three cars plunged into Florida's Anclote River and one person died when a sinkhole caused a bridge support to sway, dropping a span into the river. In the same state a sinkhole collapse in 1961 covered 40 acres and damaged five houses; in 1965 another sinkhole 24 meters in diameter damaged or destroyed four houses. A cement truck dropped 4 meters into a limestone cave-in while pouring a house foundation.

In late 1972 a resident of rural Shelby County, Alabama, was startled by a rumble that shook his house, followed by the distinct sound of trees snapping and breaking. Several days later, in the nearby woods, a crater about 140 meters long, 115 meters wide, and 50 meters deep was found (see figure 5.20). This is believed to be the largest recent sinkhole formed in Alabama and pos-

sibly one of the largest in the United States (Lamoreaux and Warren, 1973). At least 1000 sinkholes have developed in Shelby County alone in the past fifteen years, leading some hydrologists to believe that they may be the result of a lowering of the water table.

In November 1972 a house disappeared into a sinkhole in Crystal River, Florida. A county commissioner commented that a number of small sinkholes had appeared in the area before, but a few truckloads of fill had always managed to solve the problem. In this instance a family was going to move into a new house when they noticed cracks in the walls. The cracks were plastered over but reappeared. The next morning a 30-meter-deep hole opened up and the house collapsed into it along with a utility pole.

In the Roberts industrial subdivision of Birmingham, Alabama, subsidence and the formation of sinkholes have resulted in major pollution problems, in narrowly averted accidents, and in costly damages to building foundations, streets, water and sewer mains, railroad tracks, and Interstate Highway 59 (see figures 5.21 and 5.22). In 1963, damage to a sewer 3 meters in diameter resulted in the discharging of 150 million liters per day (40 million gallons per day) of raw sewage into two creeks for more than three months while repairs were being made. About thirty sinkholes have occurred beneath rails or in the right-of-way of the Frisco Railroad since 1963 (see figure 5.23). A major accident was narrowly avoided in 1964 when a sudden collapse 25 meters long and 9 meters wide beneath the tracks left the rails suspended 7 meters off the ground (Newton and Hyde, 1971).

Many of the sinkholes in the area can be related to rainfall, especially downpours. Recharge from rainfall results in fluctuations

Figure 5.20. *Giant Sinkhole in Shelby County, Alabama. This hole formed virtually overnight and is 50 meters deep, possibly one of the largest in the United States.* PHOTO COURTESY: *Alabama Geological Survey.*

Figure 5.21. *Large Sinkhole at the Construction Site of the Roberts Industrial Subdivision, Birmingham, Alabama. This area is underlain by dolomite and limestone. Over 200 collapses and areas of subsidence formed in an area of less than a half-square mile during 1963–1970.* PHOTO COURTESY: *Alabama Geological Survey.*

Figure 5.22. *Collapse of the Floor and Ceiling of a Warehouse, Roberts Industrial Subdivision, Birmingham, Alabama.* PHOTO COURTESY: *Alabama Geological Survey.*

Figure 5.23. *Sinkhole Beneath Tracks of the Frisco Railroad Near Birmingham, Alabama.* PHOTO COURTESY: *Alabama Geological Survey.*

of the water table, increased velocity in the movement of groundwater toward the center of cones of depression, and movement of water through fractures in clay to openings in underlying rocks. The result is movement of water against or through desiccated clay, which causes the collapse of existing cavities in clay or accelerates the spalling process, which then creates or enlarges cavities (see figure 5.18).

Many sinkholes occur with little or no warning, but others are preceded by certain characteristic phenomena that, when recognized, indicate potential collapse (Newton and Hyde, 1971). Among the more common phenomena are:

1. circular and linear cracks in the ground, on asphalt paving, and on concrete floors

2. depressions in soil or pavement that commonly result in ponding of water

3. slumping, sagging, tilting, or warping of highways, rails, fences, curbings, pipes, power poles, sign boards, and other vertical or horizontal structures

4. fractures in foundations and walls resulting from subsidence

5. small conical depressions in the ground that form very quickly

6. diversion or discharge of runoff or drainage into holes or fractures without rapid filling

Any construction in an area underlain by limestone, dolomite, or other calcareous rocks should be preceded by thorough surveys of the surface and subsurface geologic conditions. The distribution of all carbonate rocks, the locations of preexisting caves, sinkholes or closed depressions, and any reports of prior subsidence should be inventoried. Extensive boring or test hole programs, "downhole" television cameras, and other modern techniques can all be utilized in surveys of carbonate terrain for potential problems (Kennedy, 1968). If preconstruction surveys reveal these features or otherwise indicate an unstable foundation, a new site should be selected in the event major construction is proposed. In some cases, solution cavities have been filled with concrete or grouted to prevent collapse or further solution. In other cases, buildings have been set on deep pilings anchored in solid material. Any of these remedies, however, add to construction costs, with no real assurance that they are permanent solutions.

In areas with high water tables, solution cavities, caves, and sinkholes are filled with water and are commonly interconnected. Where communities rely on this source for their water supplies they must also be careful with waste discharge and disposal. Sewage and other chemicals and fluid wastes have been discharged into disposal wells or sinkholes and have contaminated entire groundwater reservoirs. This interconnected subsurface geologic environment is a common one in some areas of the country and should be recognized so that further groundwater contamination does not occur. The use of various tracers, such as dyes or short-lived radioisotopes, can be utilized to map the connections and flow patterns of such subsurface water systems and thereby separate them.

COLLAPSE OR SUBSIDENCE OVER MINING AREAS

Although people tend to think of mines being in unpopulated areas, many towns and cities in the United States grew up around and over mining operations. Thirty of the forty-six states where abandoned mines are known to exist have reported settlement problems related to previous mining. Of the 28,000 square kilometers (11,000 square miles) of surface land in the United States that have been undermined in search of minerals and fossil fuels, about 8,000 square kilometers (3,000 square miles) have already been affected by subsidence. The real threat is to urban areas where about 7 percent of the total subsidence occurs. Subsurface coal mining accounts for most of the problem (Candeub et al., 1973).

In the Appalachian region, where coal mining has been extensive, subsurface collapse has occurred over many abandoned mine workings. The mining operations have been stretched out here over nearly one century, under diverse ownership, operation, and control. Because the nature of mine support and the location of mine shafts and tunnels in many cases are poorly known, it is understandable how construction has occurred over areas with inadequate subsurface support.

In Wyoming the city of Rock Springs lies over a coal field (one of the largest bituminous fields in the nation) from which nearly 90 million metric tons of coal have been mined between 1868 and the present day. In the mines beneath the city the **room and pillar method** was used, a system that calls for thick supportive pillars of coal to be left at specific intervals to keep the roof of the mine from collapsing. The mines range in depth from 3 to over 90 meters beneath the surface. From inactivity, most are inaccessible and partly filled with groundwater. Subsidence began to occur in an 8000-square-meter (2 acre) portion of the city in January 1968, which led to moderate to severe dam-age to ten homes, streets, and utilities. Foundations and walls have cracked, gas and water lines have parted, and floors have been warped.

Twenty-three dwellings were damaged or destroyed in 1963 as a result of collapse over mine workings in Coaldale, Pennsylvania (see figure 5.24). In 1964 improvements valued at more than $13 million in a .33-square-kilometer area (75 acres) near Scranton, Pennsylvania, were damaged by subsidence (Flawn, 1970).

The settlement problems in Pennsylvania, and in Rock Springs, Wyoming, as in other areas are the consequence of earlier mining operations and techniques. The threat of subsidence is now more clearly recognized and continuing research and control will help ensure that a greater degree of control will be achieved in future mining operations. The problem still exists, however, of what can be done to alleviate the existing community subsidence problems that are taking their toll economically and causing human and community suffering as well.

Backfilling, or replacing the material taken from the earth with another supportive material, has long been a method of alleviating subsidence (see figure 5.25). In Coaldale, Pennsylvania, a federal-state cooperative project was set up to stabilize the area by reinforcing subsurface pillars and filling the underground openings with sand, gravel, and spoil bank fill. In Scranton, plans for control involved the drilling of some 600 holes into the underground workings and filling with the spoil from mine dumps (Flawn, 1970).

In Rock Springs a new method of hydraulic backfilling was utilized and proven to be very effective. With hydraulic backfilling, the sand or other fill material is mixed with water and pumped into the mines. In a

Figure 5.24. *House Destruction Due to Subsidence Over a Coal Mine in Coaldale, Pennsylvania. Many towns in the states where extensive coal mining has occurred are built partly over old mines whose exact location is uncertain.* Photo courtesy: *U.S. Bureau of Mines.*

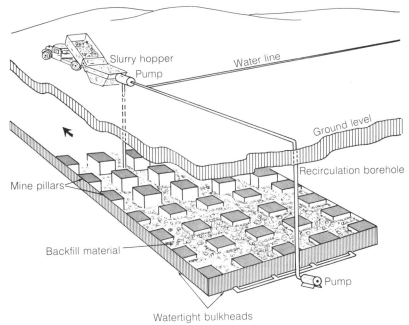

Slurry hopper
Water line
Pump
Ground level
Recirculation borehole
Mine pillars
Backfill material
Pump
Watertight bulkheads

Figure 5.25. *Hydraulic Backfilling of an Abandoned Coal Mine. Filling an abandoned mine provides support for the ground surface and also is a way to dispose of spoil or waste material left over from the mining operation. Mixing the material into a slurry or liquid form provides better filling than with dry placement.*

slurry form, the mixture is able to flow more easily than dry placement and therefore fills a much greater portion of the subsurface cavity or void.

While property damage is the primary result of subsidence, other indirect effects are varied and widespread. The threat of subsidence discourages investment and maintenance, reduces land values, complicates planning efforts and renewal programs, and may reduce the attractiveness of a neighborhood or an entire community as a place to live or work. One Pittsburg mining company that recognized the social implications of their operations guaranteed surface safety from subsidence if approximately 50 percent of the coal under a parcel of land was purchased by the property owner and left in place. This program has been one successful solution. In one ten-year period, a company with such a program provided support for 635 houses and had to repair only 2 percent of the total (Leggett, 1973).

Pennsylvania enacted legislation in 1966 designed to prevent undermining that would damage public buildings or noncommercial property. The laws regulate coal mining by requiring mining companies to leave sufficient amounts of coal in the mines for surface support. This legislation shows an overall change in attitude with an increased concern for the earth's surface and human use of it subsequent to the extraction of subsurface mineral deposits.

In addition to regulation and control over present-day mining activities, however, a complete surface and subsurface investigation for possible abandoned mine workings should precede any major construction in areas of known coal production. Plans of old mine workings are helpful but often are not available. Inaccessible subsurface mines can now be penetrated with small rotating

bore-hole cameras that can be lowered down drill holes. Photographs of the subsurface can then be used in the preparation of maps that can guide surface land use and building placement. Control of development over areas of potential subsidence can be accomplished through:

1. zoning regulations and appropriate setbacks

2. building codes to control foundation and structural design

3. on-site geologic investigations

4. land use planning to utilize endangered areas for parks, reserves, or other open-space uses

SUBSIDENCE AND DIFFERENTIAL SETTLEMENT OF ARTIFICIAL FILL

The high cost and scarcity of land in many large coastal cities, and the availability of excavated or waste material, have led to considerable filling and modification of waterfront areas for new development and construction. Milwaukee, Wisconsin, has reclaimed land from Lake Michigan. Almost 25 acres of new ground have been created on the shore of Manhattan Island, New York, by using material excavated from the site of the World Trade Center. With the exception of The Netherlands, Japan probably leads the world in the creation of this type of land for cities (Leggett, 1973). Because of its shortage of land a national program reclaiming 470 square kilometers (116,000 acres) around major ports at an estimated cost of $7 billion was announced in 1970. Although much fill material will probably be pumped from the adjacent seafloor, for years Japan has used urban waste material for filling.

Serious problems may result when structures are built on filled ground. Improper compaction or weak fill material are the factors most often responsible for subsequent subsidence or differential settlement. Many fills now consist of municipal or industrial refuse, disposed either in an open dump or as part of a sanitary landfill (see figures 5.26 and 5.27). In both cases unexpected subsidence can occur when buildings are erected

Figure 5.26. *City of Richmond Sanitary Landfill on the Margin of San Francisco Bay. The solid wastes from Richmond and many other San Francisco Bay area cities have been dumped into the bay for years.* PHOTO COURTESY: *U.S. Army Corps of Engineers.*

Figure 5.27. *Foster City, California, a Development Constructed Entirely on San Francisco Bay Fill. Many large cities on the shores of lakes, estuaries, and bays have expanded onto land filled or reclaimed from these bodies of water.* PHOTO COURTESY: *U.S. Army Corps of Engineers.*

on such material without the necessary sub-surface exploration or fill compaction.

A closely allied problem is the filling of marginal wetlands such as swamps and tidal flats to provide "usable" land for building. Today there is such a high demand for land in urban areas that reclamation of marginal land is being rapidly increased. The ecological problems involved in such filling — as, for example, the destruction of critical wildlife habitat — must be recognized, although most are beyond the scope of this book. Assuming these issues can be resolved, the prospect of subsidence or differential settlement must be dealt with when placing a fill or other heavy load on soft mud or similar material. The common properties of soil or mud found in these areas include low strength, low internal cohesion, and high water content. These characteristics cause this type of material to undergo substantial long-term settlement under sustained loads. Low shear strength can lead to plastic flow near an open face such as a channel or the edge of a fill slope.

When a load is placed upon the mud, the water in the voids of the mud is subjected to an increase in pressure. The water is squeezed out at a higher rate than under normal conditions. An analogy can be made here with squeezing water from a balloon that is punctured with many small holes. As long as the pressure is applied evenly and slowly, a certain amount of water will be squeezed out with a decrease in volume. If the pressure is great and is applied rapidly, the balloon will rupture because the water cannot escape fast enough and the entire pressure is transmitted to the water and then to the balloon. Similarly with load or fill on mud: the entire load when initially applied is momentarily transferred to and carried by the entrapped water. If the load is not too great and the application is slow, the water will manage to escape through the tiny pores and the solid grains will be forced into contact with one another to carry some or all of the load. If it is too great and is applied suddenly, high water pressures set up in the voids or pores inhibit the solid from developing the shear strength required to prevent failure (Lee and Praszker, 1969).

Early fills in many areas were placed haphazardly and it is doubtful whether any attempt was made to predict future behavior with respect to either total or differential settlement, let alone seismic response during an earthquake. All fills placed on mud are subject to subsidence. The amount and rate are unknown, however, and can only be predicted under assumed ideal conditions. Under these assumptions the mud is progressively compressed downward until stability is attained. This, however, does not account for subsidence due to lateral displacement of the mud. To a large extent it is this lateral flow from beneath fills that accounts for observed settlements greatly exceeding theoretical predicted settlements.

Subsidence and Instability of San Francisco Bay Fill

As growth has continued in the San Francisco Bay area of California in recent years, development has moved in several directions, one of which has been outward into the bay with construction occurring on artificial fill. This has now reached the point where fully one-third of the original bay has been filled (see figure 5.28). Recent legislation and the establishment of the San Francisco Bay Conservation and Development

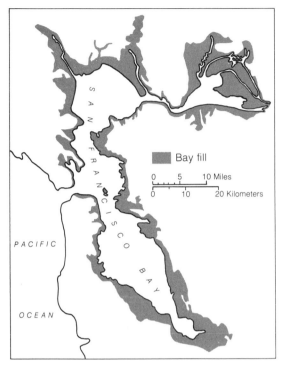

Figure 5.28. *The Extent of Filling of San Francisco Bay. During historic time, fully one-third of San Francisco Bay has been filled by either sediments added by streams or by artificial fill.* ADAPTED FROM: *U.S. Geological Survey Photo.*

Commission (BCDC) have finally come to control the haphazard and unplanned filling activity that had been going on for nearly 100 years.

Filling began in the bay as early as 1849, using dune sand with an admixture of rubbish, building rubble, old ship hulls, and occasional pilings. Differential settlement soon produced cracking, tipping, and collapse of buildings. Serious erosion from storm runoff and tidal waters was continually eating away at this early unstable fill, washing loose sand into deep water. A seawall constructed to protect the waterfront

area was constructed in 1869 and by 1917 had subsided 1.3 meters and moved laterally toward the bay as well. The fill itself along Market Street settled almost 3 meters from 1864 to 1964 (Lee and Praszker, 1969).

Geologically the bay is quite young and is floored with several kinds of bedrock covered by an unconsolidated sediment sequence consisting of an older bay mud, sand deposits, and a younger bay mud. The younger bay mud, which may be as thick as 40 meters, has long been recognized by geologists and engineers as the most troublesome in terms of soil mechanics and foundation engineering. The mud has a high natural water content, is highly compressible, and is inherently weak. These properties are to a large degree due to the clay minerals that swell and become plastic in the presence of water. The low strength, high compressibility, and variable thickness have led to failure and differential settlement of the mud under imposed loads.

During construction of the fill north of the toll plaza for the San Francisco-Oakland Bay bridge in 1947, mud was overloaded with sand fill and failed. The sand sank 6 meters and the underlying mud was forced laterally more than 150 meters into the bay (Lee and Praszker, 1969). Some of the problems that have developed on bay fill include the tilting and settling of buildings below street level, the cracking of walls and the vertical separation of buildings, and the sinking of the ground surrounding piling foundations (see figures 5.29 and 5.30). Ramps have had to be constructed to get into some buildings in the waterfront area. Portions of the highway leading to the bridge toll plaza contain over 6 meters of asphalt poured in place to counteract localized subsidence.

Figure 5.29. *Subsidence of a House Located on Old Fill Near San Francisco Bay. Note unusable doorways well below street level.* PHOTO BY: *Gary Griggs.*

Several steps can be taken to reduce differential settlement if fill is to be used, although seismic hazards are additional problems that still have to be considered. The replacement of the poorest soils with engineered fills can alleviate some of the problems although the depth of the poor soil or underlying material that can be removed is clearly a limiting factor. Surcharge or excess loadings for long periods of time prior to construction can accelerate compaction and settlement. Precautionary measures can be taken in construction, such as the use of pile supports extending down to a more solid subsurface layer or the use of a "floating foundation" that allows the building weight to be spread evenly over a large area, to minimize subsidence. It is important, however, that potential settlement problems are first identified, and a building's foundation and structural system are then designed to

be compatible with observed site conditions.

These methods, however, raise the cost of land development. In the past there has been a natural tendency for a land developer to save money and in some cases to select a soils consultant who would give him the cheapest solution. Our past experience and new legislative controls on filling should begin to suggest reasonable solutions to this problem.

To properly assess the settlement that might be encountered on soft bay mud or similar materials, it is necessary to make a thorough investigation of the geomorphic history of the site. The study should determine (1) the presence or absence of surface and subsurface sloughs, ditches, levees, and previously filled areas; (2) the length of time the area has been free of tidal action; (3) the degree of desiccation; (4) the thickness of consolidated and unconsolidated members; and (5) any other factors that would influence the behavior of the fill and improvements constructed thereon. Adequate investigation is needed to determine the expected subsidence or differential settlement within the area being filled and within specific building sites. A further analysis may also be required to ascertain where settlement can be anticipated in surrounding areas and the possible effects of such settlement.

In the San Francisco Bay region, the BCDC has established risk zones for various areas along the margins of the bay containing marshes, mudflats, or fill. Development is regulated according to the risk of the geologic unit, and the occupancy, height, and structural characteristics of the proposed building. A consulting board, composed of geologists, soils and structural engineers, and an architect, has been given the task of establishing safety criteria for the various risk zones and has the responsibility for reviewing all development proposed within the zoned areas.

Figure 5.30. *Subsidence of a House Built on San Francisco Bay Fill. Note level of driveway in contrast to the garage door level.* PHOTO BY: *Gary Griggs.*

Essay

WORLDWIDE SUBSIDENCE PROBLEMS

Subsidence of the land surface and settlement or differential settlement of large structures are ancient and recurring problems in many areas of the world. Probably the best-known example of this problem is the famed leaning tower of Pisa in Italy (see figure 5.31). The tilting of the tower is due to a 2-meter-thick layer of compressible clay only one-half meter below its massive foundation (Bolt et al., 1975). The clay is underlain by 4-meter-thick sand layer and then a thicker bed of clay. Failure of the clay beneath the shallow foundation of the massive structure must have occurred during early stages of construction. This may have been due either to shearing in the thin clay layer, or in the clay and the underlying sand, which would have caused a sudden tilting. Construction work on the tower actually stopped and started a number of times due to the subsidence and tilting. Because of the tilting the pressure on the soil on the downhill side is increased and on the opposite side decreased. The tower is inherently unstable, therefore, due to the greater settlement accompanying the higher stress. In the 800 years since construction on the tower began, the tower has continued to tilt to its present angle of nearly 5 degrees. It has also settled, more than 2 meters on the average, so that the entrance is now well below ground level.

Excessive water withdrawals from the subsurface have led to subsidence in such diverse areas as Mexico City, Mexico, and Venice, Italy. Extraction of water from sand and gravel aquifers interbedded with volcanic clay layers has produced as much as 7 meters of subsidence in Mexico City between 1900 and 1960 (Leggett, 1973). After peak subsidence rates of 30 centimeters (12 inches) per year were reached, water began to be brought into the city to reduce groundwater pumping. Recharge wells are being installed, pumping is strictly controlled, and subsidence is being halted.

Whereas settling in Mexico City has caused major structural and engineering difficulties, subsidence in Venice is even a more serious problem because the city is built at sea level. A continuing subsidence rate of about 1 centimeter per year has led to portions of Venice being submerged at high tide. A solution to the industrial demands on the underlying groundwater has not yet been found (Bolt et al., 1975).

Much of the land in The Netherlands has been reclaimed from low-lying coastal areas previously inundated by the North Sea. The soils underlying these areas have often settled with time due to the loads imposed by the large massive stone buildings. In Amsterdam a number of buildings tilt at severe angles and many have been buttressed to prevent further settlement (see figure 5.32).

Figure 5.31. *The Leaning Church Tower of Pisa. This tower is probably the world's best known example of subsidence. In the 800 years since tower construction began, its tilt has continually increased to its present angle of nearly 5 degrees.* PHOTO BY: *Gary Griggs.*

Figure 5.32. *Subsidence of Buildings in Amsterdam, The Netherlands. In an attempt to arrest subsidence and tilting, buildings in The Netherlands, where much of the land has been reclaimed from the sea, are often braced with large timbers.* PHOTO BY: *Gary Griggs.*

© 1960 Engineers Testing Laboratories Inc., Phoenix, Az.

"*. . . and we can save 700 lira by not taking soil tests.*"

PERMAFROST

Much of the earth's surface in high latitude areas is underlain by permafrost. This is not a specific material, but rather a condition of soil, sediment, or rock where temperature is continuously below 0 degrees C (32 degrees F) irrespective of the amount or state of moisture, texture, compaction, or type of material. Permafrost develops where a negative heat balance exists at the ground surface, or where the winter freezing depth exceeds the depth of summer thawing. Depths can vary from less than 30 centimeters in warmer marginal areas to over 300 meters farther north. This continually frozen ground exists over about 20 percent of the earth's land surface. Half of Canada, including almost all the Northwest Territories, Yukon Territories, and northern parts of Manitoba and Quebec, about 85 percent of

Alaska, and much of the Soviet Union are underlain by permafrost (*see* figure 5.33). Little attention or concern was given to these areas until World War II, when some military activity occurred in the Far North. Subsequently, the exploration and development of mineral resources and fossil fuels in Canada, Alaska, and the Soviet Union have led to considerable concern and research.

The term surface active layer is used to refer to the soil over permafrost that thaws annually (seasonally thawing layer). The depth of annual thaw over permafrost varies spatially and temporally in any given region. Major factors influencing thaw depth are vegetation, soil type, water content, topographic setting, and summer and annual climates (Haugen and Brown, 1971). To the north in areas of cold, continuous permafrost, the thawed soil annually refreezes to the top of the permafrost layer. Further south, where permafrost is discontinuous, the entire seasonal thaw layer does not totally freeze. Where the surface material consists of coarse-grained sand or gravel, permeability is high, drainage is good, and the surface active layer is thicker. In these situations, where water can be removed or drained, the engineering problems involved in construction should be of a minor nature. Where the surface material consists of silt, clay, or peat, permeability is low and the soil is poorly drained, which leads to a thin surface active layer. If thawing of the upper portion of this type of ground surface occurs, a fluid mass can develop that cannot drain because of the underlying impermeable frozen material.

Although the problems of permafrost and the surface active layer increase in the summer with melting, considerable change and subsequent damage can occur during the fall

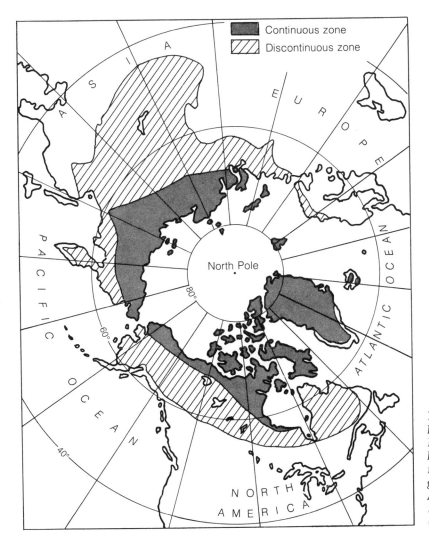

Figure 5.33. *Distribution of Permafrost in the Northern Hemisphere.* ADAPTED FROM: *Ferrians, O. J., et al., "Permafrost and Related Engineering Problems in Alaska,"* U.S. Geological Survey Professional Paper 648 (1969).

freezing period as well. As the temperature gradually drops, open water freezes first, followed by water in the more permeable sediments, such as sand and gravel. Water in the capillaries of clay may not freeze until the temperature drops to −20 degrees C (−4 degrees F). Until that temperature is reached, the water is still free to move. Since water expands when it freezes, and this water at depth is confined, tremendous pressure is exerted as ice forms. This pressure from below forces up intrusions such as dikes, which disrupt the surface and push up ridges and mounds. Rigid structures such as pavements are commonly displaced by this heaving process. When these wedges of ice melt they may cause ground collapse.

The susceptibility of permafrost terrain to

melting initiated by human activity depends upon several factors (Haugen and Brown, 1971). The thickness and general insulating quality of the organic layer and the ice content of the frozen soil just beneath it are probably the most crucial factors in such melting. Degradation of permafrost terrain can vary considerably, therefore, from place to place. Where the organic layer is thin and the ice content of the upper soil is high, such as in the Far North, the landscape is highly susceptible to almost any kind of disturbance. Further south, the organic layer is thicker, and the ice content of the soil may be found deeper, so that the likelihood of melting is more difficult to predict.

Construction in Permafrost

Although all disturbances will have a long-term effect on permafrost, some will result in more immediate effect than others. Construction activities, such as bulldozer traffic, can completely pulverize or remove the surface layer of soil and thereby expose the ice-rich soil beneath to melting (see figure 5.34). When the depth of the surface active layer is extended, as in the construction of roads, railroads, or airfields, the land may become unusable for large portions of the year (see figure 5.35). The road originally built along part of the proposed trans-Alaska pipeline route essentially became a canal with the

Figure 5.34. *Road Construction in Permafrost. The removal of the surface layer has exposed the ice-rich soil beneath to melting and turned this road into a canal.* PHOTO BY: *R. K. Haugen, U.S. Army Cold Regions Research and Engineering Laboratory.*

Figure 5.35. *Differential Subsidence Beneath Railroad Near Valdez, Alaska. Permafrost began to thaw during construction and made this railroad useless shortly thereafter.* PHOTO BY: *L. A. Yehle, U.S. Geological Survey.*

arrival of summer. The problems of permafrost were obviously not understood or were totally disregarded. The United States Army Corps of Engineers carried out a series of tests some years ago to evaluate various road and building construction techniques in this type of terrain. In preparing for construction the greater the vegetation removal and disturbance of the surface, the greater the depth of summer thaw (see figure 5.36).

With these characteristics and properties of permafrost in mind, what kinds of engineering techniques or precautions have been taken or what has been learned working in permafrost areas? For engineering purposes, frozen ground can be divided into three categories: (1) hard frozen, in which ice acts as a cement between grain boundaries; (2) plastic frozen, in which some water remains unfrozen; and (3) granular frozen, in which the grains are in mutual contact and excess ice is not present (Haugen and Brown, 1971). Where either hard of plastic frozen ground is present, the usual approach is to maintain thermal stability of the permafrost during and after construction or to thaw and remove excess water before construction is initiated. The permafrost can be preserved by the following procedures:

1. placement of an insulating pad of gravel over the building or road site

2. placement of structures on pilings anchored to frozen ground at depth, thus allowing ventilated air space between ground and building

3. artificial refrigeration of foundations

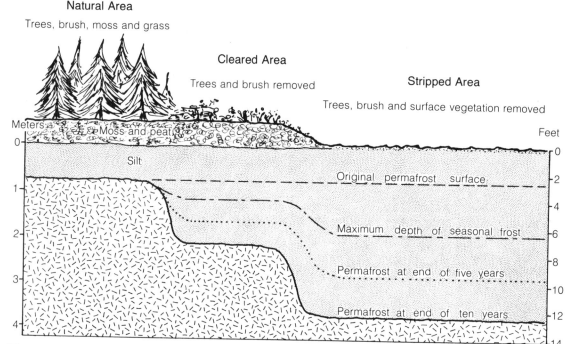

Figure 5.36. *Effects of Vegetation Removal on Permafrost. A study over a ten-year period near Fairbanks, Alaska, showed the degradation of permafrost under different surface treatments.* REDRAWN FROM: *K. A. Linell, "Frost Action and Permafrost," in Highway Engineering Handbook, K. B. Woods, ed. (New York: McGraw-Hill, 1960).*

4. installation of pipes and utilities in insulated above- or below-ground conduits

Granular frozen ground normally does not present serious engineering problems and is preferable for construction than hard or plastic frozen ground.

Adequate exploration for the selection of construction sites is the most important initial step in permafrost areas. After an appropriate site has been selected, the distribution and the properties of the frozen ground need to be thoroughly investigated and understood prior to final design and construction begin. Aerial photo-interpretation of

geomorphic features related to permafrost, the climatic regime, and geologic data from coring or drill holes are all necessary for property planning and successful construction.

Frost Heave

Frost heave of roads, highways, and railroads caused essentially by the refreezing or freezing of surface or subsurface water is also a severe problem that has been given a good deal of consideration in cold areas. In Hokkaido, Japan, mounting traffic and heavy vehicles have led to problems in this regard.

The following methods have been utilized or are being developed in Japan to minimize or eliminate frost heave of roads (Miyakawa et al., 1963):

1. replacement of frost-susceptible sub-grade soil to the freezing depth with coarse-grained materials such as sand and gravel

2. installation of heat insulation materials that are still being developed and tested for road construction

3. use of water interception, a method which involves either cutting off the groundwater supply with granular materials having no capillary action or cutting off migration of water with impermeable membranes of metal, vinyl, or asphalt

4. use of chemical additives, which would accompany the replacement of the subgrade soil and would prevent frost heave

Permafrost and the Trans-Alaska Pipeline

Recent discoveries of huge oil fields at Prudhoe Bay on the Alaskan Arctic coast have introduced permafrost to many people previously unaware of this phenomenon. The magnitude of the fields' reserves (confirmed 10 billion barrels of oil and 730 billion cubic meters of natural gas) and the increasing costs and difficulties of obtaining foreign oil have led to intensive efforts directed at moving the petroleum. Initially, three major methods of transporting the petroleum were proposed (see figure 5.37):

1. the use of heavily reinforced super-tankers that would break their way through the Arctic ice north of Canada and Alaska when conditions permitted

2. a pipeline and/or railroad route through Canada that would connect with existing pipelines in the northern United States

3. a combination of a pipeline and tankers

The method finally agreed upon uses a large pipeline that traverses most of Alaska, from Prudhoe Bay to Valdez, the northernmost ice-free port in the western hemisphere. The harbor facilities and near-shore installations at Valdez, the proposed terminus of the pipeline, were destroyed by a combination of submarine slides and waves generated by the 1964 Good Friday earthquake.

This pipeline, now under construction, will be 1280 kilometers (800 miles) long, 1.2 meters (4 feet) in diameter, and will at places either be buried in the permafrost, or laid on the ground surface, or elevated. According to preliminary estimates the initial heat in the oil plus frictional heating in the pipe are expected to maintain temperatures in the vicinity of 70 to 80 degrees C (158 to 176 degrees F) along the route at full production (Lauchenbruch, 1970).

The oil companies involved on the north slope of Alaska had invested $900 million in oil and gas leases by 1970 and were anxious from the onset to get on with the project, which would be moving 2 million barrels daily through the pipeline. The principal groups of oil companies involved initially proposed to utilize engineering practices used elsewhere in their worldwide operations and simply bury the steel pipe for all but about 130 kilometers of its route, which would cross 23 rivers, 350 streams, and several major mountain ranges and active fault

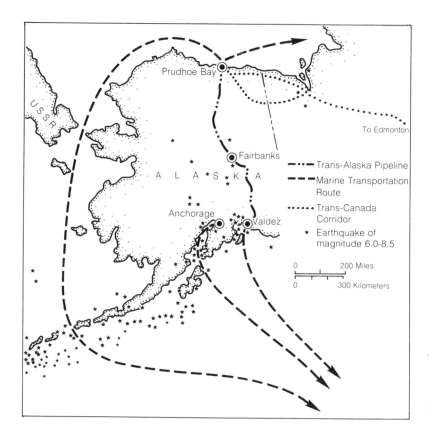

Figure 5.37. *Alternate Routes for Transporting North Slope Oil. Several different routes were originally proposed for transporting oil from Prudhoe Bay to the lower forty-eight states. Note earthquake activity throughout Alaska.*

systems. Present plans, however, involve burying the pipe for only about half of its route.

Perhaps the logical starting point in constructing oil pipelines in the Far North is to determine the thermal effects of a hot pipeline buried in permafrost. What is the amount of thawing that could occur at various times after installation for the range of climatic conditions, permafrost properties, and pipeline temperatures that would occur? What would be the effects of the thawing? Although no field testing preceded the pipeline proposal, some calculations have been made (Lauchenbruch, 1970). A 1.2-meter diameter pipeline with its axis buried

2.5 meters below the surface in typical permafrost terrain and heated to 80 degrees C will thaw a cylindrical region 6 to 9 meters in diameter in a few years in typical permafrost materials (*see* figure 5.38). At the end of the second decade of operation, typical thawing depths would be 12.0 to 15.0 meters near the southern limit of permafrost, and 10.5 to 12.0 meters in northern Alaska where permafrost is colder. Equilibrium conditions essentially would never be reached and thawing would continue throughout the life of the pipeline, but at a progressively decreasing rate. If the thawed material or the water within it were to flow, the amount of thawing could be increased several fold.

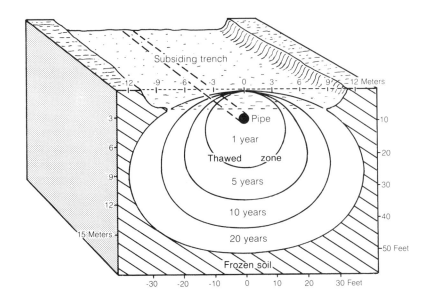

Figure 5.38. *Theoretical Permafrost Melting Profiles. Progression of thawing around a 4-foot diameter pipeline with its axis 8 feet beneath the surface and its temperature maintained at 80 degrees C.* REDRAWN FROM: *A. H. Lauchenbruch, "Some Estimates of the Thermal Effects of a Heated Pipeline in Permafrost,"* Geological Survey Circular 632 (1970).

Lauchenbruch asserts that if permafrost sediments have excess ice and a very low permeability when thawed, melting below the pipe could generate free water faster than it could filter to the surface. As a result, the thawed material around the pipeline could persist as a semiliquid slurry or slush. Under certain conditions the semiliquid slurry would tend to flow like a viscous river and seek a stable level. As an extreme example, even if these slurries occurred over distances of several kilometers on almost imperceptible slopes, the uphill end of the pipe could, in a few years, be lying at the bottom of a slumping trench many meters deep, while at the downhill end, thousands of cubic meters of mud would be extruded over the surface. The process would be self-perpetuating; where the pipeline settled to the bottom of the trench, it would accelerate thawing and flow. The pipeline would be jeopardized by loss of support in the trench and by displacement in the mudflow. It is frightening that the original pipeline planning allowed

for a sag or differential of only about 8 centimeters in a 15-meter section of pipe.

A task force appointed by the Department of Interior looked into a number of problems dealing with the proposed pipeline and its effect on permafrost and obtained a good deal of information from the Soviet Union from its permafrost experience with pipelines. The Soviets have engaged in research on the feasibility of laying natural gas pipelines in permafrost areas since World War II. Their experience convinced them that it was most desirable to avoid disturbing the thermal regime of the permafrost and where possible to construct pipelines above ground. It is less desirable, but possible, to lay pipe directly on the ground surface, with the least possible disturbance of the soil layer. The least desirable alternative is the excavation of a trench and subsequent burial of the pipeline. Despite the complex installation of various supports and difficult conditions of pipeline construction itself, the possibility of excluding the heat effects of the pipeline on frozen ground and, therefore, conservation of the best conditions to guarantee the pipeline's stability make placement above ground the most reliable decision in permafrost areas.

Construction of the Alaska pipeline was banned for about five years by court action due primarily to the uncertain environmental consequences of the project. Congress eventually authorized construction, which began in April 1974, and pipeline contractors are trying to complete the project by mid-1977.

The pipeline is being built in three modes, depending upon the environmental, topographical, and soil conditions (see figure 5.39; Allen, 1975). At Prudhoe Bay, where the oil comes out of the ground, its tempera-

ture will be as high as 82 degrees C (180 degrees F). This will drop before the oil enters the pipeline but hydraulic friction en route will maintain the temperature about 54 to 63 degrees C (130 to 145 degrees F).

In stable soil conditions the pipeline will be buried in a conventional manner. This method will be utilized in areas underlain by either bedrock, sand and gravel (which is stable when thawed), thawed soil, or where detailed exploration demonstrates that soil settlement or instability from thawing would not cause unacceptable disruption of the terrain. About 650 kilometers, or almost half, of the pipeline will be buried in a conventional manner.

The pipeline is being installed above ground wherever melting of the permafrost

Construction Methods

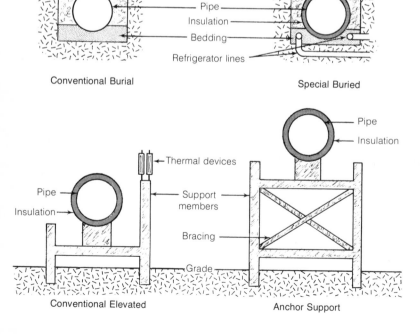

Conventional Burial

Special Buried

Conventional Elevated

Anchor Support

Figure 5.39. *Various Construction Modes for the Trans-Alaska Pipeline. Depending on the terrain and soil and environmental conditions, construction techniques will be modified.* REDRAWN FROM: *L. J. Allen, The Trans-Alaska Pipeline: the Beginning, vol. 1 (Seattle: Scribe Publishing Company, 1975).*

might cause extensive pipe deformation or create problematic soil stability conditions. The pipeline will be covered with resin-impregnated fibrous glass insulation, jacketed with galvanized steel, and then mounted on support platforms 15 to 20 meters apart. About 610 kilometers (380 miles) of the pipeline will be placed above the ground. In some instances, a special thermal device will be installed inside the support structure to keep the ground frozen.

For several shorter sections of the pipeline, where conditions are unsuitable for either conventional burial or above-ground construction, a special burial technique will be employed. By insulating the pipeline and then pumping a refrigerant through lines beneath it, the ground should be kept permanently frozen.

The temperatures and extreme working conditions along the route are taking their toll. In mid-winter temperatures can plunge to −57 degrees C (−70 degrees F); tires flatten and stick to the ground and engines must be kept running twenty-four hours per day to avoid freezing. Exposed to the wind, a person's face can turn white with frostbite in seconds. At least forty-four persons had died by mid-1975 in pipeline-related accidents. Although northern Canada and Alaska are rich in natural resources, permafrost and other hazards and environmental constraints of high latitude regions will make their extraction difficult and dangerous.

REFERENCES

References Cited in the Text

Allen, L. J. *The Trans-Alaska Pipeline: the Beginning*, vol. 1. Seattle: Scribe Publishing Company, 1975.

Bolt, B. A. et al. *Geological Hazards*. New York: Springer-Verlag, 1975.

Candeub, Fleissig and Associates. "Demonstration of a Technique for Limiting the Subsidence of Land over Abandoned Mines." Final report. Rock Springs, Wyoming, 1973.

Flawn, P. T. *Environmental Geology*. New York: Harper and Row, 1970.

Haugen, R. K., and Brown, J. "Natural and Man-Induced Disturbances of Permafrost Terrain." In *Environmental Geomorphology*. Edited by D. R. Coates. Binghamton: State University of New York, 1971.

James, L. B. "Failure of Baldwin Hills Reservoir, Los Angeles, California." In *Engineering Geology Case Histories, no. 6*. Edited by G. A. Kiersch. Boulder, Colo.: Geological Society of America, 1968.

Jansen, R. B. et al. "Earth Movement at Baldwin Hills Reservoir." *American Society of Civil Engineers Proceedings; Journal Soil Mechanics Foundation Division* 5330(1967): 551–75.

Kennedy, J. M. "A Microwave Radiometric Study of Buried Karst Topography." *Geological Society of America Bulletin* 79(1968): 735.

Lamoreaux, D. E., and Warren, W. M. "Sinkhole." *Geotimes* 18,3(1973): 15.

Lauchenbruch, A. H. "Some Estimates of the Thermal Effects of a Heated Pipeline in Permafrost." U.S. Geological Survey Circular 632, 1970.

Lee, C. H., and Praszker, M. "Bay Mud Developments and Related Structural Foundations." In *Geologic and Engineering Aspects of San Francisco's Bay Fill*. San Francisco: California Division of Mines and Geology, Special Publication 97, 1969.

Leggett, R. R. *Cities and Geology*. New York: McGraw-Hill, 1973.

Lofgren, B. E. "Land Subsidence Due to Application of Water." *Reviews of Engineering Geology* 2(1969): 271–303.

Mayuga, M. N., and Allen, P. R. "Long Beach Subsidence." In *Engineering Geology in Southern California*. Los Angeles: Association of Engineering Geologists Special Publication, 1966.

Miller, R. E. "Land Subsidence in Southern California." In *Engineering Geology in Southern California*. Los Angeles: Association of Engineering Geologists Special Publication, 1966.

Miyakawa, I.; Koyama, M.; and Takahashi, T. "Frost Heave of Roads in Hokkaido, Japan." In *Proceedings of the International Conference on Permafrost*. NAS: NRC Publication no. 1287 (1963): 497–502.

Newton, J. G., and Hyde, L. W. "Sinkhole Problems in and near Roberts Industrial Subdivision, Birmingham, Alabama." *Geological Survey of Alabama Circular 68*, 1971.

Pflaker, G. "Tectonic Deformation Associated with the 1964 Alaska Earthquake." *Science* 148(1965): 1675–87.

Poland, J. F. "Land Subsidence in the Western United States." In *Geologic Hazards and Public Problems*. Washington, D.C.: U.S. Government Printing Office, 1969.

Prokopovich, N. "Prediction of Future Subsidence along Delta-Mendota and San Luis Canals, Western San Joaquin Valley, California." Association internationale d'hydrologie scientifique, *Actes due colloque de Tokyo* (1969): 600–18.

Schiller, R. "The Growing Menace of Our Shrinking Lands." *Reader's Digest* 107, 641(1975): 124–27.

Tschebotarioff, G. P. *Foundations, Retaining and Earth Structures*. New York: McGraw-Hill Book Company, 1973.

Van Siclen, D. C. "The Houston Fault Problem." In *Proceedings of the 34th Annual Meeting, Texas Section, American Institute of Professional Geologists* (1967): 9–29.

Other Useful References

Bull, W. B. "Geologic Factors Affecting Compaction of Deposits in a Land-Subsidence Area." *Geological Society of America Bulletin* 84(1974): 3783–3802.

Legrand, H. E., and Stringfield, V. T. "Karst Hydrology: A Review." *Journal of Hydrology* 20(1973): 97–120.

CHAPTER 6

Coastal Processes

Contents

INTRODUCTION

MANY people buy or begin construction on coastal property in the spring and summer when weather and waves are calmest and the beach is high and wide. The trouble begins with the first winter storm. Large waves gnaw away at the beach and wash driftwood almost to the front door. During the big storm, strong winds send high waves to scour and undercut the concrete patio (*see* figure 6.1). Waves finally break the sliding glass doors and water surges into the living room. An all-night vigil with sand bags manages to salvage the house (Terich, 1975).

It takes such an experience for some home owners to realize the instability and dynamic nature of the coastal zone. By this time it is too late. In many cases, erosion control structures must be emplaced at considerable expense.

At present, with the increasing demands on our coasts, the processes operating there are having a greater effect on us, and we in turn are influencing coastal processes to a greater degree than ever before. Increasing population concentrations in the coastal regions indicate that the environmental stress already being felt will intensify. Planners and politicians will have to make difficult choices between complex alternatives and conflicting claims for residential, commercial, industrial, and recreational land uses, transportation, ecological and wildlife protection, resource extraction. and waste disposal (*see* figure 6.2).

Human uses of coastal areas must be decided in the context of an environment that is continually changing. Many diverse forces and processes interact along a coast to maintain a dynamic environment. Waves, tides, currents, storms, and wind are building up, wearing down, and continually reshaping this interface of land and sea. Where protec-

Figure 6.1. *Sea Cliff Erosion During a Storm. The undercutting of a sea cliff during major storms eventually led to the destruction of this house along the northern Oregon coast.* PHOTO COURTESY: *Tillamook County, Oregon, Chamber of Commerce.*

tive beaches are lacking or have disappeared, wave and surf action attack the coastline, leading to sea cliff erosion and coastal retreat.

Hardly an engineering structure is built along the coastline that does not somehow, directly or indirectly, affect the adjacent beaches. The delicate balances that exist in these environments are in many cases still not completely understood, but are being subjected to increasing pressures. Study of coastal areas will enable us to understand the effects of our past actions and will help us to avoid similar mistakes in the future. As with other geologic processes, the damages, losses, and disasters of the past have often been caused by our simple failure to understand how the processes at the earth's surface operate and what limitations they place on our actions.

Figure 6.2. *Competing Land and Water Usage in the Coastal Zone. Coastal zones are heavily utilized. This photo shows a plant that extracts magnesium from seawater, in the upper right corner. In the center is a large power plant and its oil storage tanks. On the far left are evaporation ponds for salt. There is also a state park, a marina for fishing and pleasure boats, and a marine research laboratory situated here.* PHOTO COURTESY: *California Dept. of Fish and Game.*

BEACHES AND BEACH PROCESSES

Beaches, or the deposits of sand, gravel, or other materials that flank the shoreline, are the most heavily utilized parts of the coastal environment. Although they may appear wide and stable in the summer under conditions of low, long-period waves, beaches can be eroded very rapidly when attacked by storms or heavy surf. In winter they can change their entire character or even disappear completely. Beaches commonly undergo change as a result of human activity in addition to their seasonal and storm cycles.

The beach is the buffer zone that protects the coastal areas, or sea cliffs in many cases, from direct wave attack. It makes sense then to understand the processes that control beaches and the effects of our own actions on this dynamic system. The sources and losses of beach sand, the interaction of waves, tidal action, and wind are all important in determining the size, shape, and extent of a beach and the changes that it undergoes. Along a given stretch of beach one of three conditions may prevail:

1. **Accretion** predominates over erosion and the beach is prograding or being built seaward.

2. The shoreline is stable and neither erosion nor accretion predominates, although seasonal fluctuations occur.

3. The beach is being eroded because losses of beach sand exceed supply.

We normally are only concerned with the last condition because this leads to the loss of protective beaches and consequent coastal retreat or sea cliff erosion.

Sand Sources and Beach Supply

A sufficient volume of sand or coarse-grained material is necessary for a beach to form. A source that can initially provide this material is essential whether it be upland decomposing **granite** or **sandstone**, or a near-shore carbonate reef. Sources of beach sand fall into the following categories:

Stream Runoff

The input of sand to beaches from streams is the most important process of beach nourishment. In some areas rivers are virtually nonexistent or carry little sandy material; in others, streams may empty their sediment into **estuaries** or bays and the coarse-grained material may never reach the beach. In most places, however, rivers are the major source of sand for beaches. A study of southern California beaches done years ago showed a direct correlation between stream runoff and beach conditions (Shepard, 1963). Beach size and width in an area along a mountainous coast in southern California varied directly with variations in rainfall during the preceding winter. With a high winter rainfall, streams moved a large amount of sediment to the coast (see figure 6.3). The finer grained material moved offshore while the sand was carried onto the

Figure 6.3. *Sediment Plume at the Mouth of the Yaquina River on the Oregon Coast During Winter Discharge. High winter runoff commonly produces broad beaches the following summer.* PHOTO BY: *John V. Byrne, Oregon State University.*

beaches in the subsequent months, resulting in a wide summer beach.

The two types of stream sediments contributing to coastal beaches are **suspended sediments** and **bedload** (see chapter 7). Suspended sediments are the smaller, solid particles that are turbulently suspended within the stream. The quantity of sand-sized particles in suspended sediments is highly dependent upon flow volume and velocity, inasmuch as there is usually more sand available for transport than the stream can carry in suspension. It is important to recognize that large floods can substantially increase both the amount of suspended sediment that can be carried as well as the proportion of sand in the suspended sediment. Moreover, this greater sand-carrying capacity of a stream persists for a number of years following a flood. The proportion of sand in suspended sediment varies significantly between streams. Values of 10 percent were found in southern California streams (Fay, 1972), as compared to 21 percent in the Eel River in northern California (Ritter, 1972).

Bedload is defined as the large particles that move downstream by rolling or bouncing along the stream bottom. Although few studies have been made, bedload appears to be the major mode of sand transport in arid and semiarid regions. As with suspended sediments, the proportion of sand in the bedload moving downstream is far greater during floods than during normal flow periods. These variations in stream flow introduce a seasonal factor that must be considered in evaluation of beach dynamics.

Erosion of Beach Cliffs and Coastal Formations

The erosion of the sea cliffs flanking or backing the beach is another source of beach sand. Along the coast of the Great Lakes this is a major source. The volume of sandy material contributed to a beach, or the importance of this source, is a direct function of the composition of the rock constituting the cliffs and the rate of cliff erosion. Granitic rocks, sandstones, and other coarse-grained sedimentary rocks produce a considerable amount of sand-sized material as they disaggregate (see figure 6.4); on the other hand, limestones, shales, and other fine-grained rocks produce very little sandy material as they erode and decompose. As the coastal erosion rate and grain size of the rock decrease, the importance of beach cliffs as a source of sand diminishes.

Abrasion of offshore **shoals**, reefs, or even the low tide terrace itself, if it consists of sand-sized material, can also provide material to beaches. In subtropical latitudes, Florida for example, the breakdown and erosion of coral reefs, and the redistribution of the granular material by waves can be important factors in beach construction.

Sands of the Inner Continental Shelf

Where beaches form and sand migrates in the absence of either stream or sea cliff input, there is evidence that offshore **continental shelf** sands constitute the principal sources of beach material. Beaches of the **barrier islands** along the coasts of Alabama and Mississippi west of Mobile Bay are good examples of this source (Shepard, 1963). It seems certain that the Mobile River sand load is deposited at a delta in the upper bay because the lower bay contains predominantly finer grained sediment. Cliffs are virtually nonexistent, yet the open coast here is bordered by sand islands. The islands migrate westward under influence of easterly winds, but new sand supplies keep forming

Figure 6.4. *Breakdown of Beach Cliff to Beach Sand. Here a sandy conglomerate making up a sea cliff is producing coarse-grained beach sand as it breaks down.* PHOTO BY: *Gary Griggs.*

at the eastern ends of the islands. The sand must be coming from the extensive shelf sand deposits to the south and east. A study of beach sand mineralogy in New Jersey concluded that the continental shelf must also be the sand source for the southern beaches in that state (McMaster, 1954).

Threats to the Sand Supply

The continued damming of rivers and coastal streams for water supplies has formed reservoirs that have become sediment traps. Thus the benefits of flood control and/or increased water supply have been countered by gradual loss of sand input to beaches. The nearly complete damming of the limited streams of southern California has virtually eliminated the natural sediment supply. Much of the sand presently

being contributed to the beaches is from the dredging of existing harbors and new marinas (Coe, 1966). If we are to preserve beaches, it is essential to evaluate as accurately as possible the contribution of a stream's sediment load to the coast prior to any dam construction. This aspect of dam planning, however, has never been a major consideration in the past. Measurements of sediment transport are difficult and time-consuming to make. Many factors influence the kinds of data that can be collected and extreme flood conditions, which are rarely measured, transport more sediment than normal discharge over many years (Brown and Ritter, 1971). A major difficulty here is that neither bedload nor suspended load measurements have been routinely collected as part of any past standard stream-gauging procedures, so that long-term data are lacking. There is a real need to begin the collection of these kind of data to guide our future planning.

Small settling basins can trap upstream sand almost as effectively as dams. In urbanizing areas, the attempt to remove sediments from streams has led to construction of settling and debris basins. These basins are also planned for flood control purposes to keep downstream channels clear for unimpeded passage of flood flows. They are designed to intercept sand and coarser material, but must be constantly cleaned or dredged to maintain their storage capacity. Paving of flood control channels to aid water flow during floods may also impair beach sand supply, although somewhat less than dams or settling basins.

Coastal barriers such as **seawalls** or **riprap**, erected to control erosion, reduce sand supply contributed by sea cliffs to beaches. Not only do these structures impair the breakdown and beachward movement of sandy material in the cliffs, but wave wash from a seawall also acts to accelerate beach erosion, resulting in more rapid removal of sand from the beach.

Concern for its depleted sand supply led Los Angeles County to investigate the feasibility of recapturing sand trapped behind dams and transporting it to sand-deficient beaches. This proved too costly to be feasible; however, mining of inland sand and transporting it to beaches is feasible and has been practiced in southern California for some years (Coe, 1966).

Littoral Drift

Once sand arrives at the coast and a beach forms, waves and currents provide the necessary forces to move the beach materials along the coast. Whenever waves approach the beach at an angle a **longshore current** is generated. This current pattern results in a near-shore transport of sediment known as **littoral drift**. Although all beach sand initially had to be derived from some primary source material, littoral drift is usually responsible for the largest volume of sand moving onto an individual beach.

It is essential to know the rate and direction of littoral transport, and these must be taken into account in any human intervention into the near-shore system. Decisions on sand removal or extraction operations, the planning and design of coastal engineering structures such as harbors, marinas, and ship channels, and questions of beach nourishment and sediment budgets should all be related to this system.

The direction of littoral transport can generally be determined from the beach configuration in the vicinity of either natural or artificial littoral barriers such as headlands, river mouths or inlets, **jetties**, **groins**, or **breakwaters** (see figure 6.5). With major barriers, the quantities of sand involved are large enough so that the condition observed at any particular time is probably indicative of the dominant transport direction. Because seasonal or short-term reversals in the direction of littoral drift occur, beach configurations adjacent to small structures such as groins should be observed over longer periods of time to determine predominant transport direction. The position of stream mouths and tidal inlets generally tends to migrate with time in the direction of littoral drift.

The rate of littoral transport is as important as the direction of movement in the design of coastal structures. Littoral drift varies markedly, ranging from 25,000 cubic meters

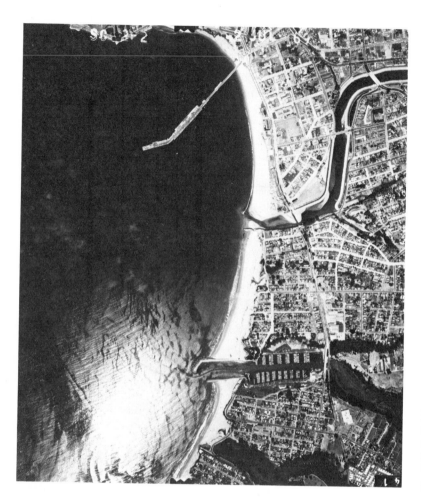

Figure 6.5. *Barriers to Littoral Drift. Littoral drift moves from top to bottom of photograph. Note beach sand buildup against the point at the mouth of the river in the upper half of the photograph and against the jetty at the small harbor mouth.* PHOTO COURTESY: *U.S. Army Corps of Engineers.*

per year at Atlantic Beach, North Carolina, to 150,000–380,000 cubic meters per year along the New Jersey shore, and up to 750,000 cubic meters per year at Oxnard, California. The amount of accretion or buildup at the upcoast side of an artificial littoral barrier, or the erosion at the downdrift side, can be a fairly accurate measurement of the drift rate along a given stretch of coast over a known period of time. The volume of material trapped by partial barriers, such as a groin, is only a partial indication of littoral drift. There is no way to determine what proportion of the drift is actually trapped by the barrier and how much may move around it. Where shoaling due to littoral drift has occurred in harbors or in entrance channels, the annual volume of sand dredged may provide a partial or minimum estimate of littoral drift.

Where no suitable littoral barriers exist for measurements of drift, or a cross-check is desirable for comparison, values of potential littoral drift can be calculated. By utilizing deep water statistical wave data, the longshore component of wave power in the surf zone can be computed (Bowen and Inman, 1966). Potential drift represents the drift that would be expected to occur under the influence of waves if an unlimited sand supply were available. The actual drift cannot exceed potential drift because this is the maximum that can be supported by existing wave conditions.

Other methods for approximating rates of littoral drift involve the use of various sediment-tracing techniques that have been employed along the coastlines of the United States in addition to those of Europe and Japan. The tracing of diagnostic heavy minerals, fluorescent or dyed sediment particles, or the utilization of radioactive materials have all been employed with varying degrees of success (Ward and Sorensen, 1970). Each method involves some problems and limitations that should be understood at the onset. The most severe limitation in the use of tracers is that of time. A study is usually made over a relatively short period. The results must then be extrapolated considerably to obtain annual approximations. The uncertainties here are obvious. Wave or surf conditions, which determine littoral drift, may change considerably throughout the year, and the short time period of study may or may not express typical conditions.

Losses of Beach Sand

It is clear that beaches are continually being supplied with sand, and littoral drift is continually transporting the material downcoast. Where is all this sand going? Should beaches either be continually increasing in size or is the sand somehow being removed? We now know that there are a number of processes operating in the coastal environment that are instrumental in removing this sand.

Rip Currents

Some sand being carried downcoast can be deflected seaward where a beach ends and a rocky promontory results in offshore **rip currents** (see figure 6.6). If the current is sufficient to carry the sand far enough offshore onto the continental shelf, it may never return. Sand can also be moved

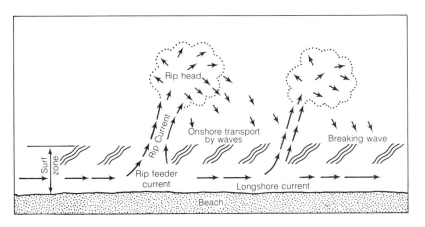

Figure 6.6. *Circulation in the Near-Shore Zone. Waves approaching the beach at an angle produce a longshore current. Periodically the buildup of water flows seaward as a rip current.* ADAPTED FROM: *Douglas L. Inmann, Fig. 42 "Near-shore Circulation System and Related Terms" (p. 78) in Submarine Geology, 2nd Edition by Francis P. Shepard. Harper & Row, 1963.*

around rocky points providing that the water depths are not great (Trask, 1955). **Wave refraction** tends to generate longshore currents that may carry sand around such promontories. Thus the sand coming down the California coast, for example, passes Point Conception, makes a bend to the southeast, and continues on downcoast to Santa Barbara.

Submarine Canyons

A very significant loss of beach sand occurs along coasts where **submarine canyons** extend across the continental shelf toward the beach. Littoral drift is commonly funneled off into the canyons, and either as a slow downslope creep, or as a more rapid movement such as a **turbidity current** or slump, the sand moves down the canyon into deeper water. The large submarine fans at the mouths of some canyons are evidence of the large volumes of sediment that have moved through the canyons in the past. Measurements and calculations have shown good correlations between rate of filling of canyon heads and amount of sediment being carried downcoast from runoff in the area between canyons (Shepard, 1963).

The southern California coast contains a number of nearly closed beach compartments or cells (see figure 6.7; Inman and Frautschy, 1966). Each compartment consists of a stretch of coastline with a series of sandy beaches terminating to the south, where a submarine canyon approaches the shore, followed by a rocky projecting point. The beach sand is in transit downcoast; supplied by coastal streams and cliff erosion, it drains off at the downcoast end of each compartment into a submarine canyon. Where the sand has been siphoned off, the beach disappears and a rocky point results. Proceeding on downcoast, another

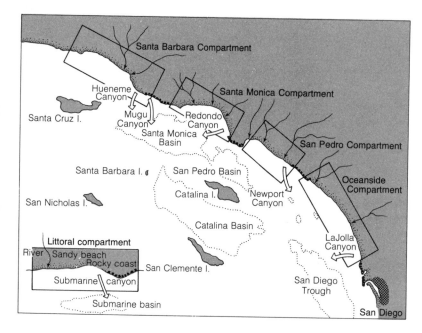

Figure 6.7. *Beach Compartments, California Coast. Individual beach compartments have been recognized along the coast of southern California. Each of these consists of (1) a river or rivers that supply beach sand, (2) a stretch of coastline with sandy beaches and down-coast drift of sand, and (3) a submarine canyon that drains off the sand to the deep sea floor.* REDRAWN FROM: *D. L. Inman and J. D. Frautschy, "Littoral Processes and the Development of Shorelines," in Coastal Engineering Proceedings of a Conference of the American Society of Civil Engineers, Santa Barbara, California, (1966).*

series of beaches appears as sand is added to the coast and a new compartment begins.

Recently methods have been developed to reduce offshore sand losses from beaches. These include engineering devices such as submerged reefs and **perched beaches** (coarse sand placed on top of an existing beach). These methods are often expensive, only marginally effective, and can sometimes create other unplanned problems.

Aeolian Transport–Dunes

Wind often blows onshore or landward and tends to build up dunes, which migrate away from the beach. As a beach widens and the expanse of dry sand expands, the losses by wind removal increase. Dunes may be seen extending inland for considerable distances at the downcurrent ends of some wide sandy beaches (see figure 6.8). Much of this sand is permanently lost to the coast. In addition to the depletion of beaches the effect of this migrating sand on human activity may be hazardous and expensive to control. Sand may cover highways, houses, vegetation — virtually everything in its path (see figure 6.9). Along the central coast of Oregon, dunes have migrated 3.0 to 4.5 kilometers (2 to 3 miles) inland and have buried and killed entire forests (Cooper, 1958). The most extensive dune sheet in this area has a

Figure 6.8. *Inland Migration of Sand Dunes, Central California Coast. Where a wide sandy beach is backed only by low-lying topography and the dominant wind direction is onshore, dunes may form that with time may migrate inland.* PHOTO COURTESY: *California Dept. of Fish and Game.*

Figure 6.9. *Damage from sand encroachment. Sand dunes near Seminole, Gaines County, Texas, have drifted to the eaves of this chicken house. Blowing sand is detrimental to both the poultry and the building.* PHOTOGRAPH COURTESY: *USDA — Soil Conservation Service.*

coastal frontage of about 86 kilometers (54 miles).

Another large dune complex is situated in northern Oregon at the mouth of the Columbia River. Active dunes in this region were threatening to overwhelm highways, towns, expensive resort homes, agricultural land, and military installations. Sand reaching the Columbia River mouth was accumulating and threatening to block passage of ocean-going vessels traveling through the river. The planting of beach and dune grasses began on the dunes in 1935 and was followed by permanent species. Dune stabilization has been achieved by careful maintenance of the plant cover (Strahler and Strahler, 1973).

Sand dunes in the Provincetown area of Cape Cod were naturally stabilized by grasses, small plants, and pine trees when settlers arrived over 250 years ago. The grazing of the dune grasses by livestock and the cutting of the forests for fuel led to the activation of the dunes and migration. Dwellings had to be abandoned and the need for continual removal of sand led to replanting of the dunes. Stabilization of most of the

dune field was achieved through legislation that prevented grazing and tree cutting. Sand in one area, however, still advances onto a highway and into a lake (Strahler and Strahler, 1973).

Dunes along many coasts can be stabilized by dune grasses and other plants. They will still have their configuration changed in response to winds and storms and may sometimes be breached so that the ocean washes through into bays. Dune stability is easily affected by human intervention, as has occurred along stretches of the New Jersey coast. Houses were built upon dunes, destroying grasses and essential retentive vegetation. Dunes were breached for beach access, groundwater was withdrawn with little control, areas were paved, and the bayshore was filled and urbanized (McHarg, 1969). The already delicate equilibrium became even more unstable.

Retribution for New Jersey's lack of planning came in March 1962 when a violent storm lashed the coastline from Georgia to Long Island. For three days one-hundred-kilometer-per-hour (60 mph) winds whipped the high **spring tides** across 1600

Figure 6.10. *Damage from Hurricane Camille in 1969 near Biloxi, Mississippi. The large waves and strong on-shore winds accompanying hurricanes cause extensive damage nearly every year to the Atlantic and Gulf Coasts.* PHOTOGRAPH COURTESY: *National Oceanic and Atmospheric Administration.*

kilometers (1000 miles) of coast. Thirteen-meter-high waves pounded the New Jersey shoreline, breached the dunes, and filled the back bays, then spilled back across the islands to the sea. When the storm subsided, the disaster was apparent. In three days 2400 houses were destroyed or damaged beyond repair; 8300 houses were partially damaged; $80 million in damage was incurred; several people were killed; and many were injured in New Jersey alone. Fires added to the destruction as roads and utilities were also destroyed.

After the disaster, bulldozers pushed the houses into the bay or into piles for burning. Sand dunes were reformed and streets uncovered as building began anew on the same sites. Foundation exposure was the most

common problem. Those houses that sat high on the dune, with the best views, found the sand swept out from under them until their pilings or foundations collapsed or were distorted beyond repair. Where in rare cases, the dune was stable and unbreached, covered only in grass, the houses endured with very minor damage from wind (McHarg, 1969).

The problems and natural processes are very clear. Dunes are an **ephemeral** geomorphic feature. Storms will occur regularly, sandbars will shift, dunes will move. Planning policy must be developed that reflects an understanding of the ecological fragility and vulnerability of the dune habitat. McHarg, in his analysis, suggests that no development, recreation, or human activity of any type occur on the primary or secondary dunes, which are the least stable and contain the most fragile vegetation. Development, if it is to occur at all should take place on backdunes, which have the advantage of protection from winter storms and may prevent backflooding of dune areas from the sea (see figure 6.11). Limited cluster development might also occur in the trough between dunes provided groundwater withdrawals would not adversely affect dune vegetation, and the dunes themselves are not breached by roads, utilities, or human trampling (McHarg, 1969). In any case, the sand dune complex, like the entire seacoast, is a dynamic environment that has to be considered carefully prior to any human activity that modifies or encroaches upon it.

COASTAL EROSION

In addition to the downcoast or longshore transport of sand, there is also a seasonal movement of beach materials offshore and onshore in response to changing wave conditions. Normally, during the summer months, the waves approaching the beach are low and have long wave lengths. These long-period waves result in onshore transport of sand because there is time for the sediment to settle out of suspension and be deposited. The summer beach is therefore high and wide and protects the shoreline or cliffs from direct wave attack (see figure 6.12A).

During winter months the prominent waves are higher and are separated by short periods. The sand on the **foreshore** is stirred up by these waves, is kept in suspension, and is moved offshore where bars and troughs commonly form. Thus the beach is cut back, in some cases leaving only bedrock or very coarse material such as gravel and cobbles (see figure 6.12B). The removal of the beach may result in the waves and surf attacking the sea cliff directly during the winter months, causing significant erosion. Marine erosion is similar to stream erosion

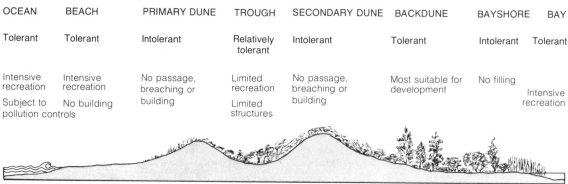

OCEAN	BEACH	PRIMARY DUNE	TROUGH	SECONDARY DUNE	BACKDUNE	BAYSHORE	BAY
Tolerant	Tolerant	Intolerant	Relatively tolerant	Intolerant	Tolerant	Intolerant	Tolerant
Intensive recreation	Intensive recreation	No passage, breaching or building	Limited recreation	No passage, breaching or building	Most suitable for development	No filling	Intensive recreation
Subject to pollution controls	No building		Limited structures				

Figure 6.11. *Tolerance of Dune Environment to Disturbance. Each dune complex consists of a number of distinct areas that vary considerably in their stability or ability to withstand encroachment or alteration. A typical dune profile above shows the backdune area to be the most suitable for human habitation.* ADAPTED FROM: *I. McHarg, Design with Nature (Garden City, N.Y.: Natural History Press, 1969).*

Figure 6.12a. *Wide Sandy Summer Beach, Santa Cruz, California. The low long-period summer waves have built up a broad sandy beach.* PHOTO BY: *Gary Griggs.*

Figure 6.12b. *Winter Beach, Santa Cruz, California. The high, short-period winter waves have moved the sand offshore, exposing the bedrock beneath the steps.* PHOTO BY: *Gary Griggs.*

in that the greatest effect commonly comes during short periods separated by much longer intervals of very slight erosion. Where beaches are nonexistent, the waves attack the sea cliff or coast throughout the year. Where extensive, wide and sandy beaches exist, the waves may never attack the cliffs directly.

Waves that reach the coastline contain enormous amounts of energy. A wave 3 meters high with a length of 30 meters exerts almost 10 metric tons of pressure per square meter. Along the coast of Scotland, waves have carried away or moved concrete blocks weighing up to 2400 metric tons (2640 tons) from a breakwater that was designed to protect the coast from wave attack. Wave impact has also thrown rocks over 30 meters into the air and broken windows in lighthouses.

Mechanisms of Erosion

Hydraulic Impact

The hydraulic action of wave or surf impact against a sea cliff can be an important erosional agent capable of dislodging large blocks of rock. This process is significant where rocks are well bedded, jointed, or fractured so that blocks can be plucked out or readily removed. The density and orientation of jointing patterns in sedimentary rocks along the coast of northern Monterey Bay, California (see figure 6.13), and in basalts along the Oregon coast (see figure 6.14; Byrne, 1966), have been extremely important in controlling coastal erosion. In addition to the hydraulic impact of the waves, the compression of air trapped or driven into the open joints or crevices in

Figure 6.13. *Erosion of a Beach Cliff Controlled by Joint Patterns in Sandstone in California. Note roots of tree wedging into joint, accelerating cliff failure.* Photo by: *Rogers Johnson.*

rock is also believed to be an important erosional factor.

Abrasion

Abrasion from the continual impact and grinding of beach material against coastal rocks is another important erosional process.

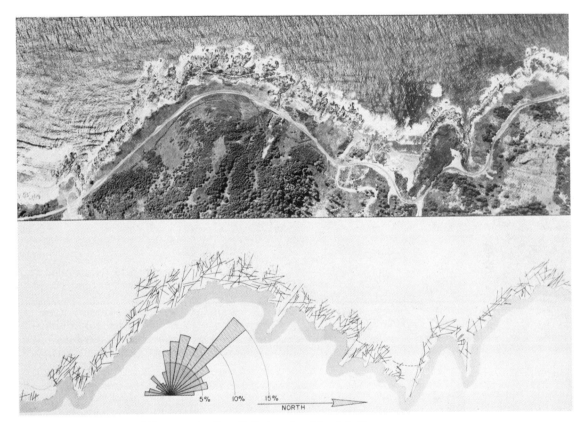

Figure 6.14. *Erosional Patterns in Basalt Determined by Jointing Along the Oregon Coast. Bottom photo shows orientation and frequency of joints.* PHOTO BY: *John V. Byrne, Oregon State University.*

The shoreline is essentially a grinding mill on a large scale. Waves, especially storm waves, throw sand, pebbles, and even large rocks against the cliffs, gradually grinding them away. Potholes in the rocky terraces and platforms of the intertidal zone are abraded by this process as evidenced by the pebble "tools" the holes contain. Other evidence for abrasion includes the presence of smooth concave surfaces cut into resistant rocks at the base of exposed sea cliffs (*see*

figure 6.15). The wave energy available, the hardness and angularity of the grinding agents or tools, and the resistance of the coastal rocks all affect the importance of this process.

Solution

Alternate wetting and drying of certain rock types in the intertidal zone can be responsible for considerable erosion. This process is

Figure 6.15. *Erosional Platform Near Santa Cruz, California. Surf action is eroding a platform at the base of a sea cliff. Photo was taken at low tide.* PHOTO BY: *Gary Griggs.*

probably the most effective in sedimentary rocks. Prolonged soaking of most shales in water causes the shale to break down into its constituent particles. The dominant process is probably not so much dissolution of cement, as it is the results of **hydration, ion exchange**, or swelling of the grains so they loosen and are washed away (Emery, 1960). The solution of calcareous cement is probably a more important process in the case of sandstones. These chemical processes tend to weaken the rocks so that hydraulic action and abrasion are more effective.

Biological Activity

Biological activity in some areas can be easily overlooked as an erosional agent. Direct mechanical boring or scraping, and indirect chemical solution, are effective mechanisms for rock breakdown and/or removal. The borings of worms, **pholads, chitons, limpets**, and sea urchins into rocks for protection and feeding all remove significant amounts of material. Borings are not restricted to soft sedimentary rocks but have been found in **gneiss, andesite,** and **chert** (Emery, 1960). In some rocks, 25 to 50 percent of the surface area has been riddled with borings, making the material much more susceptible to hydraulic and mechanical erosion.

Factors Affecting Erosion

Coastal erosion or sea cliff retreat has become a major problem in areas where roads, houses, apartments, or other structures have been built close to the edge of the sea cliff.

The rate of cliff erosion depends upon a number of factors, most of which have already been mentioned. The oceanic conditions, such as exposure to the sea, the available wave energy, and the presence or absence of a protective beach are of major importance. The rock constituting the cliffs, its hardness, bedding, and the density and orientation of jointing are controlling factors. In addition, the presence of faulting or other geologic structures and the height of the sea cliff may also affect erosion. Coastlines consisting of hard crystalline rock usually retreat or erode in an irregular fashion. Wave action etches out shear zones, joints, fractures, or zones of softer rock to form inlets and coves (see figure 6.16). More resistant rock is left as points, headlands, islands, or **sea stacks**. Coasts constructed of soft sedimentary rock or unconsolidated

Figure 6.16. *Irregular Coastline Due to Erosion in Volcanic Rocks, Oregon Coast. Surf action commonly wears away at joints, fractures, or zones of weaker rocks to produce inlets and coves in igneous rocks.*
PHOTO BY: *John V. Byrne, Oregon State University.*

material usually retreat in a more linear fashion, producing relatively straight coasts (*see* figure 6.17).

In addition to the natural oceanic and geologic factors that control sea cliff erosion, human activity in some places has either accelerated or reduced these natural rates. Loading at the edge of the sea cliff from construction of single dwellings or larger structures, if it exceeds the bearing capacity of the cliff, can lead to failure and collapse of unstable material. The alteration of normal drainage so that more water passes through the beach cliff area, whether from storm drain runoff, septic tank leaching, or swimming pool or watering activities, can also have local accelerating effects. In some instances vegetation has been planted to stabilize sea cliffs and has had exactly the opposite effect. Tree roots have grown downward into joint sets and acted as wedges to break away large blocks of material (*see* figure 6.13). Roads and parking lots constructed on the cliff edge and the subsequent load and vibration of vehicular traffic no doubt have some effect on sea cliff stability. The entire spectrum of coastal engineering structures including seawalls and riprap, groins and jetties, and breakwaters all have obvious effects on a number of coastal areas, which will be discussed subsequently.

Rates of Coastal Retreat

Keeping in mind the variations and interactions of these factors, it is easy to understand how coastal erosion or retreat rates can be so variable. Comparison of historic records and photographs with the existing coastline indicate that erosion in certain areas has been almost negligible for hundreds of years. Cliffs cut in granite in Cornwall, England, show evidence of only minimal erosion since the sea has stood at its present level (about 3000 years). On the Yorkshire coast of England, however, cliffs in **glacial drift**, an easily eroded material, have retreated at average rates of 1 to 2 meters per year. At one

Figure 6.17. *Straight Coastline Due to Erosion in Stratified Sedimentary Rocks, Half Moon Bay, California. Note undercutting and destruction of a road.* PHOTO COURTESY: *U.S. Army Corps of Engineers.*

location the coast retreated 12 to 30 meters overnight during a single storm in 1953. Surveys and records kept since the time of the Roman occupation of Britain extend back nearly 2000 years and indicate that 215 square kilometers (83 square miles) of land including at least twenty-eight towns have been lost to the sea along this coast.

The Cape Cod area in Massachusetts has been inhabited for several hundreds of years, and as a result the changes that it has undergone have been well documented (Shepard and Wanless, 1971). The Cape Cod light, for instance, because of rapid cliff recession was relocated at least three times during the nineteenth century. Storms strike outer Cape Cod from all directions and the waves have had no difficulty in eroding the unconsolidated sand and gravel deposits that make up the coast. Numerous attempts have been made to calculate the amount and rate of cliff recession at Cape Cod since the waves first began to break against the shore when the sea level rose to its present position. Erosion rates at different points vary from a minimum of 30 centimeters to a maximum of 130 centimeters (1 to 4 feet) per year. If this recession rate is projected 3000 years into the past (when the sea level probably reached its present position), a total recession of 2250 meters (7500 feet) in this area is indicated. Sand spits in this same area have migrated over 1.5 kilometers (5000 feet) in 100 years, or at an average rate of about 16 meters per year.

In the Key Biscayne area of Florida, erosion between 1884 and 1944 had caused about 150 meters of beach recession. This erosion took place principally at times of major storms, nearly half of it during a hurricane in 1926. A number of lives were lost during this hurricane as people ventured out to see the damage and were caught by rising water on a causeway between Biscayne Bay and Miami Beach.

Hurricane Eloise struck the Florida panhandle between Fort Walton Beach and Panama City on September 23, 1975 (see figure 6.18; Morton, 1976). Twelve to eighteen meters of beach **scour** and dune retreat occurred behind seawalls that failed. The most dramatic changes associated with dune retreat occurred where buildings were placed on altered dunes or on fill emplaced behind seawalls. The low seawalls failed and the undermining and removal of support beneath slab foundations led to structural failure and building collapse (see figure 6.19). Much of the structural damage in-

Figure 6.18. *Path of Hurricane Eloise Through Florida Panhandle, 1975.* REDRAWN FROM: *R. A. Morton, "Effects of Hurricane Eloise on Beach and Coastal Structures, Florida Panhandle," Geology 4(1976): 277.*

Figure 6.19. *Hurricane Damage. Collapse of a motel constructed on the beach from scour at Panama City, Florida, from Hurricane Eloise.* PHOTO COURTESY: *Robert A. Morton, Bureau of Economic Geology, University of Texas at Austin.*

volved single family dwellings or one-story motels constructed of inadequately reinforced concrete blocks.

Hurricane Eloise was not a severe storm, yet it caused considerable damage along a coastal strip about 60 meters wide. Had construction not occurred in this zone then damage would have been minimal. If alternate building designs had been utilized, much of the damage could have been prevented or minimized. The destruction was due to the failure of coastal planners, builders, and residents to recognize that (1) the beach and dunes are dynamic; (2) over the long term, the probability of recurring storm surge and wave attack is relatively high; (3) all seawalls are not designed to withstand scour and lateral loads from waves; (4) pilings provide the best foundation support in areas subject to scour; and (5) dune removal or lowered dune elevations do not provide adequate protection from hurricane forces (Morton, 1976).

On September 8, 1900, Galveston, Texas, was struck by a hurricane with tides 4.5 meters high and winds estimated at 190 kilometers per hour (120 mph). This flooded all parts of the island, caused property damage in excess of $20 million, and killed 6000 residents. The city was completely unprepared for the storm, except for a low wall, even though numerous destructive hurricanes had occurred in the past. The hurricane cut back the gulf shoreline over 50 meters. A solid concrete seawall, ultimately 6.4 meters high, was soon constructed, making Galveston the first city in the western hemisphere to be properly protected from destructive hurricanes (Shepard and Wanless, 1971). Nevertheless, Hurricane Carla in 1961 produced extensive flooding in Galveston (see figure 6.20).

Wherever houses or other structures have been built on or near the beach, very close to sea level, there is always the danger of very high tides occurring together with high storm waves and onshore winds. Protective beaches can be cut back quickly, leaving the waves to attack the houses and property directly (see figure 6.21).

Figure 6.20. *Hurricane Flooding, Galveston, Texas. A storm-surge tide moving along 21st Street in Galveston, Texas during the landfall of Hurricane Carla, 1961. A tidal-surge level of 9.3 feet above mean sea level was recorded at the seawall in Galveston during this storm.* PHOTO COURTESY: *Managing Editor, Houston Chronicle.*

Figure 6.21. *Undercutting and Subsequent Collapse of a Road Placed Too Close to a Retreating Sea Cliff, Half Moon Bay, California.* PHOTO BY: *Gary Griggs.*

Along the California coast sedimentary rocks, which commonly form the sea cliffs, are retreating at average rates of about 30 centimeters per year. Cliff erosion in resistant sandstones and siltstones at Sunset Cliffs, San Diego, California, has been occurring very slowly. The average rate has been about 1 centimeter per year for the past 75 years, and three-fourths of the area studied by Kennedy (1973) had undergone no appreciable erosion during this period. At Año Nuevo, however, approximately 80 kilometers south of San Francisco, the erosion rate has averaged 2.7 meters per year for the last 300 years — one of the highest natural rates known along this coast (see figure 6.22; Tinsley, 1972). The sea cliffs along this coastline are cut into an elevated marine terrace. At this particular location the terrace has been depressed so that the bedrock surface is nearly at sea level, leaving the unconsolidated, easily eroded terrace deposits exposed to wave and surf action.

Along much of the northern Oregon coast, landsliding has been shown to be the most important erosional process (North and Byrne, 1965). The nearly constant pounding and undercutting of the sea cliff by storm waves, the intense rainfall, and the commonly seaward dipping sedimentary rocks all contribute to this phenomena, which has affected an extensive stretch of coastline and damaged or threatened roads and structures (see figure 6.23).

The realization that the loss of real estate to the ocean is the result of the interaction of a number of geologic and oceanic factors and processes that occur naturally wherever land and sea meet is of little concern or comfort to the average citizen watching his or her backyard disappear (see figure 6.24). A new

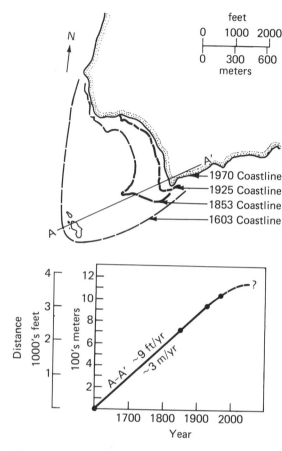

Figure 6.22. *Coastal Erosion, Point Ano Nuevo, California, 1603–1970. Old maps and charts have been utilized to determine the progressive retreat of the coastline.* REDRAWN FROM: *J. Tinsley, Sea Cliff Retreat as a Measure of Coastal Erosion: San Mateo County, California, Guidebook for Friends of the Pleistocene San Francisco Bay Area Field Trip. (Menlo Park, Calif.: U.S. Geological Survey, 1972).*

concept in real estate, "the expendable lot," has evolved as a result of the all too common residential subdivision built on rapidly retreating sea cliffs.

Figure 6.23. *Large Block Landslide Along the Central Oregon Coast. Line delineates head wall of the slide.* PHOTO BY: *John V. Byrne, Oregon State University.*

Figure 6.24. *Coastal Erosion Undercutting a Sidewalk in a Residential Area, Central California Coast. Notice how the waves have eroded the loose cliff deposits behind the protective concrete on the left side of the photograph.* PHOTO BY: *Gary Griggs.*

Essay

COASTAL EROSION ALONG THE GREAT LAKES

The coastlines of the Great Lakes are subject to severe erosion that periodically causes lake front damage. The state of Michigan, for example, has about 4600 kilometers (2900 miles) of coastline, approximately one-third of which is residential. Both long-term fluctuations in lake levels and short-term changes occur and can create coastline problems. The level of Lake Michigan has fluctuated over 2 meters since 1860 in response to prolonged periods of above or below normal precipitation. Lake levels have been rising in recent years from a low in 1964, to a high in 1973 equivalent to the level reached in 1952 (see figure 6.25). Short-term variations are produced by a combination of ice jams, winds, storm waves, and wave runup, all of which can raise local water levels as much as 2 or 3 meters. The impact of elevated water levels and storm waves on flat erodible shorelines has led to severe erosional problems. With a rising lake level, the protective beaches are narrowed, and storm waves exert more energy against the land. Where storm waves attain heights of over 6 meters, and the shoreline consists of unconsolidated **glacial till**, erosion can be rapid. In Cook County, Illinois, a section of bluff receded nearly 100 meters between 1857 and 1908 (Larsen, 1973). About one-third of the Michigan coastline is highly erodible. During the last twenty-five years, the annual rate of shore recession has averaged from 0.3 to 1.7 meters. In the worst year, 1951 to 1952, damages to shorefront property reached about $35 million (in 1970 dollars). Current high levels are also leading to severe problems. Shoreline is being eroded, houses, roads, retaining walls, and other structures have been attacked, overtopped, undermined, or destroyed by waves (see figure 6.26). As along an ocean coastline, the severity of erosion along a lake front is a function of the exposure to wave attack, the presence or absence of a protective beach, and the stability or strength of materials making up the coastline. Protective structures can be built to halt erosion, and buildings and roadways should be set back safely from the lake front where serious erosion occurs.

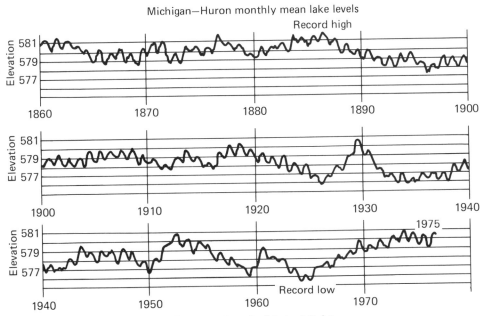

Figure 6.25. *Record of Monthly Mean Level of Lake Michigan Since 1860.* DATA FROM: *Oceanic and Atmospheric Administration, Lake Survey Center, Detroit.*

Figure 6.26. *Great Lakes Coastal Erosion Damage.* PHOTO COURTESY: *U.S. Dept. of Agriculture, Soil Conservation Service.*

225

Consultants for a proposed housing development in the Moss Beach area, about 50 kilometers south of San Francisco, were informed that (1) the erosion rate for the past 105 years was about 45 centimeters per year and could be expected to continue at that rate for the foreseeable future, and (2) the waterfront was zoned as a marine reserve and no protective engineering structures would be permitted. Developers decided to dedicate the areas closest to the ocean as a 2800 square meter (0.7 acre) common area for all residents of the subdivision. It was also noted that in about 50 years this common area will be gone and then the two homes closest to the ocean and about half of the cul-de-sac access will wash away (Tinsley, 1972). Extensive fracturing and shearing associated with an active fault in this area have created further instability in the form of slumping and landsliding, which are destroying "beach front lots" and damaging homes. About 80 kilometers further south in Santa Cruz, sea cliff erosion has undercut a large apartment building con-structed on the cliff edge (see figure 6.27). An adjacent duplex, only recently completed, was bought and sold repeatedly, and ultimately removed. Building permits for cliff construction in areas of rapid retreat are still being issued in some areas and lots are still being sold to innocent and uninformed people.

Certainly there is a strong need for some type of control and protection. An inventory of coastal erosion rates in an area of development or potential development is a first priority. The utilization of old photographs, both aerial and ground, and subdivision, street, or survey maps can usually provide this information if the coverage is good and the scales are large enough. Field surveys or studies are necessary to adequately check critical areas where geologic conditions have led to increased erosion rates. This data should then be incorporated into a coastal land use plan that would involve protective zoning, appropriate setbacks, or building bans in those areas subject to significant erosion. Real estate agents should be required

Figure 6.27. *The Undercutting of a Large Apartment Complex Due to Cliff Retreat, Capitola, California. The rate of cliff erosion was not considered at the time this building was constructed.* PHOTO BY: *Gary Griggs.*

by law to reveal geologic restrictions or problems to prospective buyers. The public would thereby be protected from purchasing an "expendable" ocean-front lot or a dwelling with a relatively short "half-life."

Another approach to planning for coastal erosion hazards follows the policy that all new developments must be sited back far enough from the cliff edge so that, under normal erosion rates, there will be no danger

to structures during an economic life of fifty years. No cliff protection is allowed, and the development must not contribute to the instability of either the cliff or beach. Using this approach, data are gathered and coastal areas are rated and then mapped for high, moderate, and low stability. An example of the type of restrictions that can be applied to each category is shown in figure 6.28. Development is excluded from areas of active

Coastal Erosion Guidelines for Geologic Stability

All developments within the immediate beach–coastal bluff area must demonstrate geologic stability of the structure for a 50-year period, must not contribute to instability of any cliff or beach, and must be consistent with other planning policies in the coastal zone.

The following definitions of coastal stability shall apply:

High stability areas (1) less than 1 foot per year historic cliff retreat,
(2) inherently stable cliff material, and
(3) not dependent upon a beach for its stability.

In high stability areas, any development proposed within the area from the toe of the bluff to a point on top of the bluff at a 1:1 (45°) slope from the toe must demonstrate stability as defined above (with a geologic engineering report).

Moderate stability areas (1) less than 1 foot per year historic cliff retreat,
(2) inherently unstable cliff material, and
(3) may be dependent upon a fronting beach for stability.

In moderate stability areas, any proposed development within the area of 2:1 (30°) slope from the toe to the top of the bluff must demonstrate stability as defined above.

Low stability areas (1) greater than 1 foot per year historic cliff retreat, or
(2) landslides or other inherently unstable material (such as beach sand or active dunes).

In low stability areas, any proposed development must be excluded from the area of 1:1 (45°) slope from toe to top of bluff, and from the area of active movement, and stability must be demonstrated for a 50 year economic life within the remaining area of 2:1 (30°) slope.

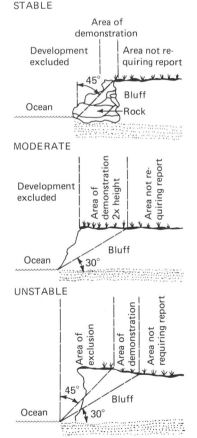

Figure 6.28. *Sea Cliff Construction Setbacks and Guidelines as Determined by Erosion Rate and Cliff Morphology.* REDRAWN FROM: *Central Coast Regional Commission — California Coastal Zone Conservation Commission, Geology: Coastal Geology and Geological Hazards (Santa Cruz, Calif. Central Coast Regional Commission, 1974).*

retreat. In marginal areas, a prospective builder must demonstrate through a geologic report that the cliff will remain stable during the fifty-year economic life of the structure. The report must discuss in detail the rate of erosion, the geometry of the cliff, the geologic properties of rock and soil materials, and the various forces acting on the cliff (California Coastal Zone Conservation Commission, 1974).

TSUNAMIS

A number of large earthquakes and explosive volcanic eruptions in the past have generated destructive seismic sea waves or tsunamis that have inundated populated coastal areas (see chapters 2 and 3). The catastrophic eruption of the Aegean island of Santorini (Thera) in about 1450 BC was accompanied by the extrusion of at least 40 cubic kilometers (10 cubic miles) of material and the generation of seismic sea waves that were propagated throughout much of the Mediterranean. The water-borne pumice and ash from the eruption was carried to elevations of 250 meters (820 feet) on the adjacent island of Anaphi, and the waves are believed to have still been 6 or 7 meters in height when they reached the coast of Israel (Galanopoulos and Bacon, 1970). The effect of such waves on early civilizations along the Mediterranean coast must have been disastrous. A great earthquake in Lisbon, Portugal, in 1755 was accompanied by destructive waves that washed ashore along the coasts of Portugal, Spain, and Morocco. Water levels in Lisbon were elevated 5 meters above high tide and about 60,000 deaths resulted from the earthquake and tsunami.

Although tsunamis are rarely noticed at sea, they are usually quite visible when they reach the coastline. As the waves reach the shallow water around islands or on the continental shelf, their speeds decrease and their heights increase to perhaps as much as 25 meters. The destructive effect of a tsunami is, therefore, controlled to a considerable degree by the offshore submarine topography and the orientation and elevation of the coastline.

Hilo, Hawaii, for example, due to its location and the surrounding coastal configuration, is a point where wave energy is concentrated or focused, producing repeated damage from tsunamis (see figure 6.29). Hilo is located at the head of a large bay that is open to the northeast. As a result, waves from a large Alaskan earthquake in 1946 were very devastating to the city (see figure 6.30). Most of a breakwater was destroyed as water rose to heights of 8 meters above normal. On the open coast water levels were raised as much as 16 meters (see figure 6.29), but because of the heavily populated coastal sections of Hilo, the waves caused the greatest damage here. Almost every house on the side of the main street facing Hilo Bay was smashed against buildings on the opposite side of the street. A steel railroad bridge across the Wailuku Estuary was broken and a span was carried 300 meters upstream. Houses were overturned; railroad tracks were ripped from their roadbeds; coastal highways were buried and beaches were washed away. The catastrophe, which came without warning, cost the Hawaiian Islands 159 lives and $25 million in property damage.

Waves from a large earthquake centered in Chile in 1960 rose as high or higher in Hawaii than those of 1946 and killed sixty-one persons. Significantly, in 1960, despite the siren warning system that had been in-

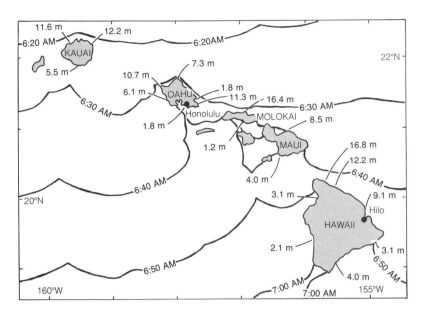

Figure 6.29. *Wave Fronts from the Alaskan Tsunami of 1946 at Hawaiian Islands. Note the refraction around the islands and exposure of Hilo. Maximum heights in meters are shown for particular locations.* REDRAWN FROM: *Geology, Resources, Society, by H. W. Menard. W. H. Freeman and Company. Copyright* © *1974.*

Figure 6.30. *Tsunami Damage in Hawaii, 1946.* PHOTO COURTESY: Honolulu Star Bulletin.

stalled by the U.S. Coast and Geodetic Survey, few people moved from the danger area. The warning was taken as a false alarm, as had occurred on several previous occasions when no appreciable waves developed (Shepard and Wanless, 1971). Although maximum damage from the 1946 tsunami occurred in Hawaii, the waves reached their greatest heights in the Aleutians. At Scotch Cap, Alaska, a reinforced concrete lighthouse 30 meters above sea level was destroyed. Along the California coast the tsunami rose over 3 meters above still water at Half Moon Bay, about 50 kilometers south of San Francisco, resulting in slight damage (Shepard, 1963).

Japan, an extremely seismic region, has suffered extensively from repeated tsunami inundation (Bolt et al., 1975). An earthquake in Japan in 1707 generated such large waves in the inland sea that more than 1000 ships and boats in Osaka Bay were swamped. In 1896 a tsunami struck the northeast coast of the Japanese island of Honshu. In the course of five minutes this wave destroyed 7,600 houses; 27,000 people lost their lives and 5,000 were injured; 18 boats and a vessel of 200 tons were hurled more than 450 meters inland from the coast.

The West Coast of the contiguous United States had little damage from seismic waves until the 1964 Alaskan earthquake. The general lack of damage is due in part to the long diagonal approach of waves to the West Coast coming from earthquakes either in the Aleutian, Middle America, or Peru-Chile Trenches. The result is a loss of energy as the waves cross the extensive stretches of shallow water of the continental shelf before reaching the shoreline. As a result of the Alaskan earthquake in 1964, however, several waves almost 4 meters in height seriously damaged the waterfront area at Cres-

cent City in northern California. One wave washed inland about 500 meters, essentially destroying twenty-nine blocks of the business district and creating losses estimated at $27 million (see figure 6.31). A number of fishing boats in the harbor were also damaged. On the Oregon coast the same tsunami caused $250,000 property loss at Seaside and Cannon Beach and drowned four persons at Depoe Bay (Shepard and Wanless, 1971).

Restrictions on future building in low-lying coastal regions that are particularly vulnerable to seismic sea waves are necessary to prevent additional or recurring damage in areas subject to inundation (see figure 6.32a and b). The first logical step in such an effort is the identification of those regions subject to tsunami runup and the frequency of occurrence and magnitude of such phenomena. Long-term records from tide gauges can be used to determine the recurrence and elevations of past tsunamis just as records of river stage can be utilized in flood frequency analysis. Over 100 years of records at the entrance to San Francisco Bay, for example, indicate that since 1868, nineteen tsunamis have been recorded with heights varying from a few centimeters to 1.2 meters (Bolt et al., 1975). Within the bay, amplitudes have been even less. The records, then, indicate the hazard in this area from tsunami inundation in the past has been relatively insignificant. Study of recurrence intervals for tsunamis in Hilo, Hawaii, in contrast, indicate that once every 100 years, on the average, a runup of 10 meters is expected. In eastern Honshu, Japan, the same runup could be expected once every 10 years.

An adequate warning system can help in eliminating loss of life from these disastrous waves. Following the destructive 1946

Figure 6.31. *Tsunami Damage to Crescent City, California, Following the 1964 Alaska Earthquake. Total damage in Crescent City reached $27 million as twenty-nine blocks of the business district were destroyed.* Photo courtesy: *U.S. Army Corps of Engineers.*

Figure 6.32a. *Tsunami Damage at Seward, Alaska, 1964. Houses and other debris were carried by waves into the lagoon area at the north end of town.* Photo by: *U.S. Army, U.S. Geological Survey.*

Figure 6.32b. *Tsunami Damage at Seward, Alaska, 1964. Heavy damage occurred in the railroad yards of Seward during the 1964 tsunami.* PHOTO BY: *U.S Army, U.S. Geological Survey.*

Hawaiian tsunami, a warning system was developed that relies on both seismographs, which are located around the Pacific, and detectors, which record the rise and fall of sea level and can pick up long-period seismic sea waves. This system is interconnected around the Pacific Basin so that it is possible to estimate the rate of approach and probable arrival time of tsunamis. Warnings then can be issued and evacuation carried out. During a 1952 seismic sea wave, the warning system was credited with preventing any loss of life and reducing property damage in Hawaii. Nevertheless, $800,000 in damage still occurred. It has become evident that a warning system in itself, no matter how efficient, is not enough. The failure to heed the warning sirens in Hawaii in 1960 and the resultant deaths are sad testimony to the problem. In addition, property damage in low-lying areas susceptible to tsunamis will continue to occur until land usage is changed and rebuilding prohibited.

STABILIZATION AND CONTROL OF COASTLINES

Coastal erosion is due to many interrelated factors and processes. The variable success of antierosion measures in the United States is partly a response to the basic complexities inherent in the dynamic nature of the erosion processes (Mitchell, 1974). Beaches, dunes, and cliffs are temporary geological features that respond to even subtle changes in the marine energy regime. The recession and **progradations** form part of the normal pattern of coastal development. Successful erosion control or coastal stabilization is often hampered by a lack of knowledge about marine processes. The long-term development of barrier islands, the sediment budgets for individual beaches, and the future of dune migration are examples of areas where knowledge is often lacking.

Because of spatial and temporal variations in coastal erosion and change it is difficult to extrapolate beyond individual study areas. Measurements taken at one location may not be at all representative of erosion of an adjacent property.

Deficiencies in existing erosion control programs constitute an additional problem. Individual property owners or even small municipalities do not normally have either the technical knowledge or financial capability to deal with the problem. Collective action and/or federal government involve-

ment has usually been the solution. The U.S. Army Corps of Engineers has been the federal agency involved and the poor planning and inadequate knowledge of this organization in the past have often resulted in projects that have created additional problems.

A number of methods and structures have been utilized to control sea cliff erosion or coastal retreat in developed areas and also to provide areas protected from wave attack (see figures 6.33 and 6.34; U.S. Army Corps of Engineers, 1966 and 1971).

Figure 6.33. *Sea Cliff Protection. A combination of riprap (right) and a homemade seawall (left) topped with a reflecting slope to prevent overtopping have been used to protect a beach front residence.* PHOTO BY: *Gary Griggs.*

Figure 6.34. *Protection of a Sea Cliff. A visually unpleasant combination of riprap, a timber bulkhead, and a concrete covering of the sea cliff have all been used by a single homeowner to protect the property.* PHOTO BY: *Gary Griggs.*

Riprap

Where protective beaches are either narrow, seasonal, or nonexistent, and property is disappearing or endangered, the emplacement of riprap (large blocks of resistant rock) at the base of a sea cliff has been a common solution (see figure 6.35). The rock mass usually absorbs the incident wave energy and may provide an effective buffer to wave attack. Wave action may erode the sea cliff around and ultimately behind riprap, however, if the riprap is placed in a discontinu-

Figure 6.35. *Protective Riprap. A retreating coastline has been protected by placing riprap at the base of the cliff.* PHOTO BY: *Gary Griggs.*

ous fashion. Although somewhat expensive, this is one of the most common methods utilized by individual property owners. With time riprap may settle or move around and then must be replaced.

Seawalls

Seawalls provide similar protection (see figure 6.36). These are normally concrete structures and vary in design depending upon local environmental conditions. They are built parallel to the beach where protection of a sea cliff, buildings, or roads from wave action is desired. The exposed face of a seawall may be vertical, convex, concave, stepped, or some combination of these. Each design has its own advantages and disadvantages, which must meet the needs of a particular site. Seawalls act both to absorb energy in the form of wave runup and to reflect wave energy directly. Reflected waves may increase erosion adjacent to the seawall, however.

Groins

Groins are shore protection structures designed to build a protective beach or prevent erosion of an existing beach by trapping littoral drift. They are usually built perpendicular to the shore and may be up to 50 meters or more in length. Because a groin poses a total or partial barrier to littoral drift, the extent to which littoral transport is altered depends upon the height, length, and permeability of the groin. Although single groins can be constructed, they are usually built in groups at some regular spacing to maintain or construct a beach along an extensive stretch of coastline (see figure 6.37).

Figure 6.36. *A Protective Curved-Face Seawall Along the San Francisco Coastline. This type of structure allows wave energy to be reflected rather than expended directly against the sea cliff or coastline.* PHOTO COURTESY: *U.S. Army Corps of Engineers.*

Figure 6.37. *Stabilization and Buildup of a Beach by a Series of Groins.* PHOTO COURTESY: *U.S. Army Corps of Engineers.*

If downcoast beaches are narrow and littoral drift rates are low, then it may be necessary to place sand between the groins artificially. Groins should not be built unless properly designed for the particular site and until the effects on adjacent beaches have been adequately considered.

Jetties

Jetties are commonly built in pairs perpendicular to the shore to protect a channel entrance from wave action, to direct and confine a stream or tidal flow to a selected channel, and ideally to prevent or reduce shoaling of a channel by littoral material. Jetties are usually of rubble mound or rock construction, but have also been built of timber or steel pilings or sheeting. The proper siting and spacing of jetties for improvement of harbor entrances are critical. The direction and strength of tidal currents, the channel section needed for navigation, the tidal volume of the harbor or basin, and the rate and direction of littoral drift all need to be given careful consideration in planning jetty construction.

Jetties, because of their length, initially impose a total littoral barrier to sand transport. Accretion equivalent to the littoral drift rate occurs updrift from the structure, and beach erosion will occur downdrift at the same rate. The total sand accumulation will depend upon the length and orientation of the jetties relative to natural forces. Planning for jetty construction where littoral drift rates are large should include some method of bypassing the sand to eliminate or reduce channel shoaling and erosion of the downdrift coast.

The jetties associated with the construction of the Santa Cruz, California, small craft harbor present a classic example of these effects (see figure 6.38). During preliminary studies for the harbor, the U.S. Army Corps of Engineers determined that the dominant littoral drift was in an easterly direction and estimated that the net annual drift rate was somewhere between 20,000 and 230,000 cubic meters. It stated that jetties would form littoral barriers, and if the net annual rate of littoral transport approached 230,000 cubic meters, erosion would be rapid and continuous downcoast. In this event, a sand bypass plant or dredging system would be needed. The decision was made to defer this, however, until its need was demonstrated.

During the first two years following harbor construction, a total of 460,000 cubic meters of sand built up along the coast next to the west jetty (Moore, 1972). The cliffs backing this beach had formerly been exposed to direct winter wave attack and erosion had removed major sections of a city street and was threatening residences atop the cliff. The littoral barrier formed by the jetty has resulted in a wide sandy year-round beach, which now protects the sea cliff immediately upcoast (see figure 6.39). The littoral drift proceeded to work its way along and around the west jetty and into the channel entrance. If a channel is to naturally maintain itself a certain relationship must exist between the tidal volume in an embayment or harbor and the cross-sectional area of the channel (O'Brien, 1931).

With a very large coastal embayment and a narrow entrance channel, there is sufficient velocity reached during ebb tide to remove the sediment and maintain the channel. However, where the tidal volume is relatively small and the entrance channel is relatively wide, as with the Santa Cruz Harbor, there simply is not enough water moving fast enough through the channel during each

Figure 6.38. *The Obstruction of the Littoral Drift of Sand by Jetties at Santa Cruz, California. The loss of sand, whether temporary of permanent, will starve downcoast beaches and usually increase erosion rates.* PHOTO COURTESY: *U.S. Army Corps of Engineers.*

Figure 6.39. *Destruction of a Road Along the Sea Cliff from Wave Erosion Prior to the Buildup of a Protective Beach. Note the jetties of the harbor that trapped the sand to form the protective beach.* PHOTO BY: *Gary Griggs.*

tidal cycle to maintain it deep enough for navigation. The shoaling and subsequent annual dredging of the entrance channel has borne this out. The channel has had to be dredged annually and in the period from 1965 to 1974, 570,000 cubic meters of sand have been removed at a cost of over $1,750,000. This sanded entrance channel has also led to an unusable or extremely hazardous harbor for three to four months each year. For two years following harbor completion sand flow was cut off altogether. In subsequent years, some sand has worked its way around the jetties to continue on downcoast, and about 30 percent of the littoral drift is trapped in the harbor each winter. This is then dredged out and added to the downcoast beach the following spring. The beaches, therefore, are partially starved during the winter months when wave action is more severe. Erosion has been occurring along the cliffs in this area for many years. Due to the emplacement of protective riprap, erosion rates along portions of this coast have declined since harbor construction. Where no protection was added, rates increased from 27 to 39 centimeters per year to 39 to 75 centimeters per year, a doubling in some instances.

Further downcoast, aerial photographs indicate a moderate to wide beach had nearly always existed at the town of Capitola prior to harbor construction at Santa Cruz. However, since that time, the beach has either been greatly reduced or is nonexistent, resulting in waves attacking the coastline directly. Cliff retreat has also accelerated in this area (see figure 6.27). Capitola is about 5.5 kilometers downcoast from the harbor and it seems likely that an interruption in littoral drift was the most significant factor in the disappearance of its

beach. Ultimately, a groin was constructed and approximately 2000 truckloads of sand were brought in to rebuild the beach at considerable expense. It seems clear here and in many other places that coastal engineering structures can have major and costly effects along extensive coastal areas.

Breakwaters

Breakwaters are structures built to protect a shore area, harbor, or anchorage from wave action. They can be either offshore or connected to the shoreline and may be constructed of rubble mound, concrete, sheet piling, or perhaps some combination of these. The most important factor in siting a breakwater is the determination of the optimum location to produce a harbor area with minimum wave and surge action over the greatest period of time throughout the year. The direction and magnitude of littoral drift are also important factors to consider in any breakwater construction.

Regardless of the placement and orientation of a breakwater, it is going to affect the coastal area in several ways. The initial or primary effect is one of greatly reducing or nearly eliminating wave action within the protected area; this produces the secondary effect of interruption of littoral drift patterns.

At Santa Monica, California, a detached offshore breakwater was constructed to provide a protected area for boat anchorage (see figure 6.40). The structure was built parallel to the coastline with the hope that littoral drift would carry the sand through the harbor to the downcoast beaches. However, a wave shadow was formed behind the breakwater and there was insufficient energy to move the sand through the harbor. The sand

Figure 6.40. *Breakwater at Santa Monica, California. Note the Widening of the Beach in the Wave Shadow Formed behind the Offshore Breakwater.* PHOTO COURTESY: *U.S. Army Corps of Engineers.*

began to widen the beach shoreward of the breakwater, which threatened to fill the harbor and to cause downcoast beach losses. As a result dredging and piping of the sand to the downcoast beaches have been required.

A completely different type of breakwater constructed at Santa Barbara, California, also affected the coastal equilibrium in several major ways. Initially (from 1927 to 1928) a detached breakwater was constructed, similar to that at Santa Monica, which experienced a very similar sand buildup. Because of shoaling, in 1930 the breakwater was extended to shore by an additional 180 meters of structure (see figure 6.41; Wiegel, 1964). The breakwater proved to be an effective trap for the sand moving eastward in the littoral zone. It first progressively filled a large embayment west of the breakwater (a college

stadium and parking lot has been subsequently built on this sand fill, which extended the beach seaward over 300 meters from the original coastline). In succeeding years the sand traveled down the breakwater, swung around its tip, and eventually began to fill the harbor. The first dredging began in 1935, five years after the breakwater had been completed. The harbor must now be dredged continually because of the littoral drift trap that was created. The sand is pumped downcoast where it again becomes part of the littoral flow. At present the dredge is committed to remove about 270,000 cubic meters of sand per year at a cost of $128,000.

Another breakwater with a different configuration was constructed from 1959 to 1961 to shield Half Moon Bay Harbor, about

Figure 6.41. *Breakwater at Santa Barbara, California. Note the Accumulation of Sand at the end of the Breakwater. The Oval Stadium and Highway to the Left of the Harbor Have Been Constructed Totally on Beach Sand Trapped by the Breakwater.* PHOTO COURTESY: *U.S. Dept. of Agriculture.*

50 kilometers south of San Francisco, from southwest swells (see figure 6.42). The breakwater construction disturbed the preexisting equilibrium conditions, which has led to increased erosion rates downcoast (Tinsley, 1972). Waves, whose energy was formerly refracted into and dissipated in the northern end of Half Moon Bay, are now reflected from the breakwater into the shoreline immediately south of the breakwater's southern terminus. Net southward transport of sand has been interrupted by movement into the harbor. The beach south of the breakwater has been narrowed so that wave action often attacks the low sea cliff directly. The sea cliffs here are only 2 to 4

meters high and consist of unconsolidated marine terrace deposits. The result has been rapid erosion of the coast here, including the destruction of a road and a bridge and the threatening of houses (see figure 6.43). The average erosion rates south of the breakwater have increased from about 30 centimeters per year (average for the period from 1914 to 1959) to values of about 150 centimeters per year in subsequent years (Tinsley, 1972). Although some riprap has been placed along the beach, it has been done in an incomplete and haphazard manner, so that wave action is now cutting away at the cliff behind the riprap.

Thorough preconstruction studies of lit-

Figure 6.42. *Breakwater at Half Moon Bay, California. Note sand buildup within harbor. Waves reflected off the breakwater in the background have led to increased erosion rates.* PHOTO COURTESY: *U.S. Army Corps of Engineers.*

Figure 6.43. *Area of Accelerated Erosion Adjacent to Half Moon Bay Breakwater. Rates of coastal retreat now reach 1.0 to 1.5 meters per year here (see figure 6.17 for an aerial view of this same area).* PHOTO BY: *Gary Griggs.*

toral drift direction and rate, wave and tidal action, and the utilization of physical models all are essential in regard to any major coastal engineering projects. By now we realize that virtually anything we construct in the littoral or near-shore zone is going to have some effect on the often delicate equilibrium that exists here, and that we have the capacity to affect significantly both depositional and erosional conditions. These environmental effects and secondary costs (i.e., dredging operations, emplacement of protective riprap, and rebuilding roads) must be considered at an early stage in any proposal and must be recognized as consequences of such projects. In some instances, modification in the orientation of structures, or their precise placement along the coast, may tend to minimize their effects. If jetties or breakwaters are constructed upcoast from rocky headlands there may be no immediate downcoast beaches to be affected. This situation exists in the case of Newport Harbor south of Long Beach, California. In other instances, the construction of a permanent sand bypass or dredging system, which will transport the littoral drift on a continuous basis across a harbor mouth or entrance channel may eliminate or minimize future problems. In any case, it is certainly time to start utilizing our past experience in future planning.

REFERENCES

References Cited in the Text

Bolt, B. A. et al. *Geological Hazards.* New York: Springer-Verlag, 1975.

Brown, W. M. III, and Ritter, J. R. *Sediment Transport and Turbidity in the Eel River Basin, California.* U.S. Geological Survey Water Supply Paper 1986, 1971.

California Coastal Zone Conservation Commission. *Geology: Coastal Geology and Geological Hazards.* Santa Cruz, Calif.: Central Coast Regional Commission, 1974.

Coe, J. J. "Searching for California's Inland Sand Sources." Paper presented at American Shore and Beach Preservation Association Meeting, San Diego, Calif., 1966.

Cooper, W. S. *Coastal Sand Dunes of Oregon and Washington.* Memoir 72. Boulder, Colo.: Geological Society of America, 1958.

Emery, K. O. *The Sea off Southern California.* New York: John Wiley and Sons, 1960.

Fay, R. C. "Southern California's Deteriorating Marine Environment." Center for California Public Affairs, Claremont, Calif., 1972.

Galanopoulos, A. G., and Bacon, E. *Atlantis: The Truth Behind the Legend.* Indianapolis: Bobbs Merrill, 1970.

Inman, D. L., and Frautschy, J. D. "Littoral Processes and the Development of Shorelines." In *Coastal Engineering.* American Society of Civil Engineers, 1966.

Kennedy, M. D. "Seacliff Erosion at Sunset Cliffs, San Diego, California." *California Geology* 26(1973): 27–31.

Larsen, C. E. *Variation in Bluff Recession in Relation to Lake Level Fluctuations along the High Bluff Illinois Shore.* Document 73-14. Illinois Institute for Environmental Quality, 1973.

McHarg, I. *Design with Nature.* Garden City, N.Y.: Natural History Press, 1969.

McMaster, R. L. "Petrography and Genesis of New Jersey Beach Sands." *State of New Jersey Department of Conservation Geology Bulletin* 63(1954): 239.

Mitchell, J. G. "Community Response to Coastal

Erosion." University of Chicago, Department of Geography, Research Paper 156, 1974.

Moore, J. T. "A Case History of Santa Cruz Harbor, California." University of California, Berkeley, Hydraulic Engineering Laboratory, Publication HEL 24-24, 1972.

Morton, R. A. "Effects of Hurricane Eloise on Beach and Coastal Structures, Florida Panhandle." *Geology* 4(1976): 277–80.

North, W. B., and Byrne, I. V. "Coastal Landslides of Northern Oregon." *The Ore Bin* 27(1965): 217–41.

O'Brien, M. P. "Estuary Tidal Prisms Related to Entrance Areas." *Civil Engineering* 1(1931): 738–39.

Ritter, J. R. "Sand Transport by the Eel River and Its Effect on Nearby Beaches." USGS, Menlo Park, California, Open File Report, 1972.

Shepard, F. D. *Submarine Geology*. New York: Harper and Row, 1963.

Shepard, F. D., and Wanless, H. R. *Our Changing Coastlines*. New York: McGraw-Hill Book Company, 1971.

Strahler, A. W., and Strahler, A. H. *Environmental Geoscience*. Santa Barbara, Calif.: Hamilton Publishing Company, 1973.

Terich, T. A. "The Retreating Shore." *Pacific Northwest Sea* 8(1975): 4–7.

Tinsley, J. "Sea Cliff Retreat as a Measure of Coastal Erosion: San Mateo County, California." *Guidebook for Friends of the Pleistocene San Francisco Bay Area Field Trip.* Menlo Park, Calif.: U.S. Geological Survey, 1972.

Trask, P. D. "Movement of Sand Around Southern California Promontories." Beach Erosion Board Technical Memorandum no. 76, 1955.

U.S. Army Corps of Engineers. *Shore Protection, Planning and Design*. Washington, D.C.: Government Printing Office, 1966.

———. *Shore Protection Guidelines*. Washington, D.C.: Government Printing Office, 1971.

Ward, M., and Sorensen, R. M. "A Method of Tracing Sediment Movement on the Texas Gulf Coast." Texas A & M University, Sea Grant Publication TAMU-SG-71-204, 1970.

Wiegel, R. L. *Oceanographical Engineering*. Englewood Cliffs, N.J.: Prentice-Hall, 1964.

Other Useful References

Bascom, W. *Waves and Beaches*. Garden City, N.Y.: Doubleday, 1964.

Byrne, J. V. "An Erosional Classification for the Northern Oregon Coast." *Annals of the Association of American Geographers* 54(1964): 329–35.

Coates, D. R., ed. *Coastal Geomorphology*. Publications in Geomorphology. Binghamton, N.Y.: State University of New York, 1973.

Dolan, R. "Man's Impact on the Barrier Islands of North Carolina." *American Scientist* 61(1973): 152–62.

Eaton, J. P.; Richter, D. H.; and Ault, W. V. "The Tsunami of May 23, 1960, on the island of Hawaii." *Seismological Society of America* 51(1961): 135–57.

El-Ashry, M. T. "Causes of Recent Increased Erosion along United States Shorelines." *Geological Society of America Bulletin* 82(1971): 2033–38.

Griggs, G. B., and Johnson, R. E. "Effects of the Santa Cruz Harbor on Coastal processes of Northern Monterey Bay, California." *Environmental Geology* 1 (1976): 299–312.

Williams, G. P., and Guy, H. P. *Erosional and Depositional Aspects of Hurricane Camille in Virginia, 1969.* U.S. Geological Survey Professional Paper 804, 1973.

CHAPTER 7

Surface Hydrology
and Flooding

Contents

INTRODUCTION

FLOODING is the most widespread geological hazard in the United States and accounts for greater average annual property losses than any other single hazard. Despite the construction of ever-increasing numbers of dams for "flood control" purposes, losses from flooding have continued to increase in recent years due primarily to expanded use, reoccupation, and development of downstream floodplains. Although dams on mainstreams and tributaries of major rivers throughout the nation stand as massive monuments to the flood control policies established in the 1930s, those policies have been seriously questioned in recent years. Overflowing rivers and streams still cause significant flooding in about half of the communities and over at least 7 percent of the total land area of the United States (White and Haas, 1975).

Virtually all stages of the natural **hydrologic cycle**, including flooding, have been significantly altered by human intervention.

Changes in land use, from agriculture to urbanization, have had profound effects on **runoff** and **erosion** of the land surface. As vegetation is removed and the soil is exposed during construction, erosion rates may increase over 100-fold. The creation of impermeable surfaces that accompany urbanization increases and concentrates runoff, leading to a greater incidence of flooding. People and their activities have obviously become major elements in the hydrologic cycle capable of altering natural systems.

Although human-induced changes to the hydrologic system produce consequences that are often interrelated, they are considered in this chapter separately. Storm water contaminants, erosion, and sedimentation are the first effects of the misuse of land, usually in the pursuit of timber harvesting and urbanization, while flooding is normally more disastrous but derives both from misuse and natural causes.

THE HYDROLOGIC CYCLE

Water is in constant motion, over and under the land surface, through the air, and within the oceans. An entire hydrologic cycle traces the flow of water in its solid, liquid, and vapor states through its various pathways and reservoirs (see figure 7.1). To give us some perspective on the distribution of water as we look at the cycle, we should know that 97 percent of the earth's water is contained in the oceans and is salty, leaving only 3 percent fresh. Of that fresh water, polar ice and glaciers contain 75 percent, groundwater constitutes about 20 percent, lakes and rivers only 4 percent, and moisture in the soil and atmosphere the remaining 1 percent. It is quite clear that our existing water resource efforts have been directed primarily toward the groundwater system and the water available in rivers and lakes, which together total less than 1 percent of the earth's available water. With so little water at our disposal, and with the growing demands on it, it is imperative that we do not misuse it.

In the study or analysis of any individual area for the purpose of planning, one begins by looking at the water balance, or budget, which involves the inflow, outflow, and storage of water. Inflow includes precipitation of all types — surface water inflow, groundwater inflow, and importation. Outflow includes evaporation and transpiration, runoff of both ground and surface water, and exportation. Storage involves surface water storage (in a lake or reservoir), groundwater storage, soil moisture, and snow or ice cover.

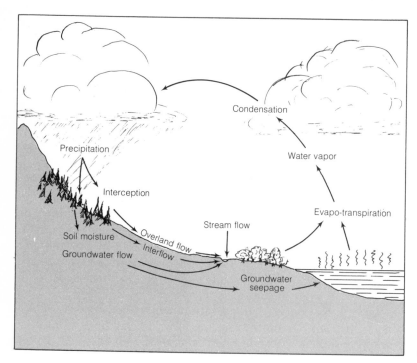

Figure 7.1. *A Generalized Hydrologic Cycle.*

Precipitation or Inflow

Precipitation is one of the most important stages of the hydrologic cycle and includes any moisture falling from the clouds, including rain, snow, hail, or sleet. The United States, on the average, receives 76 centimeters (30 inches) of precipitation per year, but 53 centimeters (21 inches) is returned to the atmosphere directly due to surface evaporation and plant **transpiration**. Thus one-third of the average total precipitation (23 centimeters) reaches the soil surface and begins to infiltrate. The rate of **infiltration** depends upon factors such as the type of soil and its permeability. When the soil becomes **saturated**, or the precipitation intensity exceeds the **infiltration capacity**, surface detention of water begins to occur with subsequent downslope flow.

Precipitation rates vary tremendously depending upon latitude, altitude, slope, orientation, and wind. Parts of the Atacama Desert in Chile, for instance, have apparently received no rainfall during historic time; parts of northern India (Cherapunjii), at the base of the Himalayas, have experienced over 2600 centimeters (1040 inches) of rainfall during a single year. Geologists, hydrologists, and engineers involved with the hydrologic cycle are concerned with how precipitation affects the flow of water, either on or below the ground surface. Precipitation is analyzed in terms of its:

1. depth: the magnitude of the rainfall

2. area: the region over which the precipitation occurs

3. duration: the time period during which the precipitation occurs

4. intensity: the rate of precipitation in centimeters per hour.

Each of these factors affects the runoff, both stream flow and groundwater, and must therefore be considered. If data from individual rain-gauging stations exist for an area, estimates can be made of the total amount of precipitation that could fall within a given period of time and the subsequent runoff that must be planned for. These determinations are not easy to make, due in part to the inadequate distribution of precipitation data, and therefore, the necessity of extrapolating over considerable areas. The most accurate method involves the plotting of contours of equal precipitation (**isohyets**) utilizing the existing precipitation data and the topography and then computing the areas involved between contours (see figure 7.2). In addition to the deficiency of precipitation data and the variability of precipitation over space, the time variations and representativeness of the data are functions of the length of available records.

For many hydrologic problems, an analysis of the temporal as well as the areal distribution of storm precipitation is necessary. Depth-area-duration compilations, for example, will indicate the maximum rainfall that has fallen during certain time periods over various size areas (see table 7.1). The average time interval between storms of a certain size, or between any events of a given magnitude, is called the **recurrence interval** for that event. If historic data exists, average recurrence intervals between major storms or major floods can easily be determined (see p. 300).

Engineers working on certain hydrologic projects, such as street drains and culverts, for example, need to know the recurrence intervals for storms of various precipitation intensities to adequately design the systems. Depending upon the magnitude of the project (e.g., a storm drain or a major dam), en-

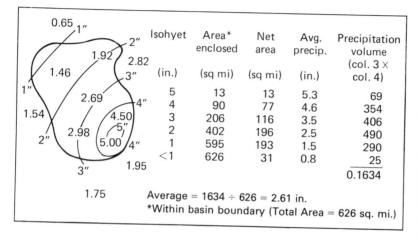

Isohyet	Area* enclosed	Net area	Avg. precip.	Precipitation volume (col. 3 × col. 4)
(in.)	(sq mi)	(sq mi)	(in.)	
5	13	13	5.3	69
4	90	77	4.6	354
3	206	116	3.5	406
2	402	196	2.5	490
1	595	193	1.5	290
<1	626	31	0.8	25
				0.1634

Average = 1634 ÷ 626 = 2.61 in.
*Within basin boundary (Total Area = 626 sq. mi.)

Figure 7.2. *Averaging Precipitation Using Isohyetal Method. Using this technique to determine the total precipitation over a drainage basin involves (1) drawing contours (isohyets) of equal precipitation, (2) determining the areas between contours, (3) multiplying the areas by the average precipitation between contours, and (4) adding these values.* REDRAWN FROM: *R. K. Linsley, M. A. Kohler, and J. L. H. Paulhus, Hydrology for Engineers (New York: McGraw-Hill Book Company, 1958), p. 35.*

Table 7.1. *Maximum Depth-Area-Duration Data for the United States (Average precipitation in inches)*

Area, sq mi	Duration, hr						
	6	12	18	24	36	48	72
10	24.7a	29.8b	35.0b	36.5b	37.6b	37.6b	37.6b
100	19.6b	26.2b	30.7b	31.9b	32.9b	32.9b	37.6b
200	17.9b	24.3b	28.7b	29.7b	30.7b	31.9c	35.2c
500	15.4b	21.4b	25.6b	26.6b	27.6b	30.3c	34.5c
1,000	13.4b	18.8b	22.9b	24.0b	25.6d	28.8c	33.6c
2,000	11.2b	15.7b	19.5b	20.6b	23.1d	26.3c	32.2c
5,000	8.1bj	11.1b	14.1b	15.0b	18.7d	20.7d	29.5c
10,000	5.7j	7.9k	10.1e	12.1e	15.1d	17.4d	24.4d
20,000	4.0j	6.0k	7.9e	9.6e	11.6d	13.8d	21.3d
50,000	2.5eh	4.2g	5.3e	6.3e	7.9e	8.9e	17.6d
100,000	1.7h	2.5ih	3.5e	4.3e	5.6e	6.6f	11.5f
							8.9f

Storm	Date	Location of Center
a	July 17–18, 1942	Smethport, Pa.
b	Sept. 8–10, 1921	Thrall, Texas
c	Aug. 6–9, 1940	Miller Island, La.
d	June 27–July 1, 1899	Hearne, Texas
e	March 13–15, 1929	Elba, Ala.
f	July 5–10, 1916	Bonifay, Fla.
g	April 15–18, 1900	Eutaw, Ala.
h	May 22–26, 1908	Chattanooga, Okla.
i	Nov. 19–22, 1934	Millry, Ala.
j	June 27–July 4, 1936	Bebe, Texas
k	April 12–16, 1927	Jefferson Parish, La.

SOURCE: R. K. Linsley, Jr.; M. A. Kohler; and J. L. H. Paulhus, *Hydrology for Engineers* (New York: McGraw-Hill Book Company, 1958), p. 44.

Table 7.2. *Typical Flood Recurrence Intervals Used in Designing Various Engineering Projects*

Project	Flood Recurrence Interval Used in Design
Streets and Storm Drains	2–5 years
Check and Coffer Dams	5–25 years
Highway Culverts	10–50 years
Small or Diversion Dams	50–100 years
Levees	100 years
Large Dams	Not specific — perhaps 200 years + a safety factor

gineers will plan for a certain recurrence interval (*see* table 7.2). Simple economics prevents every project from being built to withstand all storm conditions. There is always the possibility of an intense storm due to certain persistent meteorological conditions that can leave any area or project in ruins (*see* table 7.3 for some record rainfall intensities).

Outflow

Precipitation that does not infiltrate, or penetrate, the groundwater system begins to flow downslope and soon collects in stream

Table 7.3. *World's Greatest Observed Point Rainfalls (After Jennings*)*

Duration	Depth, in.	Station	Date
1 min	1.23	Unionville, Md.	July 4, 1956
15 min	7.80	Plumb Point, Jamaica	May 12, 1916
42 min	12.00	Holt, Mo.	June 22, 1947
2 hr, 10 min	19.00	Rockport, W. Va.	July 18, 1889
4 hr, 30 min	30.8+	Smethport, Pa.	July 18, 1942
15 hr	34.50	Smethport, Pa.	July 17–18, 1942
24 hr	45.99	Baguio, Philippine I.	July 14–15, 1911
39 hr	62.39	Baguio, Philippine I.	July 14–16, 1911
2 days	65.79	Funkiko, Formosa	July 19–20, 1913
3 days	81.54	Funkiko, Formosa	July 18–20, 1913
4 days	101.84	Cherrapunji, India	June 12–15, 1876
5 days	114.50	Silver Hill Plantation, Jamaica	Nov. 5–9, 1909
6 days	122.50	Silver Hill Plantation, Jamaica	Nov. 5–10, 1909
7 days	131.15	Cherrapunji, India	June 24–30, 1931
8 days	135.05	Cherrapunji, India	June 24–July 1, 1931
15 days	188.88	Cherrapunji, India	June 24–July 8, 1931
31 days	366.14	Cherrapunji, India	July, 1861
2 mo	502.63	Cherrapunji, India	June–July, 1861
3 mo	644.44	Cherrapunji, India	May–July, 1861
4 mo	737.70	Cherrapunji, India	April–July, 1861
5 mo	803.62	Cherrapunji, India	April–Aug., 1861
6 mo	884.03	Cherrapunji, India	April–Sept., 1861
11 mo	905.12	Cherrapunji, India	Jan.–Nov., 1861
1 yr	1041.78	Cherrapunji, India	Aug., 1860–July, 1861
2 yr	1605.05	Cherrapunji, India	1860–1861

* A. H. Jennings, World's Greatest Observed Point Rainfalls, *Monthly Weather Rev.*, Vol. 78, pp. 4–5, January, 1950.

channels. The length, width, and shape of each stream varies according to the precipitation and vegetative patterns, the slopes, soil, underlying bedrock, and geologic structure. The stream flow, or volume of water moving through the channel and its variations, both averages and extremes (i.e., floods), is of considerable importance to the field of water resources. This information is essential to the rational planning of land use along rivers and is equally important in the design of any structures (dams, bridges, reservoirs, and diversions) that directly relate to streams.

The characteristics of a drainage basin have an effect upon stream flow. Where a drainage basin contains soil that is permeable, precipitation will soak into the soil, and runoff will be less than in a basin containing relatively impermeable soil, such as clay. Topography also influences runoff, inasmuch as water flows more readily down

steep slopes than on flat topography. Runoff is further influenced by the number, type, and length of streams in a basin. Stream flow is generally enhanced if there is an adequate number of streams forming a well-integrated drainage system.

With such a variety of factors affecting stream flow, it is evident that runoff is highly variable for various geographic regions and over differing periods of time. Therefore, measurement of stream flow can be somewhat complex. However, knowledge of runoff is an essential element in water supply planning and in flood analysis, and it is important in the understanding of other environmental problems such as landsliding and erosion. Thus, it is valuable for the reader to become familiar with stream-gauging methods (see boxed essay, p. 251).

A discussion of groundwater and its importance to the hydrologic cycle will be presented in chapter 8.

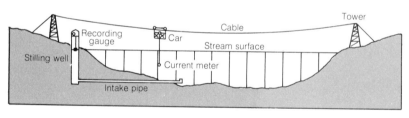

Figure 7.3. *Stream-Gauging Station. To determine the discharge of a stream, the velocity must be measured at a number of points across a section.* REDRAWN FROM: *A. N. Strahler,* Physical Geography, *John Wiley & Sons, Inc.*

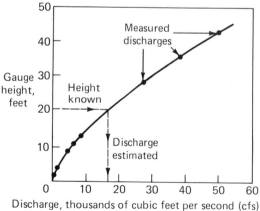

Figure 7.4. *Typical Rating Curve. A rating curve such as this one is utilized to estimate or obtain discharge values for various measured gauge heights.* ADAPTED FROM: *W. G. Hoyt and W. B. Langbein,* Floods *(Copyright 1955 by Princeton University Press): fig. 23, p. 69. Reprinted by permission of Princeton University Press.*

STREAM GAUGING AND HYDROGRAPHS

Stream gauging, or the measurement of stream flow, is a major activity of the U.S. Geological Survey. In conjunction with state and local agencies, the agency records stream flow data at over 8300 sites on major streams across the country. To calculate stream discharge, the mean velocity and cross sectional area of the stream must be determined (discharge =area × velocity). Because stream cross-sections are irregular in their depths and velocities, this involves the division of the stream crossing into a number of measured vertical sections within which velocity determinations are made (*see* figure 7.3). Average velocity can be determined for each section using a current or flow meter and then multiplied by the area to get a discharge for that section. Individual values for each section can then be summed to give a total stream discharge.

The determination of discharge by this method on a continuous basis would involve vast amounts of time. For this reason, a series of measurements are taken at each station at different discharges. These values are then related to stream height or stage (which is automatically recorded) in a stage-discharge graph, or **rating curve** (*see* figure 7.4). This is not a linear plot, as discharge volume usually increases slowly with initial increases in stage. As the channel broadens to an eventual floodplain, a small height increase brings about a huge increase in discharge. Discharge, in cubic feet per second (cfs), is calculated in this fashion and is published regularly in the U.S. Geologic Survey water supply papers for all of the nation's gauged streams. A rating curve for discharge calculations is only valid, however, as long as there are no significant changes in stream geometry at the gauging site, such as scouring during floods, deposition, or effects of water impoundment. Most stream-gauging stations are initially established in relatively stable bedrock sections of channels where possible, or where concrete weirs or low dams are constructed.

Where a well-gauged major stream and its tributaries drain a similar terrain, the relationship between average discharge and the basin area can be established (see figure 7.5). We would expect a simple relationship between an increase in basin area and the increased discharge. By working out such a relationship, a reasonable estimate of discharge could be made at any point knowing the drainage basin area upstream. Any water resource-planning or stream-engineering effort (dams, bridges, or other hydrologic structures) would need such information. The effect of human activities, such as water diversion or impoundment, on any individual gauged basin within such a watershed may well stand out as an anomaly in this basin-area–discharge relationship.

A **hydrograph**, which is simply a graph of either gauge height or stream discharge with time, is usually the most useful way of compiling information on stream flow (see figure 7.6). Height or discharge can be plotted on a daily basis, but for longer periods, monthly or even yearly means may be used instead. The use of the hydrograph determines the intervals selected and whether height or discharge is plotted. A duration or flow-duration curve can also be plotted, which would indicate the percent of time over some extended period when a given

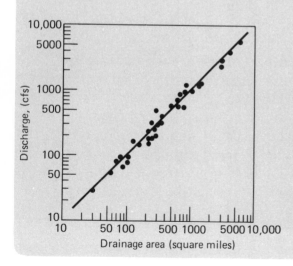

Figure 7.5. *Relation of Mean Stream Discharge to Drainage Basin Area for All Gauging Stations in the Potomac River Basin. Each point represents a gauge.* DATA FROM: *J. T. Hack, U.S. Geological Survey Professional Paper 294-B, (1957), p. 54.*

stage or discharge was equaled or exceeded (*see* figure 7.7). Such a plot is extremely useful in river bank or floodplain planning. For example, if the activity or land use being considered along a river can afford to be flooded 5 percent of the time (a parking lot for instance), it should be built 8.2 meters (25 feet) above the river bottom according to the duration curve. If a certain area or structure cannot be flooded, a wall or levee with an elevation 11.5 meters (35 feet) high must be built.

Duration curves or hydrographs plotting stage are necessary for the following:

1. proper planned utilization of rivers as navigable waterways, such that the occurrence and duration of low water periods is known

2. planning of river intakes for water companies, power companies or industry, so that the intake line is always submerged, even during periods of low water

3. the analysis of flood frequencies and magnitudes of various flood levels that are essential for planning of floodplain land use

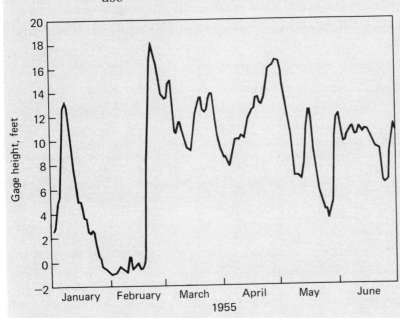

Figure 7.6. *Hydrograph Showing Mean Daily Stage of the Mississippi River at St. Louis, Missouri, January–June 1955.* Data from: *U.S. Geological Survey, St. Louis, Mo.*

Discharge plots are useful in:

1. water resource evaluations so that water availability through varying parts of the year and from year to year is known

2. an analysis of the volumes of fresh water available for dilution of the discharge of various industries or waste water treatment facilities throughout the year (it seems clear that this has been inadequately studied in the past and little consideration of dilution was ever made in many instances)

3. the analysis of minimum flow volumes necessary for cooling water necessary for power plants using fresh water

4. determination of flow necessary to recharge aquifers and sustain downstream fish and wildlife populations

5. determination of the maximum water supply available during driest weather for irrigation projects; the magnitude and duration of low flows must be known (Kazmann, 1972)

It is easy to see the need of industry, agriculture, and municipalities for this kind of long-term hydrographic data if we are to do any rational planning for the future.

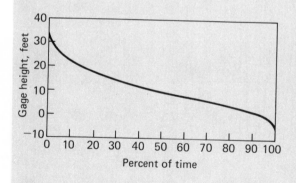

Figure 7.7. *Duration Curve for Gauge Height. Curve shows percent of time that any gauge height was exceeded for lower Mississippi River at St. Louis, Missouri. Curve is based on stream records from 1861 to 1960.* DATA FROM: *Daily stages recorded from 1861 to 1960 by U.S. Corps of Engineers, St. Louis, Mo.*

Storage

In a hydrologic system the storage factor is indicated by the equation:

RATE OF OUTFLOW = RATE OF INFLOW ± CHANGE OF STORAGE

Storage occurs in a number of places within a watershed. Major quantities of water infiltrate into the soil and underlying rock formations and are stored in groundwater aquifers. This concept will be discussed more fully in chapter 8. Water is also stored in river channel banks, natural lakes, and in artificial reservoirs. As indicated in the equation above, lake and river storage serve to regulate flow and distribute it more evenly during periods of varying runoff. Flood control reservoirs are designed to take advantage of this principle. Similarly, river channels accumulate enormous volumes of water during periods of high runoff and flooding. For example, during the January 1937 Ohio River flood, the volume of storage in the channel system was approximately 55 million acre-feet — a quantity of water twice the capacity of Lake Mead, a large reservoir on the Colorado River (Leopold, 1974). Much of this water drains back into the river and flows downstream during subsequent periods of lower flow.

HYDROLOGIC IMPLICATIONS OF LAND USE

With any type of land use alteration, we can expect changes in the natural hydrologic cycle to occur. Cutting of timber, overgrazing by animals, or removal of vegetation and creation of impermeable surfaces during urbanization, for example, all have pronounced and well-documented effects on the hydrologic system (see figures 7.8 and 7.9). Decreased infiltration, increased runoff, and accelerated erosion are some of the more

Figure 7.8. *Rilled and Gullied Land Due to Overgrazing. Note less disturbed area beneath oak tree.* Photo courtesy: *U.S. Forest Service.*

Figure 7.9. *Road Cut Erosion. Gullying in vertical cuts in unconsolidated alluvial materials is due to vegetation removal and road construction.* PHOTO BY: *Gary Griggs.*

obvious hydrologic results of these changes in land use patterns. Timber harvesting and urbanization produce the most serious and long-lasting changes.

Timber Harvesting

Hydrologic Effects

During timber harvesting certain logging practices and associated road construction drastically alter many aspects of the normal hydrologic cycle and thereby modify runoff patterns and rates of soil erosion and sediment transport.

The Eel River Basin, a heavily logged area in northern California, has the highest recorded average annual suspended sediment yield per square kilometer of drainage area of any river of its size and larger in the United States. The yield in tons per square kilometer is more than four times that of the Colorado River and over fifteen times greater than that of the Mississippi River. Erosion in the Eel River Basin is, therefore, a major watershed management problem. The potential for erosion and **siltation** is an important land use constraint and must be considered before logging, road construction, or any activity resulting in vegetation removal is undertaken.

The greatest stream sediment loads and the highest rates of erosion in northern California and southern Oregon occur in areas characterized by abundant landslides and/or rapid downhill soil creep (*see* figure 7.10; Janda, 1972). In this particular climate and physiographic setting, even during in-

Figure 7.10. *Mass Wasting Adjacent to the Middle Fork of the Eel River, Northern California.* PHOTO BY: *W. M. Brown III, U.S. Geological Survey.*

tense rainstorms, the natural storage and high infiltration capacity of the forest floor make overland water flow a rare occurrence. Most of the water in small tributaries, therefore, probably has flowed through the forest floor instead of over it. Much of the sediment in larger streams is derived from erosion of previously deposited surficial sediments and soils in and immediately adjacent to stream channels.

Mass wasting processes and stream or **fluvial erosion** are often accelerated by poorly designed roads and timber harvesting. The improper design of road fills and culverts at stream crossings also leads indirectly to increased fluvial erosion. A culvert may be plugged by sediment or debris, resulting in upstream water impoundment. Either through saturation and slumping, or overtopping and subsequent gullying, the road fill then fails. The downstream erosion and sedimentation can quickly alter and destroy the natural equilibrium stream course

and its vegetation and wildlife (*see* figure 7.11).

Qualitative and semiquantitative aspects of the impact of logging and road construction on rates and types of erosional processes are visually apparent. Convincing quantitative documentation of the impact of land use changes on stream sediment loads in areas already characterized by excessively high natural loads requires careful study. Recent comparative studies indicate that sediment yields from unroaded, clear-cut, cable-yarded drainage basins (those utilizing overhead cables to remove timber rather than conventional tractors) may initially be up to 8 times greater than yields from uncut control basins. Under favorable circumstances the impact may diminish rapidly with time. The impact of the more extensive tractor-yarded logging operations is undoubtedly far greater. The dumping of debris from road construction into stream channels and the landsliding associated

Figure 7.11. *Excessive Sediment Deposition Along Lower Redwood Creek, Northern California. Improper land use upstream has led to excessive erosion and subsequent deposition.* PHOTO COURTESY: *U.S. Army Corps of Engineers.*

with forest roads can initially increase stream sediment loads up to 100 times relative to unroaded clear-cut areas. The effects associated with forest access roads and roaded clear-cut areas can persist or even become accelerated with time (*see* figure 7.12; Janda, 1972).

It is clear that logging and associated road construction practices in the Coast Ranges of the western United States have increased erosion rates and stream sediment loads. Silted fish spawning grounds, silt-filled pools, and log jams are visually obvious environmental damage from the increased sediment load (*see* figures 7.13 and 7.14). More quantitative data are necessary to document these harmful effects to control specific practices such as road design, road maintenance, and yarding procedures during logging operations.

Erosion Control Methods

Studies done to date, however, have shown that certain road construction methods and logging practices can reduce erosional damage significantly. Roads should be planned using whatever topographic, soil, geologic, and vegetative maps are available. Unstable and steep areas should be avoided; or if not avoidable, extra care in erosion control must be taken. Roads should be located on benches or ridges, as far away as possible from stream channels. Where streams must be crossed, bridges and culverts should be used and located where channels do not need to be altered. Installation of such structures during summer low flow periods has the least impact on fish and wildlife habitat. After an area is logged, it is imperative that water bars or ditches be installed to divert

Figure 7.12. *Watershed Destruction. Total destruction of a stream course has occurred as a result of poor timber harvesting and road construction practices.* PHOTO COURTESY: *California Dept. of Fish and Game.*

Figure 7.13. *Watershed Destruction. Destruction of a stream bottom from the deposition of logs and debris.* PHOTO COURTESY: *California Dept. of Fish and Game.*

Figure 7.14. *Logging Road Construction Destroying a Stream Bottom. Buffer zones must be left along streams to protect them from logging activities.* PHOTO COURTESY: *California Dept. of Fish and Game.*

runoff to undisturbed forest areas. Seeding of fast growing grass or brush is necessary on all road cuts and disturbed soil areas (California Water Resources Control Board, 1973).

Generally selective cutting of timber results in less erosion than clear cutting. Use of balloons, helicopters, or overhead cables to transport logs from the harvest area to a pickup point has been shown to cause significantly less soil disturbance than tractor yarding. These methods are more costly, however, and in many cases, can be justified

only with the large timber volume derived from clear cutting.

Urbanization

As urbanization occurs, the hydrology of an area undergoes some important changes. Of most immediate concern is the creation of impermeable surfaces, such as roofs, streets, sidewalks, and parking lots, which reduce the amount of water infiltration and greatly increase runoff. Estimates indicate that larger suburban parcels have significantly less impermeable surface (25 percent for a .33 acre lot, about 1350 square meters) than small urban lots (80 percent for a 540 square meter developed lot, about .12 acre; Leopold, 1968).

Urbanization also is commonly accompanied by the construction of storm sewers, which collect runoff from these impervious areas and discharge it directly to stream channels. With the creation of impermeable surfaces and the emplacement of storm drains, the frequency of floods (see figure 7.15) and also the size of flood peaks increase (see figure 7.16; Leopold, 1968).

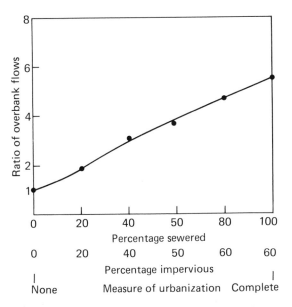

Figure 7.15. *Increase in Flood Frequency from Urbanization. For a 1 square mile drainage area, the curve shows the increase in overbank flows per year (expressed as ratio of overbank flows after urbanization to those before urbanization) for varying degrees of urbanization.* REDRAWN FROM: *L. B. Leopold, "Hydrology for Urban Land Planning," Geological Survey Circular 554 (1968).*

Storm Water Contaminants

Urban runoff can also contribute to a variety of problems including a direct pollution of receiving waters and impairment of sewer and catch basin functions. A recent study concluded that runoff from city street surfaces is highly contaminated (Sartor et al., 1974). Moreover, chemical tests indicated runoff from the first hour of a moderate to heavy storm in a typical city would contribute considerably more contaminants to watercourses than would the same city's untreated sewage during the same period of time. Major types of pollutants in runoff include common sand or silt, petroleum products, phosphates, nitrates, heavy metals such as zinc and lead, organic pesticides, and coliform bacteria. There is little question that street surface contaminants warrant serious consideration as a major pollution source having potentially significant detrimental effects.

Because storm water runoff has just recently been recognized as a serious environmental problem, workable solutions are just beginning to emerge. Off-channel facilities such as check dams and artificial ponds, treatment lagoons, infiltration basins, and roadside swales are all capable of partially controlling silt and other pollutants from urban runoff. In a recent study of storm runoff problems in Columbia, Maryland, many small reservoirs dispersed throughout the community were proposed to control urban pollution. Water collected and stored in these basins was to be treated to drinking water standards and then released for use by the city. Not only would part of the facilities' cost be offset by use of the water, but the reservoirs could also serve recreation needs and would reduce peak runoff rates (Detwyler and Marcus, 1972).

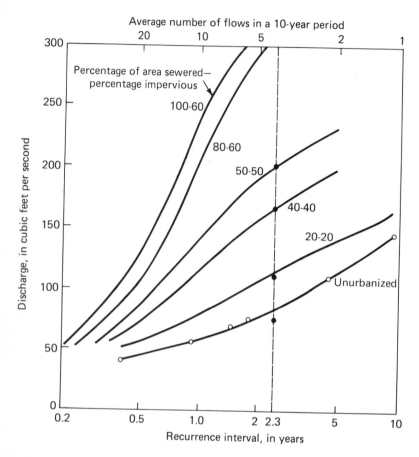

Figure 7.16. *Increase in Size of Flood Peak from Urbanization. For a 1 square mile drainage basin the curves show the increase in discharge for varying degrees of urbanization.* RE-DRAWN FROM: *L. B. Leopold, "Hydrology for Urban Land Planning," Geological Survey Circular 554 (1968).*

Several of the methods mentioned above also serve to recharge groundwater reservoirs in urbanized areas, which due to the impervious surfaces, no longer receive an adequate amount of infiltration. In Fresno, California, a large number of gravel packed wells receive runoff from streets and have proven successful in disposing of storm drainage to the subsurface. Nassau and Suffolk counties on Long Island, New York, have over 2000 recharge basins that for years have successfully collected all the storm water runoff from large and small subdivisions and allowed it to infiltrate the underlying aquifers.

A new process has recently been developed that could contribute significantly to the solution of urban runoff problems. This development is porous asphaltic pavement, which allows percolation of storm water into the underlying soil, rather than overflow into storm facilities and receiving watercourses. Testing conducted under an Environmental Protection Agency (EPA) contract at the University of Delaware indicates that porous pavement substantially reduces storm water runoff and could measurably increase groundwater recharge. The construction of porous pavement over an impervious membrane offers a potential

Figure 7.17. *Urban Development Waterproofs the Land Surface. In this aerial view of Wheaton, Maryland, roofs and pavement have made most of the surface impervious to water.* PHOTO COURTESY: *U.S. Department of Agriculture.*

method for storing storm water and slowly releasing it for subsequent discharge. Other benefits noted from the experiment include improved automobile skid resistance, low maintenance cost, preservation of roadside vegetation (roots under impermeable pavement suffer from water drought), and preservation of natural drainage patterns. Porous asphaltic pavement also acts as a filter and can remove most of the contaminants from storm runoff (Thelen et al., 1972).

Erosion and Sedimentation Problems

Every year over 400 square kilometers (1 million acres) of land in the United States are converted from agricultural use to urban use. As the population has increased and become more urban, the open land surrounding

existing cities has been sacrificed to support ever-increasing numbers of houses, shopping centers, industrial parks, schools, and highways (see figure 7.17). The alteration of the land usually involves the removal of all or most of any existing vegetation and the exposure of the bare soil to the elements. When exposed to rainfall, earth laid bare by construction erodes easily at rates about 10 times greater than land cultivated in row crops, 200 times greater than land in pasture, and 2000 times greater than land in timber (Vice et al., 1969). Sediment yields from urbanized or developing areas range from 350 to more than 35,000 tons per square kilometer per year (see figure 7.18; Wolman and Schick, 1967). It is not uncommon for such land to remain bare for as long as six months during construction of a single residence

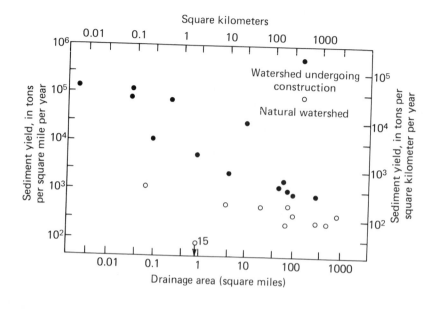

Figure 7.18. *Effects of Construction on Sediment Yield. Data are from the Baltimore and Washington, D.C. metropolitan areas.* ADAPTED FROM: *M. G. Wolman and A. P. Schick, "Effects of Construction on Fluvial Sediment, Urban and Suburban Areas of Maryland,"* Water Resources Research 3 (1967): 451–64.

and to be exposed for more than one year during construction of large housing and apartment developments.

The effect of the progressive alteration of the land surface on increased erosion rates has been well documented by Wolman (1967) for a small area near Washington, D.C. A very low erosional rate of 0.2 centimeters per 1000 years under original forest condition was increased slowly to about 10 centimeters per 1000 years as extensive farming with cultivation began to occur in the middle to late nineteenth century (*see* figure 7.19). As some of the area returned to forest and grazing land, this rate was reduced by half, only to increase dramatically to approximately 10 meters per 1000 years as construction activities began to occur. Erosional rates of 100,000 tons per square kilometer were reached for very small areas during the 1960s, as urbanization occurred, accompanied by surface vegetation removal, asphalting and creation of impermeable sur-

faces, and concentrated runoff. With complete urbanization, rates dropped again to about 1 centimeter or less per 1000 years, mainly due to the lack of soil exposure.

A recent study of a small basin draining

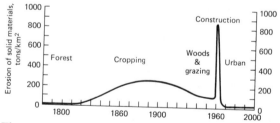

Figure 7.19. *Progressive Change in Erosion Rates with Changing Land Usage. As forests were cleared for farming and grazing, and land was gradually used for construction and urbanization in the Middle Atlantic region, erosion rates were greatly altered. Note the lack of erosion with total urbanization when virtually all the soil is covered.* REDRAWN FROM: *M. G. Wolman, "A Cycle of Sedimentation and Erosion in Urban River Channels,"* Geografiska Annaler 49 (1967): 385–95.

into San Francisco Bay showed greatly accelerated erosion, increased flood frequency, and storm runoff as the small watershed was given over to urban development. The study of the Colma Creek Basin indicates that sediment yields in areas under construction increased as much as eighty-five times the rates under natural conditions (Knott, 1973).

Lake Bancroft, in northern Virginia, originally served as a water supply reservoir for the surrounding area. As urban development forced its abandonment as a water supply source, the lake became the focal point of a high priced suburban development. Continued urban expansion resulted in even greater accumulation of sediment from the construction and development. Eventually more than 200 thousand tons of sediment were washed into the lake, almost a ton of silt for every 100 square meters in the basin under development (40 tons per acre). Dur-

ing peak development, local residents spend over $200,000 to dredge sediment from the lake to maintain its recreational and aesthetic qualities (U.S. Geological Survey, 1972).

In Fairfax County, Virginia, 0.8 square kilometers (197 acres) undergoing highway construction over a three-year period contributed 34,000 tons of sediment to the local stream. Although the highway construction was limited to 11 percent of the basin, the sediment contributed by the area was 94 percent of the total sediment yield during the three-year period (Thomas and Schneider, 1970). Similar problems occurred during state highway construction in loose, very erodible silt-rich sand in the central California coastal area. Slope failures due to oversteepening combined with heavy rainfall led to serious erosion and siltation problems (see figure 7.20).

The erodibility of the soil, the steepness

Figure 7.20. *Excessive Runoff and Siltation Associated with Highway Construction. Steep slopes, unconsolidated sand, and lack of replanting led to this problem.* PHOTO BY: *Gary Griggs.*

and length of the slope, the intensity of rainfall, vegetation removal, and construction methods all affect the amount of erosion on a site and, therefore, the sediment production to the drainage systems. Rilled and gullied slopes (see figure 7.21), gullied waterways and channels, washed out roads and streets (see figure 7.22), undercut pavements and pipelines (see figure 7.23), clogged storm sewers, flooded basements, and debris-laden work areas (see figure 7.24) all slow down construction and increase costs to the public and private sector.

All eroded soil or surficial material even-

Figure 7.21. *Erosion and Gullying of an Unpaved Road That Is Part of a New Subdivision, Pleasant Hill, California. Where land has been cleared for development, winter rains can often produce this situation.* Photo courtesy: *California Div. of Mines and Geology.*

Figure 7.22. *Washed Out Road Due to Poor Compaction During Construction of a Mobile Home Park.* Photo by: *Gary Griggs.*

Figure 7.23. *Excessive Runoff from Urban Area Eroding Fill Around a Culvert.* Photo by: *Gary Griggs.*

Figure 7.24. *Effects of Poor Road Building. Improper road construction for a mountain development has led to deposition of debris and washed out culverts.* Photo courtesy: *California Div. of Mines and Geology.*

tually enters a stream channel, where it continues to be harmful in a number of ways (Guy, 1970). The deposition of coarse sediment may reduce flow capacity or completely plug natural or constructed channels or drains. Increased sediment deposition in stream channels, estuaries, or other water bodies may cause serious biological damage, such as suffocation of fish and the destruction of gravel spawning grounds for certain fish. Water treatment costs for domestic and industrial uses are increased. Reservoir storage and channel conveyance for water supply are lost. Each year approximately 1 million acre-feet of sediment are deposited in the nation's reservoirs. This total annual storage loss only amounts to about 0.2 percent of the initial capacity, but depletion rates for the more abundant smaller reservoirs (capacities of 100 acre-feet or less) are much greater than those for larger reservoirs (see table 7.4).

Sedimentation and Flooding

Periodic flooding and accompanied sediment deposition have become common problems in semiarid southern California due to urbanization (Rantz, 1970). When the area was first settled, the channels on the alluvial fans and valleys below the surrounding mountains were often poorly defined and semipermanent. During flood periods the waters spread over wide floodplains. On many square kilometers of alluvial fans, old river bottoms and floodplains, where water once flowed relatively unconfined, there are now homes, highways, railroads, towns, and cities (see figure 7.25). The flood hazards are not confined to water runoff alone but include debris brought down the surrounding steep mountain slopes and deposited on streets and developed property (see figures 7.26 and 7.27). The problems have also been aggravated by

Table 7.4. *Summary of Reservoir Capacity and Storage Depletion Data*

Reservoir Capacity (Acre-Ft.)	Number of Reservoirs	Total Initial Storage Capacity (Acre-Ft.)	Total Storage Depletion (Acre-Ft.)	(%)	Individual Reservoir Storage Depletion		Average Period of Record (yrs.)
					Average (%/yr.)	Median (%/yr.)	
0–10	161	685	180	26.3	3.41	2.20	11.0
10–100	228	8,199	1,711	20.9	3.17	1.32	14.7
100–1,000	251	97,044	16,224	16.7	1.02	.61	23.6
1,000–10,000	155	488,374	51,096	10.5	.78	.50	20.5
10,000–100,000	99	4,213,330	368,786	8.8	.45	.26	21.4
100,000–1,000,000	56	18,269,832	634,247	3.5	.26	.13	16.9
Over 1,000,000	18	38,161,556	1,338,222	3.5	.16	.10	17.1
Total or average	968	61,239,020	2,410,466	3.9	1.77	.72	18.2[1]

[1] The capacity-weighted period of record for all reservoirs was 16.1 years.

ADAPTED FROM: F. E. Dendy, "Sedimentation in the Nation's Reservoirs," *Journal of Soil and Water Conservation* vol. 23, No. 4 (1968), pp. 135–137.

Figure 7.25. Wrecked Railroad and Automobile Bridges from Flooding in Big Tujunga Wash, North Hollywood, California, 1938. PHOTO COURTESY: U.S. Forest Service.

Figure 7.26. Erosion and Deposition Following Brush Fire Near Santa Barbara, California, 1932. Just over 1.16 inches of rain fell in thirty hours, and at no time was the rainfall severe. Erosion started eighteen hours after the storm began and took twenty hours for this amount of debris to be deposited. PHOTO COURTESY: U.S. Forest Service.

Figure 7.27. Deposition of Debris in an Urban Area Following Brush Fire and Subsequent Flooding, Southern California, 1938. PHOTO COURTESY: U.S. Forest Service.

increased runoff and erosion in mountain areas where erodible soils have been laid bare as a result of forest and brush fires of both natural and human origin (*see* figures 7.28A and 7.28B).

Permanent solutions to these problems after urban sprawl has occurred are difficult (Rantz, 1970). The surrounding mountains are geologically young and break down rapidly under prevailing environmental conditions. Reservoirs and debris dams fill with sediment and rapidly lose their useful capacity (*see* figure 7.29). Although seasonal precipitation is not high, rainfall from a single storm is often excessive. Past floods have not been indicative of maximum flood potential, especially in light of the creation of more impermeable surface with increased urbanization. As population has grown in greater Los Angeles (from about 2.5 million in 1935 to about 7.0 million in 1970) pressure has been exerted on planners by developers. More building sites for more houses and attendant industrial and commercial activity are being sought. No effective regional planning authority exists for the heavily urbanized areas because of local jurisdiction in the incorporated communities. Local planning, although sometimes based on regional concepts, is not uniform in quality. Difficulties of planning on a regional scale to provide a logical pattern of

Figure 7.28a. *Aerial View of Foothill Area of Angeles National Forest Above Glendora, California. Note fire-scarred hillsides from forest fire, August 1968.* PHOTO COURTESY: *U.S. Forest Service.*

Figure 7.28b. *Local Flood Control Debris Basin Fills Quickly After Heavy Rain Storm Near Glendora, California, January 1969.* PHOTO COURTESY: *U.S. Forest Service.*

Figure 7.29. *Mono Debris Dam in Los Padres National Forest, California, Filled with Sediment.* PHOTO COURTESY: *U.S. Forest Service.*

zoning for protection of residents and property from sedimentation and flood hazards are obvious.

The following measures have been utilized in an attempt to remedy the flood situation in the Los Angeles Basin (Rantz, 1970):

1. construction of reservoirs designed to store as much flood flow as possible

2. construction of debris basins to catch debris and sediment

3. diversion, where possible, of sediment-laden waters into areas where sediment can be deposited and excess water can percolate into the groundwater reservoir

4. realignment, enlarging, and paving of permanent channels to convey excess runoff to the ocean and prevent erosion

Intense storms in January 1969 produced rainfall greater than had ever been recorded at the stations surrounding Los Angeles. The surrounding mountains recorded 24 to 121 centimeters (9.4 to 47.6 inches) of precipitation during an eight-day period. The death toll from the storm and flooding was ninety-two persons. Ten thousand people were driven from their homes and physical damage was estimated at $62 million. Yet, due to the flood and sediment control basins, $1.2 billion in damage was prevented. Debris basins in Los Angeles County trapped an estimated 1.5 million cubic meters (2.0 million cubic yards) of debris. Of the sixty-one debris basins only seven were completely filled, and only three had debris pass over their spillways into downstream drains. The improved channels, in general, contained the flood flows within their banks or levees.

The greatest damage occurred in foothill areas surrounding the Los Angeles Basin. The combination of intense precipitation and, in places, ground laid bare by brush fires or construction, led to mudflows and torrents of sediment- and debris-laden water. Rocks and sediment covered streets, sidewalks, lawns and flowed into homes (see figure 7.30).

Housing has been moving into the upland hillside, mountain, and canyon areas. Hillside developments in these areas are extremely difficult to protect from these types of disasters because of countless small watercourses, each of which requires individual control measures for protection against inundation and debris damage. Roads and construction have aggravated already unstable areas and when saturation by heavy rainfall occurs, the slides and mudflows are inevitable.

Methods of Erosion and Sedimentation Control

Erosion and sedimentation can be controlled effectively, and at reasonable cost if study and planning precede any kind of land surface alteration and if some basic principles are followed. The damage and cleanup costs involved when no planning has been done, or when no preventative measures have been taken, are usually never calculated or else considered too late to be of any consequence. With the passage of the National Environmental Policy Act in 1969 and the requirement of environmental impact statements for construction projects involving federal financing, problems of erosion and sedimentation will now have to seriously be considered. (See chapter 11, p. 441.)

Many methods have been used to control erosion and sedimentation caused by urban

Figure 7.30. *Effects of Urban Encroachment in Alluvial Fan Areas of Southern California Following Brush Fires and Heavy Precipitation.* PHOTO COURTESY: Los Angeles Times.

development (U.S. Department of Agriculture, 1970). These range from temporary procedures to permanent installations, from selecting the proper season for construction to the building of engineering works. The methods are based on some principles that can be universally applied: (1) selection of an environmentally appropriate site for the project, (2) reducing the area and duration of exposure of soil to erosion, (3) mechanically retarding the runoff or trapping the sediment removed from a site.

Site Selection.   The initial selection of land in an area where drainage patterns, slopes, and soils are favorable for the intended use, and the fitting of any development to the site are preliminary planning decisions that can greatly minimize future problems. Planners, developers, and builders need to make use of soil and other resource data in making zoning decisions, in selecting land for development, and in preparing site develop-

ment plans. Although soil surveys have benefited farmers for years, highway engineers, land assessors, banks, and land planning agencies in increasing numbers are now realizing their importance and value (U.S. Department of Agriculture, 1970). Knowledge of the erodibility and depth of soil and terrain steepness can assist in designing highway routes that reduce erosion problems. Detailed soil surveys describe the characteristics and properties of each soil type — its texture, slope, depth, erodibility, and permeability. Other useful information such as the presence of impervious or porous layers may also be included. On the basis of these factors, the limitations and suitabilities of the soils can be delineated. Variations in permeability determine percolation rates and will affect suitability of soils for septic tanks. Expansive clays or those with high **shrink-swell ratios** and seepage problems may impose limitations on other soils for certain uses.

Essay

SOILS

Soils are the result of physical and chemical weathering of some basic rock material. The most important variables involved in soil formation include parent material, climate, topography, time, and the **biosphere**. Although parent rock, whether residual or transported, is the basis of all soil, the same rock, granite for example, will produce different soils under different weathering conditions. Climate then determines the processes of soil development.

Rock breakdown or weathering is a composite of mechanical disintegration and chemical decomposition. These two types of breakdown are intimately related to the two principal climatic factors, temperature and precipitation. Mechanical breakdown is a result of expansion and contraction, heating and cooling, and freezing and thawing, which simply break rocks up into smaller pieces. Chemical decomposition, more important overall in the weathering process, involves oxidation, solution, **hydrolysis**, and **carbonation** of the rock material. These chemical reactions and physical processes are all affected by temperature and the amount of water that percolates through the soil.

Time is also important inasmuch as the development of a mature soil profile is a very slow process. Development of acid soils in forested regions, for example, where sandy well-drained material exists, may only take 100 or 200 years. In the mid-continent region, on the other hand, it may take 5000 years to dissolve 30 centimeters of limestone, and it takes 9 meters of limestone to form 30 centimeters of soil. In other words, it would take 150,000 years of weathering to form 30 centimeters of soil (Cargo and Mallory, 1974).

Soils with time generally form in layers, which give rise to a distinct soil profile (see figure 7.31). The uppermost, or A, horizon is most familiar to us, and the one usually called soil. Rainfall percolates through this zone, dissolving and leaching rock and soil material as it moves. Organic matter, such as leaves and roots, break down and accumulate in this zone, whereas soluble chemical constituents are removed.

The lowermost, or C, horizon grades downward into unweathered rock. This zone is actually a layer of partially decomposed bedrock or parent material, which is being mechanically and chemically broken down. Between the two lies the B horizon, which is a transitional zone. The material that has been leached from the soil above, and the residue left from weathering of the underlying bedrock, constitute this layer. Under humid conditions iron may be deposited in the B horizon; where the climate is dry, calcium carbonate is usually deposited.

Our total dependence on this thin layer of soil, the ease with which it can be totally eroded, and the incredibly slow rate of formation are all sufficient reasons to take every conceivable measure to protect this nonrenewable resource.

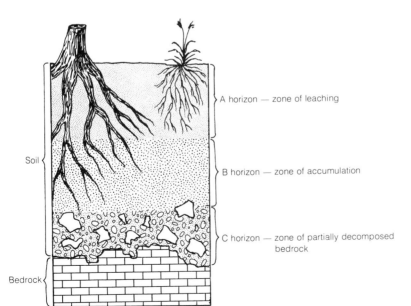

Soil

Bedrock

A horizon — zone of leaching

B horizon — zone of accumulation

C horizon — zone of partially decomposed bedrock

Figure 7.31. *Soil Development. A generalized soil profile showing the soil horizons proceeding from fresh bedrock to the ground surface.*

Where development is proposed, those areas unsuitable for it or sensitive to alteration should remain as open space or low use areas. On sites where the cost of controlling erosion may be high because of the measures needed to overcome site limitations, alternate uses or a plan more compatible with the land should be considered. Clustered development, for instance, fits buildings and streets to the natural characteristics of the land and consolidates building construction. On sloping land, buildings should be sited only on the more level areas. The steep, more erodible land should be left undisturbed (see figures 11.7 and 11.8).

If zoning based on the potential erodibility of the landscape, considering land slopes, rainfall intensities, and related factors, is developed, these problems can be controlled at the onset. For example, disposing of runoff can be a problem for some downslope areas. Drainage from building sites and upslope watersheds must be controlled and disposed of in a safe manner.

Offsite measures need to be implemented that prevent damage to downstream land and property from either erosion or sediment.

Reducing Soil Exposure. Proper planning can ensure that large areas are not cleared of surface cover and left bare for extended periods of time. In both highway construction and house subdivision development, it is not uncommon for large tracts of cleared land to lie exposed to intensified erosion through entire winters (see figure 7.32). In areas of high seasonal precipitation all major grading and/or land-clearing operations should coincide with periods of minimum rainfall. Where development of large blocks of land or extensive stretches of highway can be carried out in smaller workable units, construction can be completed more rapidly and bare soil need not be exposed for long periods. Grading should always be held to the minimum that makes the site suitable for its intended purpose without appreciably

Figure 7.32. *Exposure of Bare Land to Heavy Winter Rains, Lake Tahoe Area, California and Nevada.* Photo courtesy: *California Div. of Mines and Geology.*

increasing runoff. In many instances, heavy cutting, filling, or reshaping of the natural topography is performed to increase the amount of usable land on a parcel. This type of heavy grading almost invariably increases erosion hazards and should be accompanied by well-timed use of appropriate erosion control measures.

Vegetative measures provide temporary cover to aid erosion control during construction and permanent cover to stabilize the site after project completion. Establishing and maintaining good plant cover is easy in areas of fertile soil and moderate slopes. The plants endemic to the area can ordinarily be used for long-term stabilization. They should be selected on the basis of adaptation to soil and climate, ease of establishment, suitability for the specific site use, longevity

or ability to reseed, and ease of maintenance. For example, grasses used for waterway stabilization must be able to withstand periodic submergence and provide a dense cover to prevent channel scouring.

Other sites, because of exposed subsoil, steep slopes, or other conditions, may erode very severely and require special treatment (see figure 7.33). Straw or other mulch is commonly used to protect exposed slopes initially, especially if the time is inappropriate for seeding. Mulch serves to reduce rain impact and runoff, and, therefore, to increase infiltration or percolation into the soil. It also serves to retain soil moisture and hold seed and fertilizer in place. **Hydromulching**, which involves the application of a mixture of seed, fertilizer, and mulch, is becoming more widely utilized as

Figure 7.33. *Highway Failure. Failure of this steep highway bank despite vegetative cover is due to very weak soils.* PHOTO BY: *Gary Griggs.*

a rapid labor-saving soil protection measure.

A number of fibrous or mesh materials also have value in erosion control as well as in plant stabilization. Jute netting, a coarse, open mesh, weblike material, can be applied directly on the soil to protect seeded areas and hold soil and mulch in place. Meshing can also be used along the sides of channels and in repairing outlets and diversions where gullies have developed.

Structural Measures for Runoff Control. Structural measures for runoff and subsequent control involve reshaping land to intercept, divert, convey, retard, or otherwise control runoff. Terraces or benches can be fitted to the natural terrain and used to break long slopes and slow the flow of runoff. Such a bench or ridge at the top of a slope may also be used to divert runoff that could cause serious erosion away from a slope (see figure 7.34). Erosion can be controlled on unsurfaced roads or slopes by installing low barriers or water bars to divert runoff to roadside drainage ditches. This is commonly required on logging roads in the western United States. Most drainage ways or watercourses undergo significant change as a result of surrounding construction or urbanization. The stabilization of channels with riprap or concrete may be necessary to control increased runoff from construction sites.

The construction of temporary sediment traps along watercourses can effectively retain material that might be carried off during construction. Sediment traps are small reservoirs that retain suspended and coarse material rather than allowing it to move into streams and rivers. Such traps can be eliminated later when landscaping is established and they are no longer needed.

All these efforts should rightly be demanded by project planners, but unfortunately, in the interests of economics and time, they have too often been eliminated or minimized. In this case, the responsibility falls in the hands of the planning staff and

Figure 7.34. *Terracing and Drainage Used to Stabilize Highway Cuts.* PHOTO BY: *Gary Griggs.*

planning commission of the city or county. At this point, the necessary environmental constraints must be considered and the appropriate conditions placed on any proposed development. Environmental impact statements can be used to elucidate problems that might occur from a project and to propose erosion control methods to deal with these problems. Planners need to be able to integrate this type of analysis into the project before approval is given.

FLOODING

Water, in large amounts, has become essential to modern civilization. At times, however, because of the locations where people have chosen to settle and to farm, and often because of their own actions, they find themselves with an abundance of water, over which they have no control, and which does considerable damage. Hundreds of floods, small and large, occur annually in the United States and will continue to occur. From time to time meteorologic and hydrologic conditions combine to produce huge floods of unprecedented magnitude. These are natural phenomena to which we must learn to accommodate ourselves.

To some people floods may be only a source of inconvenience, to others they may be an immediate danger to life and property. To farmers they mean erosion and loss of topsoil on barren or newly planted hillside fields, and inundation and flooding of valley bottoms or floodplain lands (see figure 7.35).

Figure 7.35. *Overflow of the Salinas River in Central California onto Its Floodplain.* PHOTO COURTESY: *U.S. Army Corps of Engineers.*

To engineers floods mean measurements of stages and discharges under the worst possible conditions, the study of ways to reduce damage, and the design and construction of dikes, floodwalls, channels, dams, and reservoirs. To legislators floods often represented a peg on which to hang measures apparently designed not so much as to reduce flood damage as to stabilize the economy, promote the general welfare, or in some instances, curry political favor. To all of us as citizens they are a natural phenomenon on which we have spent billions of dollars under our so-called flood control policies (Hoyt and Langbein, 1955).

In hydrologic or geologic terms, a flood exists when the level of a stream is such that it overflows its normal channel and spreads out on its floodplain. Nearly all large or mature streams have such a floodplain or low-land area that has formed by the periodic overflow of the stream, deposition of its sediment, and the subsequent readjustment or realignment of the stream course. As with other geological or hydrologic phenomena, floods only become a hazard because people insist on getting in their way.

Since streams usually only occupy a small portion of their fertile floodplains, people from the earliest civilizations have chosen to inhabit, cultivate, and later, urbanize these areas. Probably one-third of the world's people obtain their food from soils developed on such alluvium. Many small communities are built entirely on river floodplains (see figure 7.36) and large parts of some major cities are also well exposed to flooding. New Orleans is the largest city in the United States built entirely on a floodplain, and as a result, the city has had to

Figure 7.36. *Construction on a Floodplain. Aerial photograph showing the encroachment of the city of Capitola, California, onto the floodplain of Soquel Creek. Broken line delineates floodplain.* PHOTO COURTESY: *U.S. Army Corps of Engineers.*

ward off the Mississippi floods with levees since its early history.

Flood Damage

Floods are probably the most costly of all natural hazards in terms of loss of life, property, and land (see table 7.5). There has also probably been more money spent for various projects labeled as "flood control" than for all other geologic hazards combined. Floods are not restricted to particular geographic areas but can occur almost anywhere. They can range in magnitude from submerged suburban intersections to overbank discharge of major rivers, which can destroy millions of dollars worth of property and take many lives. Damage is primarily a result of the erosive capacity of flood waters, the

Table 7.5. *Major Worldwide Floods and Their Damages*

Year	Month	Location	Lives Lost	Property Damage (millions of dollars)
1973	March–June	Mississippi River		$1200.0
1972	June	Eastern United States	113	$3000.0
1972	June	Black Hills, South Dakota	242	$163.0
1970	May	Oradea, Rumania	200	225 towns destroyed
1969	August	James River Basin, Virginia	154	$116.0
1969	Jan.–Feb.	Southern California	60	$399.2
1968	August	Gujarat, India	1000	
1967	Jan.–March	Brazil (Rio de Janiero and Sao states)	600+	
1966	November	Arno Valley, Italy	113	Art treasures destroyed in Florence
1964	December	California and Oregon	40	$415.8
1963	October	Belluno, Italy	2000+	Vaiont Dam overtopped
1962	September	Barcelona, Spain	470+	$80.0
1955	October	Pakistan and India	1700	$63.0
1954	August	Kazuin District, Iran	2000+	
1953	Jan.–Feb.	Northern Europe	2000+	Coastal areas devastated
1951	June–July	Kansas and Missouri	41	$1000.0
1951	August	Manchuria	5000+	
1911		Yangtze River, China	100,000	
1889	May	Johnstown, Pennsylvania	2000+	
1887		Honan, China	900,000+	

DATA FROM: H. W. Menard, *Geology, Resources, Society.* (San Francisco: W. H. Freeman and Company, 1974); B. A. Bolt et al. *Geological Hazards* (New York: Springer-Verlag, 1975).

impact of the water itself on structures (see figures 7.37 and 7.38), the deposition of sediment and debris as flood waters recede (see figures 7.39 and 7.40), and the contamination of water systems.

The damage to commercial, industrial, or residential buildings (see figure 7.41), to agricultural land, to transportation systems (see figure 7.42), and to utilities is direct and can be calculated in dollars and cents. Other

Figure 7.37. *Structural Damage as a Result of Flooding, Northern California, 1965.* PHOTO COURTESY: *U.S. Army Corps of Engineers.*

Figure 7.38. *Destruction of Highway and Bridge During Flooding of Salinas River, Central California.* PHOTO COURTESY: *U.S. Army Corps of Engineers.*

Figure 7.39. *Destruction of a Home and Sediment Deposition Due to Flood Waters, Weott, California.* PHOTO COURTESY: *U.S. Army Corps of Engineers.*

Figure 7.40. *Deposition of Flood Debris in Harbor at Crescent City, California.* PHOTO COURTESY: *U.S. Army Corps of Engineers.*

Figure 7.41. *Damage to Store Front and Automobiles During Flooding.* PHOTO COURTESY: *U.S. Army Corps of Engineers.*

Figure 7.42. *Railroad Cars Overturned by Force of Flood Waters, Northern California.* PHOTO COURTESY: *U.S. Army Corps of Engineers.*

types of damage are more indirect and include losses of businesses or services, decline in property values, and related effects. Intangible losses, such as loss of life and aesthetic damage, do not lend themselves to economic analysis. Damages, therefore, are varied and diverse. They extend from the initial soil erosion losses from upstream farmland to the overtopping of banks and flooding of entire cities built far downstream on floodplains.

As stream velocity and discharge increase during flooding, the stream's ability to erode land and to transport debris increases. Erosive power, in fact, increases exponentially with increased discharge, resulting in huge sediment loads. The bed and banks of the stream will be eroded and the size and amount of material carried will increase considerably. The suspended load of a river may be increased during flood stages to 10,000 times that which may be carried during periods of low flow (see figure 7.43 Brown and Ritter, 1973). Floods on the Eel River in northern California during the winter of 1964–1965 discharged 145 million metric tons of suspended sediment in a thirty-day period. This is equivalent to 51 percent of the total suspended load computed for the previous ten years. Major floods are, therefore, extremely important in terms of erosion and sediment transport.

Causes of Floods

A number of conditions or factors can produce or contribute to a flood. These include:

1. heavy precipitation
2. snowmelt
3. ice jams
4. dam failure

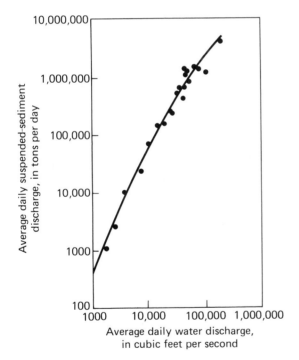

Figure 7.43. *Sediment-Transport Curve for Middle Fork Eel River Near Dos Rios, California, October 1957–September 1967.* REDRAWN FROM: *W. M. Brown and J. R. Ritter, "Sediment Transport and Turbidity in the Eel River Basin, California," U.S. Geological Survey Water Supply Paper (1968).*

Precipitation Intensity and Duration

The intensity and duration of precipitation are probably the most common single factors producing floods. When the infiltration capacity of soil, or maximum rate at which water can be absorbed by it, is exceeded, runoff increases to flood proportions. The combination of frost or frozen ground with high precipitation can also accelerate runoff and produce flood conditions. There is no question, however, that the alteration of the landscape, whether from logging, overgrazing, farming, mineral extraction, or construction of one kind or another has involved the partial or complete removal of the vegetative cover and subsequently has greatly increased the rate of runoff. This has led to greater flood damage due to higher runoff volumes and the increased erosion and deposition caused by them.

Snowmelt

Over much of the Midwest and Far West, extended periods of freezing temperatures lead to snowfalls in excess of 5 meters, which is equivalent to 1.25 meters of water. Rapid melting of snowpacks usually occurs during the spring months and floods of this origin, therefore, occur later in the year. They are commonly felt in downstream areas far removed from the areas of actual snowmelt, such as along the lower Mississippi River.

Ice Jams

In cold regions where rivers freeze solidly to considerable depths, such as Canada, Alaska, and the Midwest, the annual spring breakup of ice may commonly be followed by flooding. Large blocks and sheets may become jammed at channel constrictions, abrupt bends, or bridge openings, damming the river, with resultant overflow and flooding.

Dam Failure

In addition to natural flooding, the failure of human engineering works, either during periods of high stream flow, or independent

of it, for whatever reason, has given rise to monumental floods (see figure 7.44). The failure of the St. Francis Dam along the Santa Clara River in southern California (Clements, 1969), and the Vaiont Reservoir failure in Italy, which caused the death of more than 2000 people (Kiersch, 1965), are examples of flooding due to human error. Most recently, a failure in the $2.8 billion California water project aqueduct near Lancaster led to the flooding of 6 square kilometers (1500 acres) of farmlands, in addition to roads and houses.

On February 26, 1972, a coal waste dam on Buffalo Creek, in Logan County, West Virginia, improperly built, maintained, and inspected, completely collapsed and triggered the most tragic flood in the state's history. Precipitation in the area for the three days prior to failure was an event that could be expected once every two years. Water was reported within 30 centimeters of the crest of the dam four hours prior to the flood. Failure probably occurred through foundation deficiencies, causing sliding and slumping on the front face, aggravated by the waterlogged condition of the dam and the lack of overflow control. Approximately 500,000 cubic meters of water in a 3- to 6-meter-high flood wave washed through the 24 kilometers of Buffalo Creek Valley, at average speeds of about 8 kilometers per hour. In three hours, at least 118 lives were lost, 500 homes were destroyed, 4000 people were left homeless, and property and highway damage reached $65 million. Five causes led to dam failure:

1. The dam was not designed or constructed to withstand the potential amount of water that could be impounded. It was primarily a waste pile that grew from routine dumping.

Figure 7.44. *Levee Overtopping and Resultant Flooding, Sacramento River Delta.* PHOTO COURTESY: *U.S. Army Corps of Engineers.*

2.   The dam had inadequate water level or overflow controls.

3.   The sludge beneath the dam was an inadequate foundation and led to slumps and subsidence.

4.   The great thickness of the dam in relation to height, without engineering compaction, led to decreased stability.

5.   The coal waste that constituted the dam disintegrated rapidly, was noncohesive, and did not compact uniformly (Davies et al., 1972).

Numerous coal waste banks exist in the Appalachian Mountains, many which impound water. Although extensive inspection of the banks was carried out in 1966 by the U.S. Bureau of Mines subsequent to a failure in Wales, coal banks can grow and change rapidly where mining is ongoing. The Buffalo Creek waste dam was inspected in 1966. With 900 tons of waste being dumped daily, however, the bank grew 15 meters in elevation in a single year and a large pool had formed behind it. The instability and potential hazard of dams constructed of this material are clear.

Reducing Flood Damage

Despite the huge losses of life and property that have resulted from flooding, settlement in river valleys has not been discouraged. In fact, evidence indicates that occupancy of floodplains in many parts of the world is increasing, and at a greater rate than overall population increase. Flood losses, therefore, will no doubt continue to mount. What action has been or can be taken then either to reduce or to minimize flood damage? The efforts, at present, have been expended in two different directions:

1.   Water control includes any attempt to limit erosion and runoff in drainage basins, or in stream channels, through various structural modifications, as well as projects to store flood waters in reservoirs, for example.

2.   Floodplain management, based on the recognition of flood hazards, includes such nonstructural measures as the initiation of floodplain zoning and appropriate land use restrictions, in combination with flood forecasting and emergency protection measures.

Water Control

Watershed Management.   All measures of flood protection have their limits, including those involving "land treatment." The goal in such watershed management is to reduce or delay surface runoff before it gets to a river channel and either initiates or contributes to a flood. Efforts in this area have involved soil protection, planting or replanting vegetation, and small water control structures such as ponds and check dams. To the extent that the infiltration capacity of the soil can be increased, surface runoff and soil erosion can be decreased. However, the relationship between rainfall and runoff is very complex, and on the whole, floods seem to flow out of woodlands as well as farms. Any excess rainfall, over that which the land can absorb, will run off and soon reach a stream channel. Experience seems to indicate that both small streams and small floods can be modified to some degree by improved land practices in tributary or upstream areas.

Great floods in downstream areas, however, occur independently of vegetation and land surface.

Dams and Reservoirs.   The construction of dams and the utilization of reservoirs are very old and very direct ways of providing flood protection for downstream areas. About 30 percent of the reservoir capacity of the United States is dedicated to such flood control. To provide full control a storage reservoir should have enough capacity to retain excess runoff for a flood of a given magnitude (see table 7.2), and then the ability to release the stored water downstream after the flood has subsided (see figure 7.37). Mismanagement can lead to unavailable storage capacity at the time of maximum runoff or to too hasty release of water adding to downstream flooding. Reliable runoff forecasting and experienced management are necessary, because reservoirs can always be expertly handled in retrospect. Although higher lands are kept dry throughout a flood, farmers and home owners along a reservoir may see these structures as offering "controlled flooding" rather than flood control (Hoyt and Langbein, 1955).

Controversies have long existed over the value of more smaller dams and reservoirs in upstream locations as contrasted with fewer large dams in downstream areas. Both have advantages and limitations. Small reservoirs spread the protection to upstream areas and are less likely to take agricultural or forest land out of production or to displace large numbers of people. the U.S. Soil Conservation Service, the U.S. Forest Service, and various conservation organizations have long held this view. Larger dams are more effective in controlling large downstream floods, and, therefore, in protecting populated areas. They can usually be used to generate hydroelectric power as well. Their reservoirs, however, often require the displacement of entire communities, can lead to the loss of extensive areas of productive farmland, and usually destroy numerous fish and wildlife habitats. The main advocate of these larger dams in the United States has been the U.S. Army Corps of Engineers. Such large-scale projects have in the past been rationalized on the basis of a benefit-cost ratio of greater than 1.0, over some extended period. Unfortunately, the indirect costs, such as the loss of scenic or natural areas with values that cannot be quantified, have never been considered in such economic analyses.

An example of such large-scale responses to flood control is a dam site selected by the U.S. Army Corps of Engineers on Soquel Creek, a small stream along the central California coast (see figure 7.36). Although flood damage over the years had been relatively minor, the water supply system was ample to meet projections of future use, and the adjacent Monterey Bay is recreational area, the corps proposed a "multiple purpose reservoir" on the creek for flood control, water conservation, and recreation. The reservoir would have flooded about 3.4 square kilometers (845 acres) of redwood forest and some farmland. The following economics were reported by the U.S. Army Corps of Engineers (1966):

Costs:*	Dam and improvements	$28,400,000
	Annual upkeep	1,070,000
Annual Benefits:	Flood control	180,000
	Higher land utilization	40,000
	Water supply	680,000
	Recreation	760,000
	Annual Benefit	$1,660,000

* An interest rate of 3⅛ percent was utilized, resulting in a much more favorable benefit-cost ratio than actual market conditions would have produced.

A benefit-cost ratio of 1.2 based on a 100-year record was calculated. The loss of the redwood forested valley, with its biological habitats and aesthetic value, was not considered. The actual need for the dam could not be justified by any means other than a 100-year excess of calculated benefits over calculated costs. This kind of accounting and project rationalization fortunately is now coming under control.

Large reservoirs do offer a high degree of flood protection, and may be used for power generation and recreation as well. They also, however, represent enormous artificial concentrations of water that are vulnerable to the kinds of geologic hazards previously discussed. Vaiont Reservoir in northern Italy, the St. Francis Dam and Baldwin Hills Reservoir in southern California, and the Teton Dam in Idaho are examples of past failures. In the case of large dams, the secondary effects of construction may ultimately make such a project very questionable.

The examples of embarrassing environmental and fiscal problems posed by dams and reservoirs of all sizes and shapes throughout the United States are widespread. The U.S. Army Corps of Engineers' Fishtrap Dam and Reservoir project in eastern Kentucky is a good example. The $54 million dam was built primarily for flood control. Yet officials allowed free-wheeling coal mining and haul road construction to continue in the area. The resultant sediment accumulation is now expected to impinge on the reservoir's flood control capacity by 1985, sixty-five years earlier than planned. A proposed swimming and fishing area had to be abandoned due to deposition of a 2-meter-thick layer of sediment in one year. An estimated $1.2 million will be necessary to reclaim the damaged land and provide additional sediment control (Odell, 1975).

Essay

CONSEQUENCES OF THE ASWAN DAM, EGYPT

The construction of the Aswan Dam on the upper Nile River in Egypt is an excellent example of the incompletely understood direct and indirect consequences of the construction of a large dam. It was expected that the dam would conserve and control the fertile waters of the Nile, that its hydroelectric power would promote rapid industrialization, and that it would quickly pay for itself. Major miscalculations and side effects of dam construction have all but destroyed this initial optimism.

The annual Nile floods have now been controlled. These flood waters previously had three indispensable functions: (1) they flushed away salts from the soil that would otherwise choke plant life; (2) they swept away the snails carrying Bilharzia larvae, which causes a debilitating or fatal intestinal infection in humans; and (3) they left behind a new layer of fertile soil.

Built without sluices, the dam is trapping all the Nile's silt in its reservoir. The loss of this sediment load has led the clear river water to scour its bed, undermining numerous bridges and dams downstream. To prevent this, the Ministry of the High Dam proposed plans to build ten barrier dams between Aswan and the sea at a cost of $250 million — one-quarter of what the high dam cost.

Due to the decline in sediment discharge and, therefore, delta outbuilding, the Mediterranean Sea is eroding the delta and advancing inland, threatening many square kilometers of rich agricultural land. Approximately one-third of the delta front is actively eroding, some sections as rapidly as 2 meters per year.

Deprived of the Nile's rich sediment, much of the land already has depleted nutrient levels and needs artificial fertilizer. Two-thirds of the annual fertilizer usage is needed to make up for lost fertility and mineral content since Nile silt stopped coming. A number of minerals will also have to be added to the soil with time.

The marine food chain in the eastern Mediterranean has also been broken. The loss of nutrients formerly carried with Nile sediment has reduced plankton populations. As a result sardines, mackerel, and crustaceans have either died off or been driven away. Approximately 30,000 Egyptian fishermen have lost their livelihood.

Perhaps the saddest truth is that, despite all the money that has been spent, there is now not enough water behind the dam, and there may never be. The reservoir capacity, planned to meet demands of Egypt and Sudan in a succession of dry years, was supposed to be reached in 1970. By 1971 the reservoir was only half full and most authorities feel the level may not rise much more in the next 100 years. Two factors have combined to produce large water losses: infiltration of lake water into the underlying permeable Nubian Sandstone and evaporation. The assumption was apparently made during planning stages that the fine clay coming into the reservoir would eventually seal any porous rock underlying the reservoir. Most of the sediment, however, has been settling in the deep center of the basin. Porous sandstone capable of absorbing endless quantities of water is exposed along the new lake's 500-kilometer-long western bank. It is uncertain how much water is escaping or how it can be halted.

It was recognized that the reservoir was to form in one of the hottest and driest places on earth and that losses due to evaporation would be tremendous. However, the calculations failed to account for the high wind velocity over this vast expanse of water. The measured evaporation rate, about 15 billion cubic meters per year, is 50 percent greater than estimated and is half the amount that used to be "wasted" by the Nile flowing unused to the sea.

The one undisputed benefit of the dam has been the conversion of 2,800 square kilometers from flood to canal irrigation, so that several crops can be produced on the land each year. The drawback has been that the canals are providing habitats for the snails that carry the Bilharzia larvae. A healthy human scarcely need set foot in the water to pick up the disease. Because of the absence of indoor plumbing, Egyptian peasants use canals as both toilet and bathing facilities and with the increase in canals and loss of the cleansing flood waters, infection rates have increased considerably. This practice is not likely to stop.

This dam and reservoir, which were symbols of national pride for Egyptians, have proved to be an ecological disaster.

A unique and costly problem plagues the huge Gavins Point Dam on the Missouri River near Yankton, South Dakota. The dam, constructed in 1957, was part of another U.S. Army Corps of Engineers' vast flood control project on the Missouri River, but was constructed on the wrong site. Originally the dam was to be built above the mouth of the Niobrara River, but politicians downstream persuaded the corps to build it near their town about 65 kilometers east of the original site. As expected, the dam prevented a lot of silt from being carried down the river with the spring rains and melting snow. The buildup of silt in the reservoir was far more rapid than anticipated, however. The result has been a sharp rise in the groundwater table at the town of Niobrara 56 kilometers upstream. Water has been seeping into basements of homes in the town of 500, up to 1 meter deep in places. Hundreds of cottonwood trees have rotted and mosquito-breeding swamps thrive where fields of corn once grew. The corps, realizing its mistake, gave the residents a choice — build a dike around the town, go elsewhere to live, or move the town to higher ground. At a cost of $15 million, and the discomfort and displacement of the residents, the town is being moved to an adjacent, less fertile, upland area. How can a government agency like the U.S. Army Corps of Engineers continue to make such mistakes? How much longer are we going to

have to pay the environmental and economic costs of such unplanned disasters?

Channel Alterations. In the lower reaches of rivers, relief is commonly low and floodplains are very large and usually extensively farmed or urbanized. Various types of "channel improvements" or alterations are commonly made in these downstream areas for flood protection.

Channel Clearing and Dredging. By enlarging the discharge capacity of a stream channel, either by cleaning out vegetation or by widening, deepening, or realigning and straightening the channel, flood stages can be decreased to some degree. This work is usually carried out in conjunction with levee construction. A reduction in channel length serves to steepen the average channel gradient and to deepen the channel. A deeper channel allows flood discharges to pass through without rising to levels that would overtop the banks or levees. Stream meanders can also be cut off or straightened to facilitate channel flow and to decrease overbank flooding. Conspicuous examples of cutoff and straightening procedures may be seen along the Mississippi River from below Cairo, Illinois, to the mouth, a total river length of 1350 kilometers (850 miles). In this reach, 530 kilometers (331 miles) of river channel were shortened by 185 kilometers (116 miles) in the 1930s with appreciable benefits. Cutoff maintenance is a continual battle, however, and is only successful to the extent that cutoffs can be stabilized. Those whose job it is to master large meandering rivers come to realize that theirs is an art more than a science (Hoyt and Langbein, 1955).

Diversions and Bypasses. The diversion of flood waters is a common practice in many areas, notable examples being found along the lower Sacramento River in California and along the lower Mississippi River. At certain stages, floodways can be opened to divert some portion of the stream flow through another channel to lower the overall flood stage. Along the lower Sacramento River there are five overflow weirs that are connected to five bypasses. The U.S. Army Corps of Engineers estimates that flood damages prevented by these and other improvements along the lower Sacramento River amounted to $860 million in the period from 1950 to 1969 (U.S. Army Corps of Engineers, 1969).

Along the lower Mississippi River diversions of flood waters at critical times are made through the Birds Point–New Madrid Floodway below Cairo, Illinois, through the Boeuf Floodway between the Arkansas River and Ouachita River, through the Atchafalaya River, and through the Bonnet Carre Spillway into Lake Pontchartrain. The Mississippi floodways, which were constructed by building a pair of levees across the natural floodplain, are placed in operation when the main river reaches a critical stage. They are simply a way to return a portion of the floodplain to the river, which had been cut off by levee construction (Hoyt and Langbein, 1955).

Channel Stabilization. In the southern California area, rainfall and consequently runoff are sometimes irregular. Nearly all the rain for a year may fall within a week's time. The steep slopes of the surrounding mountains, the recurring brush fires that bare these slopes, and the nearly complete urbanization and suburbanization of the flat alluvial plain areas have led to routine flooding of street intersections and low-lying areas. Mudflows commonly occur with the high discharge so that debris is often left scattered through commercial and residential areas. As a result, most of the natural stream channels have been either lined with riprap or concrete to control the flow and any possible flooding (see figure 7.45). Although expensive and aesthetically unpleasant, these washes, or "rivers," such as the Los Angeles River, have usually been effec-

Figure 7.45. *Former Creek Channel Stabilized by Concrete.* PHOTO BY: *Gary Griggs.*

tive in containing high runoff. In some areas, old automobiles have been systematically anchored along stream margins to prevent bank erosion and confine the flow of high water. Although inexpensive and functional, the results are visually not very pleasing (*see* figure 7.46).

Dikes and Levees.   Dikes, levees, or flood-walls are perhaps the most common methods of directly controlling channel overflow. Throughout history, as populations have grown and along with them the need for food, levees and dike systems have been built and expanded to protect fertile floodplain land that has become increasingly more valuable. The construction of these walls eliminates natural overbank flow and floodplain storage, and therefore tends to increase downstream flood peaks.

When flood waters overtop a dike or levee, disaster usually occurs because a large area is quickly inundated with little warning, causing a great amount of personal and property damage. During the worst flood in fifty years in the Pacific Northwest, the Columbia River, flowing at a rate of over 1,000,000 cubic feet per second (cfs), broke through a railroad fill serving as a dike and completely inundated Vanport City, Oregon. The town was totally destroyed, inundated with 3.0 to 4.5 meters (10 to 15 feet) of water. Nineteen thousand residents were left homeless. Living in an area protected by levees therefore involves a certain degree of risk.

Failure of the levees and floodwalls that flank almost 3600 kilometers (2250 miles) of the Mississippi River during periods of high discharge has inundated hundreds of square kilometers of farmland. During the severe 1973 spring floods over 1200 square kilometers of land were inundated (*see* figure 7.47). Over 10,000 homes were destroyed in the delta section alone, and dam-

Figure 7.46. *Stream Bank Stabilization with Automobile Bodies, Corralitos Creek, Central California.* PHOTO BY: *Gary Griggs.*

ages reached $140 million (see figure 7.48). These severe floods had their beginnings in the mild, wet fall and winter of 1972. Reservoir levels and tributary streams were well above normal throughout the basin when heavy spring rains, which continued through March and April of 1973, began to fall.

The flooding on the main stem was the highest ever observed over a 595-kilometer

Figure 7.47. *Mississippi River Flooding. Inundation of farmland along the Mississippi River near Wilson, Arkansas, during April 1973 floods.* PHOTO BY: *James Sanders, U.S. Army Corps of Engineers.*

Figure 7.48. *Mississippi River Flooding. Flooding along Mississippi River near Caruthersville, Missouri, March 1973.* PHOTO BY: *Charles Franks, U.S. Army Corps of Engineers.*

(370-mile) reach of the river (Chin et al., 1975). New records were also set for consecutive days above flood stage for most main-stem gauging stations from southern Iowa to Louisiana. For example, the Mississippi River remained above flood stage for seventy-seven days at St. Louis, Missouri, ninety-seven days at Chester, Illinois, sixty-three days at Memphis, Tennessee, and eighty-eight days at Vicksburg, Mississippi. The total sediment discharge to the Gulf of Mexico from March through June was approximately 240 million tons.

Although flood control facilities on the Mississippi River are designed to control extreme flood conditions, during a flood of this magnitude a great deal of effort is continually necessary to keep facilities intact and operating. In 1973, despite the facilities and the effort, damage was still severe. North of St. Louis thirty-nine levees were breached or overtopped, flooding many square kilometers of farmland and unprotected communities. Extensive tributary flooding occurred in Arkansas, Tennessee, and Mississippi. Over 53,000 square kilometers (13,000,000 acres) of land were inundated (Bolt et al., 1975), over 10,000 homes were destroyed in the delta section alone, and damages reached $1.2 billion.

Floodplain Management

Flood Forecasting.   A great amount of life can be saved and property protected through an adequate flood forecasting system. Entire towns can be evacuated and valuable property, goods, or livestock can be moved to higher ground if ample warning occurs that flood conditions are imminent. The River and Flood Forecasting Service, operated by the U.S. National Weather Service, has sev-

eral offices around the country that issue river and flood forecasts to their various districts. Cooperation and coordination have also been developed between concerned organizations and agencies such as the U.S. Coast Guard, the National Guard, the U.S. Army Corps of Engineers, and the American Red Cross to aid in evacuation of people and removal or protection of property.

Through long-term analysis of stream flow data, the National Weather Service has compiled information and prepared detailed graphs illustrating the probabilities of certain water levels or flood conditions for individual rivers (see figure 7.49). The plots can then be utilized in making land use decisions in these areas.

Floodplain Planning.   Despite preventative measures floodplain damages have increased each year as a direct consequence of more intensive land use and population growth. This trend is partly a result of the obvious value and usefulness of floodplain lands (flat and fertile land with available water), but it is also a consequence in many areas of the failure of the average person, whether a home builder, real estate salesman, or merchant, to actually understand the geologic and hydrologic significance of a floodplain. Simply stated, these attractive, fertile floodplains have been constructed through time by natural overbank flow. People first, therefore, have to be aware that a floodplain and, therefore, a flood hazard actually exist.

An analysis of long-term discharge records for two river basins in Pennsylvania indicates the general relationship between the frequency and intensity of floods (see figure 7.50). Several definitions will make this figure and the concept of frequency of

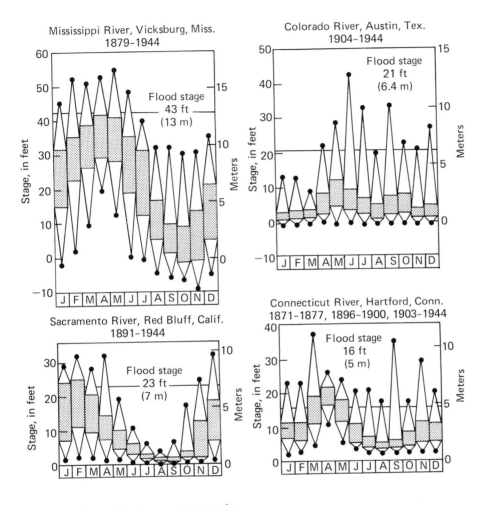

Key to flood expectancy graphs

During 25% of the years
of record the maximum
monthly stage fell in
this range

● Maximum of record

▢ Upper quartile

During 50% of the years
of record the maximum
monthly stage fell in
this range

Lower quartile

During 25% of the years
of record the maximum
monthly stage fell in
this range

● Lowest monthly
maximum of record

Figure 7.49. *Flood Expectancy Graphs for Four United States Rivers. Note the different months during the year when major flood peaks occur for different parts of the nation (broken line indicates flood stage for each river.)* REDRAWN FROM: *A. N. Strahler,* Physical Geography, *John Wiley & Sons, Inc.*

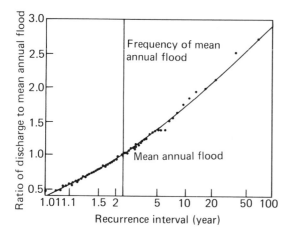

Figure 7.50. *Regional Flood Frequency Curve for Two River Basins in Pennsylvania. Using such a graph, the recurrence interval for various discharges or floods, relative to the mean annual flood, can be determined. For instance, a flood twice as great as the mean annual flood occurs only once in eighteen years on the average.* REDRAWN FROM: *Fluvial Processes in Geomorphology, by L. B. Leopold, M. G. Wolman, and J. P. Miller. W. H. Freeman and Company. Copyright © 1964.*

flooding (see boxed essay on flood frequency analysis) more meaningful. First, the annual flood is simply the highest discharge for any given year. The arithmetic mean of these values over a period of years is called the mean annual flood. In figure 7.50 we can see that the mean annual flood has an average recurrence interval of 2.3 years; that is, once every 2.3 years, on the average, the highest flow of the year will equal or exceed the value of the mean annual flood. Floods with twice the discharge volume of this mean annual flood occur every 18.0 years on the average. In contrast, studies of many rivers have revealed that, with few exceptions, the bankful discharge (or the level at which the stream will just start to top its banks) has a recurrence interval of 1.0 to 2.0 years with an

average of 1.5 years. This means that once every 1.5 years, or two years out of three, on the average, the highest flow of the year will equal or exceed bankful conditions (Leopold, 1974). A river flowing at bankful will not be at overflow level everywhere along the channel, however, because there are always slight variations in height of banks or depth of channel.

If accurate, large-scale topographic maps are available for an area, delineation of stream or river floodplains is relatively easy to accomplish and should be carried out in all areas where this has not already been done. Historic records of river stages during past floods can be utilized to determine the recurrence intervals for floods of various magnitudes. With aerial photographs and topographic maps, the areas that could be inundated in the event of a flood of a given stage or recurrence interval can then be delineated (see figure 7.51). Potential risk for various portions of the floodplain can then be determined. Finally, zoning or land use criteria can and should be established to control human activity.

However, floodplain zoning, like seismic zoning, is difficult to accomplish because floodplains are already heavily used, and because of the likelihood that damages will be recovered through federal relief funds. Floodplain zoning does not necessarily mean evacuation, but it does involve serious consideration of flooding as a guide to land use planning.

Agricultural land (grazing land in particular), recreational areas, even parking lots, all represent logical uses of floodplains. The construction of homes, clearly, is an example of unwise usage of the floodplain. Because of the effects of flooding on structures and their occupants, and the moral obligation for rehabilitation and relief imposed

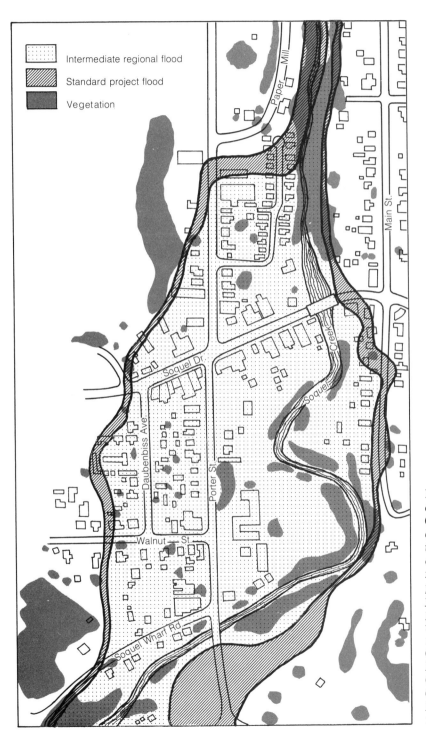

Figure 7.51. *Development on Floodplain, Soquel Creek, California. The areas delineated on the map are those that would be inundated by (1) the intermediate regional flood (approximately the 100-year flood), and (2) the standard project flood (approximately the 200-year flood).* RE-DRAWN FROM: *U.S. Army Corps of Engineers, "Flood Plain Information, Soquel Creek, Santa Cruz County, California" (San Francisco District, Corps of Engineers, 1973).*

Essay

FLOOD FREQUENCY ANALYSIS

The objective of a flood frequency analysis is to determine how often, on the average, we can expect to have floods of various magnitudes. The recurrence interval is the average time span between natural events of a given size (for instance, the 50-year flood, the 100-year storm, and the like). In the case of floods it is derived from stream discharge data. A frequency analysis can be performed very easily:

1. Obtain the stream flow records for a selected gauging station for all the years of record.

2. Select a high discharge rate, or stage height (such that no more than three or four peaks a year, on the average, exceed this) and tabulate all of the discharges over the period of record that are greater than this volume or height.

3. Rank the discharges in decreasing order.

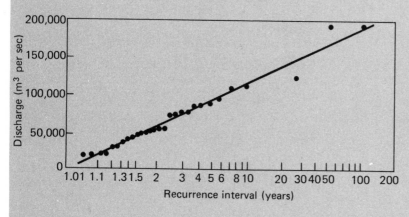

Figure 7.52. *Flood Frequency Curve for Eel River at Scotia, California. This graph, based on data collected from 1932 to 1959, indicates how often on the average a given discharge will occur.* DATA FROM: *U.S. Geological Survey, Menlo Park, California.*

4. Utilizing the formula $T_r = \dfrac{n+1}{m}$ where T_r = average recurrence interval in years, n = number of years of record, m = rank of individual discharge in the series, take selected discharges, or stages, and determine recurrence intervals for each.

5. Using these values, construct a graph of recurrence interval versus discharge (see figure 7.52). Plot recurrence interval on a log scale. Depending upon the number of years of record, the graph can be extended to estimate the discharge or stage of the 50- or 100-year floods.

The relationships between flood magnitude and frequency are a function of the size of the drainage basin, runoff characteristics, climate, and the human alterations of the system (dams, diversions, and the like). The determination of the recurrence intervals for floods of a certain magnitude is of critical importance to the design of any engineering structures that deal with runoff and also to the rational planning of floodplain land usage.

Two additional terms related to flood magnitudes are often used in flood planning. The **intermediate regional flood** is another phrase used to describe the 100-year flood. Because 100 years of data are not usually available, an extrapolation of the recurrence interval graph must be made considering the regional hydrologic characteristics as well as the individual watershed. The **standard project flood** is a major flood that can be expected to occur from a severe combination of meteorological and hydrologic conditions. The conditions are those that are considered reasonably characteristic of the geographic area involved. Such a flood might have a recurrence interval of several hundred years or more. It is called a standard project flood because flood control projects are normally designed for this magnitude of flood.

upon the remainder of the affected community, and the entire nation for that matter, there is no justification for the construction of schools, hospitals, or housing developments in hazardous floodplain areas. A number of communities are buying stream frontage for parks and recreational uses and, at least at this point, restricting any further building on areas prone to flooding. Although floodplain zoning regulations may exist, their usefulness or enforcement may be limited or nonexistent.

There are many towns and cities that have never learned their lesson. Albuquerque, New Mexico, on the Rio Grande, is a good example of a city that has expanded onto the floodplain of a passive river, and on the beds of "dry" washes or **arroyos** that drain the adjacent mountains. Many areas of the Southwest are exposed to cloud bursts and flash floods that can turn arroyos into raging torrents within hours. Many of Utah's cities are built on flood hazardous alluvial fans and at the mouths of canyons through which perennial streams drain. Because residents have become so firmly rooted, relocation is more and more remote, and cloud bursts and their destruction have simply become accepted as one of the hazards or costs of living in some of these areas.

Protective zoning of remaining unpopulated floodplain areas, elimination of disaster redevelopment funding and government-financed housing developments in such areas, and the protection, flood proofing, or relocation of high occupancy structures seem to be the most logical approach to this problem.

National Flood Insurance Program

In December 1973 Congress passed the Federal Flood Disaster Protection Act. A series of catastrophic floods in 1972 prompted the federal government to consider substantial revisions of the original (1968) Flood Insurance Act. These floods, resulting in extraordinary human suffering and property loss, indicated that even when federal insurance is purchased, it did not cover actual damages because of unreasonably low statutory limitations on property coverage. Furthermore, homes that were swept away in floods were often rebuilt again in the same flood-prone area due to inadequate land use controls.

The 1973 program was initiated to provide flood insurance at rates made affordable to property owners through a federal subsidy. It applies not only to areas subject to normal flooding, but also to flood-related mudslide and shoreline erosion hazards. Eventually it is expected that this insurance will replace disaster assistance. A community is first notified by the U.S. Department of Housing and Urban Development (HUD) that it is flood prone. HUD uses the area covered by a 100-year recurrence interval flood in its consideration of what is flood prone. The community then must enter the program within a certain time period or face denial of federal government assistance and federally insured loans or mortgages for projects located in these areas.

One of the more far-reaching provisions of the act is that communities, upon entering the program, must adopt and enforce stringent land use controls that will regulate new building construction in the flood-hazard area. Land use requirements have been established by HUD in two phases. At the time a community enters the program, it must, at a minimum, set up a review procedure for building permits, subdivisions, and water and sewage systems in flood-hazard areas to ensure that all new construction is consistent with the need to minimize flood damage. More restrictive land use measures are required when HUD provides the community with detailed maps and flood data. At this stage it must prohibit any development in the floodway that could inhibit flood flows and must require the elevation or flood proofing of all structures in areas subject to flooding that are outside the floodway but within the floodplain.

Further provisions of the act encourage the development of comprehensive flood-plain management plans in flood-prone areas. The primary objectives of these plans should be the long-term reduction of flood losses through open-space tax incentives, acquisition of development rights, initiation of emergency preparedness programs and flood warning systems, and the establishment of detailed flood-proofing requirements for existing structures in flood-prone areas. In combination with these approaches, a management program could also investigate the construction of flood control structures (levees, dams, channelization) where they might be consistent with environmental objectives and economic constraints.

REFERENCES

References Cited in the Text

Bolt, B. A. et al. *Geological Hazards.* New York: Springer-Verlag, 1975.

Brown, W. M. III, and Ritter, J. R. *Sediment Transport and Turbidity in the Eel River Basin, California.* U.S. Geological Survey Water Supply Paper 1986, 1971.

California Water Resources Control Board. *A Method for Regulating Timber Harvest and Road Construction Activity for Water Quality Protection in Northern California.* Publication no. 50, 1973.

Cargo, D. N., and Mallory B. F. *Man and His Geologic Environment.* Reading, Mass.: Addison-Wesley Publishing Company, 1974.

Chin, E. H.; Skelton, J.; and Guy, H. P. *The 1973 Mississippi River Basin Flood: Compilation and Analyses of Meteorological, Stream Flow and Sediment Data.* U.S. Geological Survey Professional Paper 937, 1975.

Clements, T. "St. Francis Dam Failure of 1928." In *Engineering Geology in Southern California.* Los Angeles: Association of Engineering Geologist Special Publication, 1969. Pp. 89–92.

Davies, W. E.; Bailey, J. F.; and Kelly, D. B. *West Virginia's Buffalo Creek Flood: A Study of the Hydrology and Engineering Geology.* U.S. Geological Survey Circular 667, 1972.

Detwyter, T. R., Marcus, M. G. et al. *Urbanization and Environment.* Belmont, Calif.: Duxbury Press, 1972.

Environmental Protection Agency. *Processes, Procedures, and Methods to Control Pollution Resulting from Agricultural Activities.* Washington, D.C.: U.S. Government Printing Office, 1973.

Guy, M. P. *Sediment Problems in Urban Areas.* U.S. Geological Survey Circular 601-E, 1970.

Hoyt, W. G., and Langbein, W. B. *Floods.* Princeton: Princeton University Press, 1955.

Janda, R. J. Testimony concerning sediment load in northern California and southern Oregon presented in public hearings before the California North Coast Regional Water Quality Control Board, Eureka, Calif., July 19, 1972.

Kazmann, R. G. *Modern Hydrology,* 2nd ed. New York: Harper and Row, 1972.

Kiersch, G. A. "Vaiont Reservoir Disaster." *Geotimes* 9,9(1965): 9–12.

Knott, J. M. *Effects of Urbanization on Sedimentation and Flood Flows in Colma Creek Basin, California.* U.S. Geological Survey Open File Report, Technical Report no. 6 (1973).

Leopold, L. B. *Hydrology for Urban Land Planning: A Guidebook on the Hydrologic Effects of Urban Land Use.* U.S. Geological Survey Circular 554, 1968.

——————. *Water: A Primer.* San Francisco: W. H. Freeman and Company, 1974.

Odell, R. "Silt, Cracks, Floods, and Other Dam Foolishness." *Audubon* 77,5(1975): 107–14.

Rantz, S. E. *Urban Sprawl and Flooding in Southern California.* U.S. Geological Survey Circular 601-B, 1970.

Sartor, J. D.; Boyd, G. B.; and Agardy, F. J. "Water Pollution Aspects of Street Surface Contaminants." *Journal of the Water Pollution Control Federation* 46,3(1974): 458–67.

Thelen, E. et al. *Investigation of Porous Pavements for Urban Runoff Control.* Washington, D.C.: U.S. Government Printing Office — EPA, 1972.

Thomas, H. E., and Schineider, W. J. *Water as an Urban Resource and Nuisance.* U.S. Geological Survey Circular 601-D, 1970.

U.S. Army Corps of Engineers. *Survey Report for Flood Control and Allied Purposes, Soquel Creek, Santa Cruz County, California.* San Francisco: Corps of Engineers, 1966.

——————. *Water Resources Development in California.* San Francisco: Corps of Engineers, 1969.

U.S. Dep't. of Agriculture–Soil Conservation Service. *Controlling Erosion on Construction Sites.* Agricultural Information Bulletin 347 Washington, D.C.: U.S. Government Printing Office, 1970.

U.S. Geological Survey. *Erosion and Sedimentation.* Washington, D.C.: U.S. Government Printing Office, 1972.

Vice, R. B.; Guy, H. P.; and Ferguson, G. E. *Sediment Movement in an Area of Suburban Highway Construction, Scott Run Basin, Fairfax County, Virginia, 1961–64.* U.S. Geological Survey Water Supply Paper 1591-E, 1969.

White, G. F., and Haas, J. E. *Assessment of Research on Natural Hazards.* Cambridge, Mass.: The MIT Press, 1975.

Wolman, M. G. "A Cycle of Sedimentation and Erosion in Urban River Channels." *Geografiska Annaler* 49-A(1967): 385–95.

——————, and Schick, P. A. "Effects of Construction on Fluvial Sediment, Urban and Suburban Areas of Maryland." *Water Resources Research* 3,2(1967): 451–62.

Other Useful References

Bue, C. D. *Flood Information for Flood Plain Planning.* U.S. Geological Survey Circular 539, 1967.

Dendy, F. E. "Sedimentation in the Nation's Reservoirs" *Journal of Soil and Water Conservation* 23,4(1968): 135–37.

Hinson, H. G. *Floods on Small Streams in North Carolina: Probable Magnitude and Frequency.* U.S. Geological Survey Circular 517, 1965.

Judge, J. "Florence Rises from the Flood." *National Geographic* 132(1967): 1–43.

Leopold, L. B.; Wolman, M. G.; and Miller, J. P. *Fluvial Processes in Geomorphology.* San Francisco: W. H. Freeman Company, 1964.

Schneider, W. J., and Goddard, J. E. *Extent and Development of Urban Flood Plains.* U.S. Geological Survey Circular 601-J, 1974.

Shaeffer, J. R.; Ellis, P. W.; and Spieker, A. M. *Flood-Hazard Mapping in Metropolitan Chicago.* U.S. Geological Survey Circular 601-C, 1970.

Strahler, A. W., and Strahler, A. H. *Environmental Geoscience.* San Francisco: Hamilton Publishing Company, 1973.

U.S. Environmental Protection Agency. *Processes, Procedures and Methods to Control Pollution Resulting from Construction Activities.* Environmental Protection Agency Publication 430/9–73–007. Washington, D.C.: U.S. Government Printing Office, 1973.

CHAPTER 8

Groundwater

Contents

INTRODUCTION

IF we cooperate with nature to efficiently utilize the groundwater supplies in the United States, the potential volumes available are vast. The estimated fresh water stored beneath the ground is 50 times the annual surface runoff and 150 times as large as the existing surface storage (Kazmann, 1972). Our increasing population and demand for water, however, may make even this source insufficient in time.

These groundwater reservoirs, or aquifers, have many advantages over those at the surface: no construction costs, no silting up, essentially no evaporation, and no hazards created. Existing below the ground surface, they do not take up or destroy any valuable surface area. Groundwater reservoirs must be carefully protected, however, and watched closely so that they are not damaged. As pressure has increased to control or eliminate various waste water discharges to rivers, lakes, and the marine environment, the attractiveness of subsurface disposal has increased. Underground disposal or accidental discharge of toxic and long-lived waste fluids, such as radioactive material, can damage or totally destroy a groundwater reservoir.

Overdraft, or the continued lowering of the **water table** due to excessive withdrawal, can eventually damage an aquifer. A decline in water quality can occur, perhaps due to saltwater intrusion; excessive pumping can also result in ground subsidence (see chapter 4). Because of the magnitude of our groundwater supplies, our dependence upon them, and the threats of contamination and overwithdrawal, it is essential to have some knowledge and understanding of this subsurface water system, how it operates, and what kind of precautions are necessary for its protection.

THE GROUNDWATER SYSTEM

The groundwater zone is that volume at some depth, or in some cases at the surface, where all the voids in the rock or soil overburden are filled with water. This water is recharged or replenished either by the downward infiltration of surplus soil water due to gravitational movement or at times by percolation from streams and lakes. This water is referred to as **meteoric water**, or that derived originally from precipitation. **Connate water**, which is water trapped in the pore spaces of sedimentary rocks at the time of their formation, and **juvenile water**, that added to the earth's surface from magmatic or volcanic sources (commonly as hot springs), also occur in small quantities in the subsurface. Both connate and juvenile water,

however, often contain undesirable salts or minerals and are therefore of poor quality. Groundwater in the San Joaquin Valley of California, for instance, commonly contains considerable boron brought to the surface from connate water at great depths (Linsley, Kohler, and Paulhus, 1958).

Two concepts of fundamental importance in dealing with groundwater are those of porosity and **permeability**. Porosity is the total volume of pore or void space in a rock and is determined by the type of rock or material, its grain size, sorting, and fracturing (see figure 8.1). Porosity, expressed as percent of void space, controls the volume of water that any earth material can hold. Lithified sedimentary rock, such as shale and limestone, commonly may have porosities of 5 to 10 percent, whereas uncon-

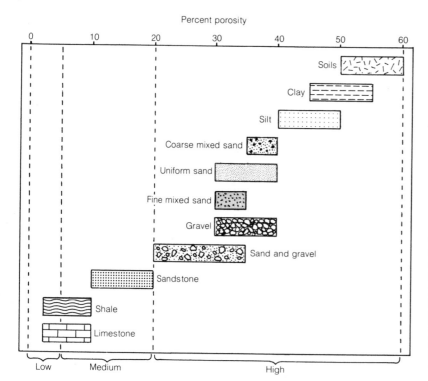

Figure 8.1. *Variations in Porosity. The grain size, shape, arrangement and amount of cementation or solution will all affect the porosity or pore space in a material.* BASED ON: *Fig. 7.5 from R. C. Ward,* Principles of Hydrology *(*MAIDENHEAD, *Berkshire, England: McGraw-Hill (UK) Ltd, 1975), p. 191.*

solidated material, alluvial sands and gravels, can have porosities up to 35 or 50 percent, and can therefore store large quantities of water.

The permeability of a material is its ability to transmit fluids through its pore spaces. Although porosity is an important factor in determining permeability, highly porous materials are not necessarily highly permeable. To be permeable, a formation must not only have many open spaces between the constituent particles, but these spaces must also be sufficiently large and interconnected to allow flow. For example, some clays may have total porosities as high as 75 percent, but their spacing may be so small and unconnected that water cannot easily pass through. On the other hand, sands may have a lower total porosity than clay because of unequal particle size and sorting. However a sandy material will normally have a higher permeability because there are larger and more connected spaces between the sand particles than between the clay particles. Usually permeability is expressed as the rate of discharge per unit area under controlled hydraulic conditions.

Aquifers

A geologic material or rock unit that both contains and yields significant amounts of water is called an aquifer. Aquifers can perform three principal functions. They act as:

1. filters

2. pipeline or transmitting devices

3. reservoirs (Kazmann, 1972)

Aquifers are natural filters of impurities both at the intake point and as groundwater moves through water-bearing formations.

Suspended material is filtered out by surface soil and underlying formations; and as groundwater moves through aquifers, dissolved constituents are removed by **adsorption**, ion exchange, or other processes. However, this filtering mechanism is not always effective. Leaking septic tanks and sewage lines, unsealed landfills, and sewage disposal sites can allow pollutants to pass directly into groundwater reserves where the water table is near the surface, the soil mantle is thin, or the aquifer is very permeable. Further, some contaminants such as salts, petroleum products, and certain dissolved chemicals pass relatively intact into and through an aquifer even after percolating through a considerable thickness of soil.

Aquifers transmit water from intake area to subsurface locations or to and from surface lakes, streams, and wetlands. This surface-subsurface interconnection is important in that many aquifers depend upon surface water bodies for all or part of their recharge. Conversely, streams and ponds depend upon aquifer-fed springs and seeps for supplementary water to provide **base flow** during dry periods. Thus pollution or depletion of surface water can reduce the quantity and quality of groundwater.

Aquifers are also natural storage reservoirs for groundwater used for drinking, irrigation, and industry. Groundwater provides about 20 percent of the United States water supply. Many urban communities and most rural communities depend upon this source for their water supply.

The most productive aquifers consist of clean or well-sorted, coarse-grained sand or gravel — the coarser and cleaner deposits being the most productive. Mixtures of sand and gravel, clean finer-grained sand, and mixed alluvial deposits, in that order, are progressively less productive although they still may be good aquifers. Added factors of importance include the areal extent and thickness of the aquifer. In consolidated rocks, secondary factors, such as **cementation** in sedimentary rocks, solution in limestones and calcareous rocks, fracturing or jointing in **igneous rocks**, are all important in affecting aquifer potential. For instance, although we would not normally think of granite as an aquifer, localized areas of intense fracturing may contain adequate water for individual wells.

Most of the highly productive aquifers are unconsolidated gravels and sands found along watercourses and in floodplains, abandoned river beds, alluvial valleys, and coastal plains. Most of the water supply for industrial areas in the East and Midwest is pumped from such deposits, as is irrigation water from California's San Joaquin Valley — one of the most productive aquifers in the country. Cavities in carbonate rocks, such as limestone, formed by groundwater dissolu-

tion often provide excellent reservoirs. Examples of such aquifers are found in central Tennessee, the Shenandoah Valley of Virginia, and parts of Georgia, New Mexico, and Missouri. Other types of consolidated rocks such as gypsum, sandstone, **conglomerate**, and some volcanic rocks can yield significant quantities of water depending upon local geologic properties.

Water, then, enters an aquifer from percolation and recharge and flows slowly under the influence of gravity (see figure 8.2). At some depth beneath an aquifer there is impermeable rock that serves to confine the groundwater. Impermeable layers or strata that cannot transmit water but that can contain it are called **aquicludes**. Groundwater or an aquifer that is not overlain by impermeable material is designated as **unconfined**. The upper surface of the groundwater zone in this situation is simply the water table. The elevation of the water table may vary considerably, commonly following the to-

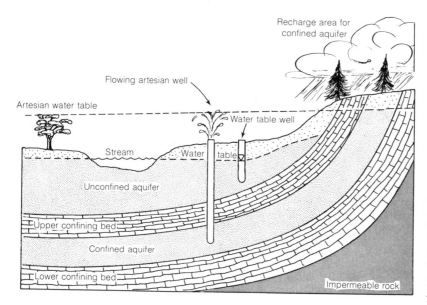

Figure 8.2. *Groundwater System. Depending upon the presence or absence of an overlying confining layer, a groundwater aquifer can be either confined or unconfined. A well tapping a confined aquifer is called an artesian well, which will often flow freely at the surface.*

pography to some degree, but the pressure at this level is always equal to atmospheric pressure. A water table is **"perched"** if the aquifer lies over impermeable strata and the water is not able to percolate through the soil to the normal water table lying below.

In **confined aquifers**, the groundwater is overlain by relatively impermeable layers and is under greater than atmospheric pressure. Thus a well that penetrates a confined aquifer is called an **artesian well**, and the water in it will rise above that of the water table (see figure 8.2). If the pressure is great enough, or the location of the well is right, water may actually flow out of the artesian well at the surface. A close look at the relationships in figure 8.2 shows why a clear understanding of the subsurface geology is essential in the selection of a good well site. By mapping the surface geology and the attitudes of the bedding units, accurate cross-sections can be plotted that will delineate the subsurface structure, and the relative depth and position of existing aquifers.

Without human interference, a groundwater basin fills with water and discharges the excess in several ways. Streams that intersect the water table and receive flow from the groundwater are called **gaining streams**. Those that recharge the groundwater are known as **losing streams**. Whether a stream is a gaining or a losing stream depends on the season of the year and the water level (see figure 8.3).

Where an aquifer underlain by an impermeable layer intersects the earth's surface, a spring or a line of springs may form. Such springs may constitute the headwaters of a creek and may be year-round or, in some cases, seasonal. Most natural springs are relatively small and marked at the surface by denser or greener vegetation than the surrounding areas.

Changes in Groundwater Level

Groundwater levels fluctuate considerably, both on an annual basis and over longer periods of time. Variations in the amounts and rates of recharge and discharge are the major factors influencing these levels. The direct effects of precipitation and river stage on the level of the water table have both been studied and documented (Linsley, Kohler, and Paulus, 1958; Strahler and Strahler, 1973). Normally we would expect the water table, if an aquifer is recharged primarily by precipitation, to reflect the seasonal increase in intensity and duration of rainfall by rising in the winter and subsequently dropping in the summer and fall (see figure 8.4). Water levels in wells adjacent to major rivers show

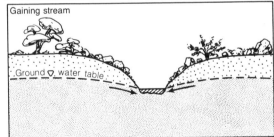

Figure 8.3. *Losing and Gaining Streams. A stream that recharges the groundwater table is called a losing stream, whereas one that is fed by the groundwater is called a gaining stream. Some streams will change seasonally, being losing streams in winter and gaining streams during dry summer months.*

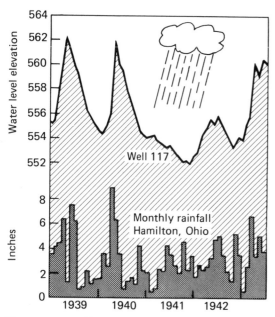

Figure 8.4. *Effect of Seasonal Rainfall on Groundwater Table. Note close correlation between periods of high rainfall and level of water table.* ADAPTED FROM: *F. H. Klaer and R. G. Kazmann, "A Quantitative Study of the Well Fields of the Mill Creek Valley Water Supply Project, Butler County, Ohio," U.S. Geological Survey, Open File Report (1943).*

marked annual fluctuations that correlate with seasonal river stages (*see* figure 8.5).

In addition to the natural fluctuations in the water table and the **drawdown** or lowering of the water table by pumping from wells, a number of construction or engineering projects can either raise or lower the water table significantly, often with detrimental effects. From past experience, and a knowledge of the permeability of the rock units involved, a geologist can evaluate the effect of a major road cut or dam on an aquifer and at least consider its suitability prior to the initiation of any major project.

In the construction of Glen Canyon Dam in 1959 on the Colorado River, the permea-

bility of the Navajo sandstone surrounding the reservoir site was not adequately evaluated. Although the U.S. Bureau of Reclamation calculated a 15 percent "**bank storage** factor" (the volume of water that would percolate into the surrounding rocks) before construction, a check in 1965 indicated 1,900,000 acre-feet, or 30 percent of the total amount of water stored in Lake Powell behind the dam, had seeped into the porous lake bottom and sides since the dam had been completed. This water is essentially in dead storage, that is, it is not readily available for use (Sierra Club, 1965).

The creation of deep cuts or extensive fills can have major effects on both the groundwater and the surface water regime. Extensive filling, or the emplacement of abutments, retaining walls, or sheet pilings can obstruct groundwater flow, raise the water table and drown vegetation, and alter runoff and recharge characteristics of an area. Extensive highway cuts can behead aquifers present in soils and shallow bedrock and thus destroy shallow local water supplies (see figure 8.6). Cutting deep into an aquifer can lower the water table by allowing the groundwater to flow out along the excavation or face of the cut. This discharge permanently lowers the water table to some new equilibrium level, resulting in a reduction of the volume of stored water, a decline in well yield, and even dry wells (Parizek, 1971). Slope failures may also occur along such cuts due to seepage pressure and unconfined slopes.

A deep cut for a new county highway in the Santa Cruz Mountains passed directly through the major aquifer in the area. The excavation penetrated to the underlying confining layer and as a result the water now drains from the aquifer onto the road. The drop in the water table and decline in dis-

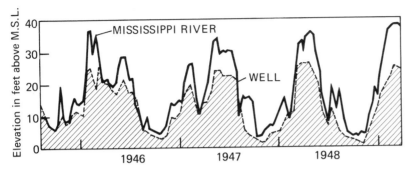

Figure 8.5. *Effect of Changing River Level on Groundwater Table. As water levels in the Mississippi River have fluctuated, corresponding changes have occurred in an observation well 2 miles away.* ADAPTED FROM: *R. R. Meyer and A. N. Turcan, Jr., "Geology and Ground-Water Resources of the Baton Rouge Area, Louisiana," U.S. Geological Survey Water Supply Paper 1296 (1953), p. 20.*

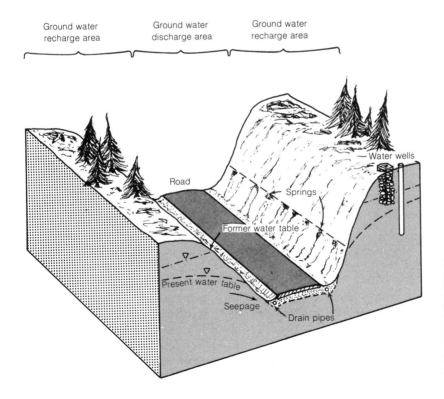

Figure 8.6. *Effect of a Road Cut on the Water Table and Adjacent Wells. Making a major cut through an aquifer will (1) lower water table, leading to dry wells; and (2) lead to seepage and, therefore, instability along the road cut.*

charge from adjacent wells led to a court case against the county.

The water table can be raised by the creation of a large reservoir behind a dam. If the material beneath and adjacent to the reservoir is permeable, water will migrate into this permeable layer, eventually raising the surrounding water table. This may have a positive effect for some farmland, or, on the other hand, it may create a saturated lowland area with little use, possibly permanently altering the preexisting land use patterns.

The Gavins Point Dam on the Upper Missouri River is a case in point (see chapter 7). Due to the improper location of the dam and the excessive siltation in the reservoir, the water table upstream began to rise considerably. As a result, water has seeped into basements, cornfields have turned to swamps, and an entire town will have to be displaced. On the other hand, although Glen Canyon Dam on the Colorado River is an extreme example of percolation of water into surrounding geologic strata, no adverse effects from it have yet been detected.

Effects of Impervious Surfaces

In addition to altering the groundwater regime, human activities can also impede the process of groundwater recharge. Many aquifers are recharged from streams or other surface water bodies. However, an aquifer may receive a substantial portion of its water from a readily identifiable surface recharge area. Development of such an area, and the accompanying coverage of this surface with impervious material, will prevent recharge of the underlying aquifer by physically sealing the subsurface from percolation. This can result not only in a depleted aquifer and, in time, severe impairment of the aquifer's ability to yield stored water, but also in accelerated surface runoff and increased flooding (see chapter 6).

Methods have recently been recognized that protect recharge areas where established communities are expanding and threatening to reduce aquifer recharge. One is the development of a porous asphaltic pavement that allows percolation of precipitation into the soil (see chapter 6). Communities can also zone recharge areas for low density development to reduce the amount of impervious surface per unit area. The township of Southampton on Long Island, after suffering water shortages, studied the areal geology and determined an important recharge region was an undeveloped glacial **moraine** on the north side of the island. It subsequently zoned the 70 square kilometers in this area for very low density residential development (Thurow, 1975).

UTILIZATION OF GROUNDWATER

Water Wells

Digging a well provides access to the groundwater reservoir, which can then be utilized by a pumping system. For the needs of one family or a small farm, where water is plentiful, a shallow well is usually sufficient. Drilling a well, especially a deep or a dry well, can be expensive, time-consuming, and frustrating. Before selecting a site and embarking on such a project, a survey of the surrounding surface and subsurface water availability (information on well depths, groundwater levels and their fluctuations, and the nature of nearby streams) and the surface and subsurface geology can prove extremely useful. If a shallow well is sufficient, whether vertical or horizontal (in the case of side hill wells) it can be (1) hand dug; (2) bored — using some type of power-driven auger; (3) jetted — where the cutting action is supplied by a directed stream of water; or (4) driven — where a series of connected lengths of perforated pipe are driven into the ground below the water table.

Where the groundwater level is deep or has been lowered, or where large industrial or agricultural demands exist, deeper wells are necessary. These wells are drilled utilizing specialized machinery and may extend hundreds or even thousands of meters into the subsurface. Hydraulic rotary drilling is probably the most common method used in unconsolidated or weakly consolidated material. This involves the rotation of a hollow drilling bit weighted with a heavy collar. Water, or a clay slurry, is usually pumped down the drilling pipe and flows through the outside of the hole, serving to lubricate the bit and extract rock chips. Following the drilling of a well, a casing or pipe is usually installed to prevent the cave-in of the walls and the possible contamination of the groundwater by inferior quality surface water. At the base of the well, in the vicinity of the aquifer, the casing will be perforated, and commonly surrounded or packed with gravel to increase the permeability and improve the flow of water into the well.

The rate at which water can enter a well and be pumped out is a function of the thickness and extent of the aquifer and its permeability. Flow rates of wells are usually stated in gallons per minute or gallons per day and can vary considerably. A flow of several gallons per minute is minimal to continually support a small family. The irrigation of gardens and livestock requires somewhat greater water production, perhaps two to five times this volume, depending upon the number of acres or animals.

Aquifer permeability determines the rate of flow of groundwater, which is usually quite slow, perhaps 1 to 2 meters per day, even in a good aquifer; flow rates of 10 to 20 meters per day may occur in coarse gravels (Strahler and Strahler, 1973). As pumping rates increase, some point is reached when the water is being removed from the aquifer faster than it can enter. At this point, the water level in the well and the surrounding water table begins to drop, eventually creating a cone of depression (see figure 8.7). This drawdown creates a steepened gradient in the water table, which initially leads to an increased rate of flow into the well. Well testing commonly involves a determination of the pumping capacity of a well expressed as the ratio of discharge to drawdown, or how many gallons per minute can be pumped out per foot of drawdown. The increase in flow rate only holds for a limited amount of drawdown, as cones of depression

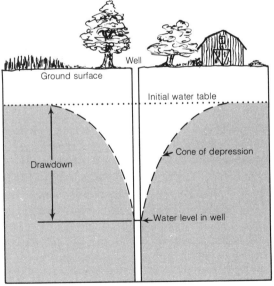

Figure 8.7. *Water Table Drawdown and Cone of Depression Produced by Pumping of a Well. The size of the cone is related to well pumpage and the permeability of the aquifer.*

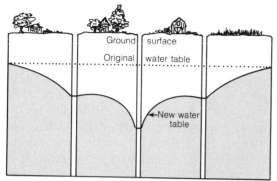

Figure 8.8. *Water Table Drawdown Due to Overlapping Cones of Depression. Many wells in close proximity will create an overall drop in the water table due to overlapping drawdown.*

can become quite large (kilometers across for large wells) with no increase in flow rate. The common end result of numerous wells with high pumping rates is an overall lowering of the water table due to the overlapping of these cones of depression (*see* figure 8.8).

Artificial Recharge

To either raise the level of groundwater, compensate for a declining water table, or halt subsidence, the natural infiltration of precipitation or surface water can be augmented by artificial or induced recharge. By spreading surface runoff or water released from reservoirs over the ground surface in basins, infiltration can be increased (*see* figure 8.9). The success of recharge basins is a function of the area covered with water, the length of time water is in contact with the soil, and the permeability of the surface and subsurface material. Natural gravel basins constructed in a nearly dry stream channel in the Campbell area of the western Santa Clara Valley in California have proven to be effective recharge basins. Water is released from an upstream reservoir and slowly works its way through a series of individual

basins. These recharge efforts, in addition to controls on groundwater withdrawals and increased importation, have succeeded in controlling the subsidence that was occurring in the valley (*see* chapter 5).

In other locations, water may be spread out over extensive flat areas or distributed into a series of ditches or furrows. Experience has shown that alternate spreading and drying periods are more efficient than constant coverage in promoting infiltration. In addition, recharge rates generally decrease as soil particle size decreases and are directly proportional to the depth or **head** of water.

Where surface recharge by spreading is not feasible due to low permeability, impermeable subsurface strata, or lack of adequate area for spreading grounds, recharge through shafts or wells may be practical. The hydraulics of recharge wells are opposite to those of a discharge well in that a mound of groundwater builds up around the well. The recharge rate is dependent upon the rate at which water moves out into the aquifer away from the well. The capacity of individual recharge wells is generally small, and the cost relatively high, compared with the cost of spreading basins, unless producing wells are used for recharge on a seasonal basis (Linsley, Kohler, and Paulhus, 1958). The quality of water being pumped down recharge wells must be carefully controlled. Silt and clay can clog an aquifer, and bacteria, viruses, or various chemical constituents may degrade the subsurface water quality. Therefore, recharge efforts utilizing treated waste water need to be very thoroughly studied and controlled because of their potentially high concentrations of suspended material and bacteria.

The results of research and field experience in this area are encouraging. A five-year

Figure 8.9. *Groundwater Recharge Basin. Water is diverted from the concrete lined Rio Hondo River Flood Control Channel (diagonally crossing the picture). The combined flood waters and purified wastewaters flow into the large, flat, pond-like basins. A basin four feet deep will trickle all its water through the soil and into the underground water basins in about four days. This is a part of the 450 acre complex. The basins are shown in the upper right and center of the photo.* PHOTO COURTESY: *Dept. of County Engineer, Los Angeles, California.*

project in Orange County, California, indicated that reclaimed **secondary effluent** was acceptable for domestic use after it traveled through 165 meters of a confined aquifer (Baier and Wesner, 1971). Bacteria, viruses, and toxic materials were consistently absent, although work continues on removal of undesirable taste, odor, and dissolved inorganic material.

Effects of Excessive Withdrawal

The continued withdrawal of groundwater at rates significantly greater than natural recharge usually leads to a number of undesirable consequences. This practice, commonly called **"mining,"** is a serious problem in areas such as the Southwest where water demands from communities and agriculture greatly exceed recharge from annual precipitation and the limited stream flow.

The concept of **safe yield** is one that has been used in the past in regard to excessive groundwater withdrawals but has never been totally understood. Safe yield is defined in various ways (Kazmann, 1972), but it is essentially the rate or amount of groundwater that can be extracted from a basin without producing negative effects. Unfortunately, due to annual variations in recharge, the unknowns in the subsurface geology and groundwater flow, and the increasing individual thirst for water, this quantity is very elusive. Damage to an aquifer or adverse effects of excessive pumping are usually recognized too late to correct the situation.

Both water quality and quantity can be affected by overdraft. Commonly the water table will decline over an entire basin or aquifer, thus affecting extensive areas. One immediate result is the need to dig deeper wells to reach the lowered water table. Digging many new and/or deeper wells is an expensive undertaking. The drop in the water table will also affect the flow of small streams that are sustained in the drier parts of the year by the groundwater flow or base flow. These streams, as well as some ponds and lakes, may dry up altogether with a significant decline in the water table. In such instances aquatic life and riparian vegetation may be totally destroyed.

Water of inferior quality may enter an aquifer, either from underlying layers or from the margins of a groundwater basin, and lower the overall quality. Deeper water may be of connate origin, for example, and contain high concentrations of dissolved salts. The subsidence of the ground surface as a result of collapse of the soil structure due to groundwater withdrawal and the effects of this on human activities and structures are additional side effects that have already been discussed (see chapter 4).

Groundwater in many aquifers must be classed as a nonrenewable resource like any other mineral deposit, because the rate of withdrawal greatly exceeds natural recharge

rates. A good example is in the high plains of northwestern Texas (Kazmann, 1972). The major aquifer in the area, the Ogallala Formation, is a sandy unit 60 to 90 meters thick. The number of irrigation wells in the region has increased from less than 300 in 1935, to 3500 in 1944, and 44,000 in 1960. Acreage under irrigation increased from 40,000 acres (178 square kilometers) in 1935 to more than 4,000,000 acres (17,800 square kilometers) in

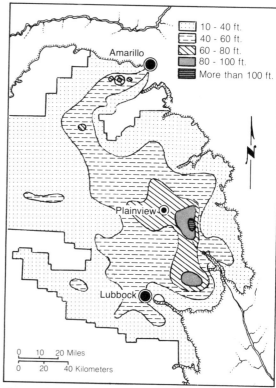

Figure 8.10. *Water Table Decline in High Plains Area of Texas, 1938–1962. The drop in the groundwater level in this area resulted from the removal of approximately 50 million acre-feet of water from the aquifer.* ADAPTED FROM: *Fig. 5.22, p. 214 (Based on data from High Plains Underground Water Conservation Dist. #1, Lubbock, Texas) in Modern Hydrology, 2nd Edition by Raphael G. Kazmann. Harper & Row, 1972.*

1960. The recharge of the aquifer, primarily from rainfall, averages between 50,000 to 100,000 acre-feet per year. Almost all the water is being taken from aquifer storage. The water table has been lowered, over 30 meters in places (see figure 8.10), and there is clearly a finite limit to the 200,000,000 acre-feet of storage in the aquifer. Approximately 50,000,000 acre-feet have already been withdrawn. The people living in the area have foreseen the outcome of the mining operation and through political organization and pressure, have had the Texas Water Plan modified to provide for water importation. The lower Mississippi River, about 1000 kilometers away, would presumably be the source of their water. Annual withdrawal volumes could reach 14 million acre-feet per year. The project would be financed, in most part, by funds collected in taxes by the federal government — a capital expenditure informally estimated at $10 billion, some $2500 per irrigated acre (Kazmann, 1972). This high cost, the long distances involved, the uphill pumping required, and the opposition of Louisiana have discouraged such a massive project to date.

Sea Water Intrusion

In coastal areas, fresh water aquifers are commonly in contact with the ocean and exposed to sea water intrusion (see figure 8.11). If the water table within the aquifer is above sea level, the intrusion of salt water is repelled and little or no contamination occurs. However, the increased demand for

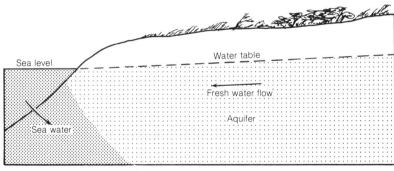

(a) Undisturbed coastal aquifer

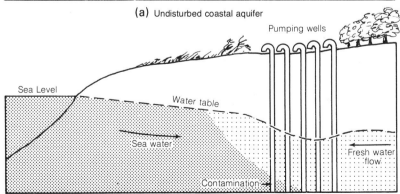

(b) Aquifer with sea water intrusion

Figure 8.11. *Seawater Intrusion in a Coastal Aquifer. (a) An undisturbed aquifer in contact with the ocean and having a water table sloping seaward. In this case the groundwater gradient serves to keep seawater repelled (b). A coastal aquifer with heavy withdrawals of water leading to water table decline. The landward slope of the water table has led to seawater moving into the aquifer.*

groundwater in many fertile coastal plain areas, both for agricultural and domestic uses, has led to the lowering of the water table and the consequent intrusion of a wedge of sea water into certain aquifers so that salt or brackish water begins to appear in wells (*see* table 8.1). Continuous heavy pumping of groundwater from the lower Salinas River Valley adjacent to Monterey Bay on the central California coast since the early 1940s has depleted the groundwater storage in that area. Water levels have been lowered to a point where sea water encroachment has occurred, leading to degradation of the groundwater quality. This de-

cline in water quality has led to reduced yields of crops irrigated by degraded water and increased costs of drilling new wells to replace wells producing saline water. Over 100 wells within one small part of this area (around Castroville — see figure 8.12) were capped or abandoned from 1943 to 1968 due to excessive chloride content (California Dep't. of Water Resources, 1970).

Groundwater accounts for more than 95 percent of the water used in the Salinas Valley. The increasing demand for high quality water for both agricultural and domestic uses has made the search for a solution to the sea water intrusion problem a major objective of the local and state water resources agencies. Sea water intrusion has also been a major problem in the Sacramento River Delta area and along the southeast side of San Francisco Bay (*see* figure 8.13). In the delta, the diminishing discharge of the Sacramento River, due to various water removal and control projects, has accelerated the problem.

Long Island on the East Coast has also experienced sea water intrusion problems. Groundwater development on Long Island has progressed through several distinct phases (Heath et al., 1966). Initially the hydrologic system was in equilibrium, with the natural long-term average groundwater recharge equal to the discharge. The interface of salt and fresh water was stable in the glacially deposited sediments underlying the island. As population grew, larger and deeper public supply wells were drilled, individual wells and septic tank systems were replaced by sewer systems with ocean discharge, the withdrawals increased, and recharge decreased. The water table declined and sea water began to move landward. Kings County and Queens County, the most developed parts of Long Island, experienced the most pronounced intrusion. Water levels

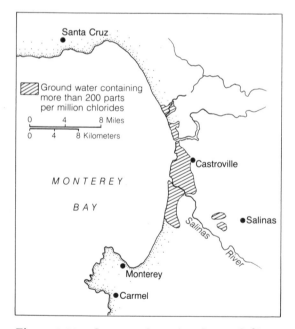

Figure 8.12. *Seawater Intrusion, Lower Salinas Valley, California. Excessive pumping for agricultural uses has led to extensive seawater intrusion and the abandonment of many wells due to their high chloride content.* DATA FROM: *California Dept. of Water Resources, Sacramento, California.*

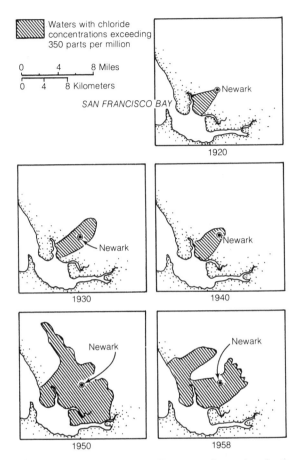

Waters with chloride concentrations exceeding 350 parts per million

0   4   8 Miles

0   4   8 Kilometers

SAN FRANCISCO BAY

Newark

1920

Newark

1930

Newark

1940

Newark

1950

Newark

1958

Figure 8.13. *Progressive Seawater Intrusion in the Shallow Groundwater Aquifer, East Side of San Francisco Bay, California.* DATA FROM: *California Dept. of Water Resources, Sacramento, California.*

have begun to recover with substantial reductions in consumptive groundwater uses and the importation of water from upstate New York.

There are several ways in which sea water intrusion can be controlled. By reducing the pumping in a coastal aquifer, the overdraft causing intrusion can be eliminated, permitting the groundwater levels to rise above sea level and to reestablish a seaward gradient. This is the most obvious solution but one that is difficult to get users to begin voluntarily. Rearranging the basin pumping pattern by moving wells landward may raise water levels on the coast enough to repel sea water influx.

The artificial recharge of an intruded aquifer, from either spreading areas or recharge wells, is another method. The water levels and appropriate gradients could then be properly maintained. This method was used in the coastal plain of the Los Angeles Basin, where sea water intrusion had become a serious problem due to groundwater withdrawals of up to 90,000 acre-feet per year. Treated Colorado River water was introduced by means of a series of recharge wells. A groundwater or pressure ridge of fresh water was formed between the area of withdrawal and the sea, which halted the intrusion. It is also possible to recharge an intruded aquifer with waste water treated to advanced secondary or tertiary levels. This type of recycling may well become a common method in the future for solving water shortage problems. In some areas economic incentives have to be provided or legislation implemented that will require or encourage recharge of groundwater by major users.

The emplacement or construction of a subsurface barrier that would reduce the permeability of an aquifer sufficiently to prevent inflow of sea water might be a workable solution in the case of a shallow aquifer. The construction of dikes, or the injection of grout, asphalt, or some other material may produce such a semi-impermeable barrier. For such a method to be feasible, the outcrop area of the aquifer exposed to sea water cannot be large.

Saline pollution also occurs in arid or semiarid regions where high irrigation water

Table 8.1. *Representative Examples of Saltwater Intrusion in the United States*

Location	Nature of Problem	Corrective Measures	Outlook
Gulf Coast			
Alabama — Mobile	Lateral movement from Mobile River caused by intensive pumping	Pumping curtailed; deeper wells for fresh water	Shallow aquifer unchanged
Louisiana — Baton Rouge	Movement of saline water (or possibly industrial waste) into water well field owing to intensive pumping	None yet	Reduced pumping; supplementary water supply studies
Texas — Galveston, Texas City	Upward and/or downward intrusion of saline water into producing aquifers because of heavy pumping	Pumping moved inland; surface supplies developed; desalting being considered	Continuing problems
East Coast			
Connecticut — New Haven and Bridgeport	Intrusion from tidewater in harbors caused by intensive pumping	Pumping relocated landward; alternative supplies used	Further pumping curtailment; use of alternate supplies
Florida — Dade and Broward Counties	Infiltration of tidal water from canals constructed to drain inland areas and lower water table	Canal construction controlled; salinity control structures installed to keep out saltwater and raise fresh water level	Continued management, surveillance, and study
New Jersey — Newark-Passaic, Sayreville	Intrusion from tidal estuaries into producing aquifers, aggravated by intensive pumping, harbor and canal dredging, and disposal of industrial and municipal wastes	Pumping relocated; use of alternate supplies	Serious until control measures established; continuing study
West Coast			
California — Ventura County — Oxnard plain	Intrusion caused by intensive pumping	Experimental facilities in operation for control with a pumping trough	Economic pressure will force a solution
Santa Clara County	Intrusion from San Francisco Bay caused by intensive pumping	Pumping curtailed; artificial recharge of aquifer	Managed ground-water basins
Washington — Tacoma area	Intrusion from ocean due to pumping	Wells moved inland	Continued intrusion

Adapted from: The Task Committee on Saltwater Intrusion, "Saltwater Intrusion in the United States," *Journal of the Hydraulics Division, American Society of Civil Engineers* 95(1969): 651–59.

demand decreases surface water flushing, thereby allowing accumulation of harmful salts. Downward percolation of salts into the groundwater reservoir, coupled with repeated evaporation and circulation of the same water, concentrates these salts in the aquifer. As a result, fresh water aquifers become brackish. This process is occurring presently in the Imperial Valley near the California-Mexico border, resulting in reduced crop yields and the gradual withdrawal of lands from agricultural usage.

GROUNDWATER CONTAMINATION

Incidences of contamination of groundwater are numerous and usually are simply the result of ignorance or lack of understanding. Wastes that are intentionally and unintentionally spread over the ground surface, dumped into pits, or injected into the ground can cause serious problems. Because groundwater generally moves very slowly, the wastes may go undetected for some time. Even when discovered, it may take years to correct the situation (Pettyjohn, 1972).

Groundwater pollutants or contaminants may consist of any of the following:

1. microorganisms, including pathogenic viruses and bacteria

2. organic matter, primarily from domestic sewage

3. chemical wastes, from agricultural or industrial operations, and leaching from sanitary landfills

4. nuclear wastes, generated by atomic power plants and weapons fabrication

One of the greatest concerns for groundwater contamination is the fate of domestic sewage, whether from individual septic tanks or municipal disposal systems (see chapter 9). This concern is recent, however, in that septic tanks have been used extensively in urban subdivisions in the past twenty-five years on a scale sufficient to serve more than 30 million persons (McGauhey, 1968). Many people, however, tend to ignore the contamination of groundwater where human wastes are involved, even though the relationships between these wastes, pathogenic organisms, and diseases are well known. Where shallow wells and septic tanks exist side by side in permeable alluvial material, the interconnection via the groundwater system is very obvious. Incidences of well water contamination with bacteria and detergent from septic tank effluent are common in certain areas.

Common disposal methods for solid waste, either open dumps or so-called sanitary landfill operations, may also be sources of groundwater contamination (see chapter 9). Municipal waste, which can consist of organic garbage, sewage sludge, and rubbish and trash of endless varieties, is commonly dumped or buried, usually in low areas, and is then subject to percolating rainwater which filters through the debris. The water can transport the various ions or substances that have been leached from the waste disposal site and eventually, if conditions are favorable, add these to the groundwater reservoir. If subsurface geology and hydrology are studied prior to the development of such a site, a favorable location can usually be selected where possibilities for groundwater contamination are very low.

Injection wells for the disposal of various chemical wastes have been utilized in the past, often with inadequate understanding of the subsurface environment. The thickness and distribution of various rock units, their permeabilities, as well as the existence of joints, fractures, and faults must be understood before disposing of any toxic material in the subsurface. These factors are all important in determining the directions and rates of fluid movement beneath the ground. Where carbonate rocks occur, and have been subjected to solution, the existence of caves and sinkholes often allow for very rapid movement of subsurface water. This can lead to free interchange of fluids over considerable distances and therefore serious subsurface waste disposal problems.

Probably the greatest potential threat to the groundwater system is the contamination by long-lived radioactive isotopes produced from either weapons fabrication or nuclear reactor operations (see chapter 9). Radioactive wastes are distinguished from all other kinds of chemical wastes by the lack of a treatment method to counteract their innate biological harmfulness. An essential requirement, therefore, for a nuclear power industry is a system for the safe management and disposal of radioactive wastes. Unfortunately, a permanent solution has not yet been found. At the Hanford Nuclear Center adjacent to the Columbia River in the state of Washington, and at the National Reactor Test Station at Idaho Falls, Idaho, intermediate level radioactive wastes are either stored in earth ponds or discharged underground through wells into the subsurface body of circulating groundwater (NAS-NRC, 1969). When such wastes are dangerously radioactive for many years, this is an unacceptable and unsafe solution for their disposal.

Individual studies of groundwater pollu-

tion incidents reveal that contamination can occur in diverse geological regions in many different ways and by many different contaminants (Pettyjohn, 1972). Heavy metals, toxic chemical wastes, oil field brines, agricultural chemicals, pathogenic viruses and bacteria, and most recently, long-lived radioactive wastes are all potential threats to the groundwater system whose storage and/or disposal has to be carefully controlled and monitored. No surface or subsurface waste disposal operations should be initiated when doubts exist as to the nature of the rocks at depth and their permeability, the positions of aquifers, and any possible interconnections between the two. Aquifers can easily become contaminated, and once introduced, the contaminants may be very costly and difficult to remove, if they can be removed at all.

REFERENCES

References Cited in the Text

Baier, D. C., and Wesner, G. M. "Reclaimed Waste Water for Groundwater Recharge." *Water Resources Bulletin* 7(1971): 991–1001.

California Dep't. of Water Resources. "Sea Water Intrusion in the Lower Salinas Valley." Progress Rep't, 1968–1969, DWR Rep't. 151, 1970.

Heath, R. C.; Foxworthy, B. L.; and Cohen, P. *The Changing Pattern of Groundwater Development on Long Island, New York.* U.S. Geological Survey Circular 524, 1966.

Kazmann, R. G. *Modern Hydrology.* New York: Harper and Row, 1972.

Linsley, R. K.; Kohler, M. A.; and Paulhus, J. L. H. *Hydrology for Engineers.* New York: McGraw-Hill Book Company, 1958.

McGauhey, P. H. "Man-Made Contamination Hazards." *Groundwater* 6(1968): 10–13.

NAS-NRC. *Resources and Man.* San Francisco: W. H. Freeman and Company, 1969.

Parizek, R. R. "Impact of Highways on the Hydrogeologic Environment." *Environmental Geomorphology.* Edited by D. R. Coates. Binghamton: State University of New York, 1971.

Pettyjohn, W. A. *Water Quality in a Stressed Environment.* Minneapolis: Burgess Publishing Company, 1972.

Sierra Club. "Dams in Grand Canyon — a Necessary Evil?" *Sierra Club Bulletin* (Aug. 1965).

Strahler, A. N., and Strahler, A. H. *Environmental Geosciences.* Santa Barbara, Calif.: Hamilton Publishing Company, 1973.

Thurow, C.; Toner, W.; and Erley, O. *Performance Controls for Sensitive Lands: a Practical Guide for Local Administrators.* Washington, D.C.: U.S. Government Printing Office, EPA, 1975.

Todd, D. K. *Groundwater Hydrology.* New York: John Wiley and Sons, 1966.

Other Useful References

Hackett, O. M. *Groundwater Research in the United States.* U.S. Geological Survey Circular 527, 1966.

Lohman, S. W. et al. *Definitions of Selected Groundwater Terms: Revisions and Conceptual Refinements.* U.S. Geological Survey Water Supply Paper 1988, 1972.

McGuinness, C. L. *Scientific or Rule-of-Thumb Techniques of Groundwater Management: Which Will Prevail?* U.S. Geological Survey Circular 608, 1969.

McGuinness, C. L. "Groundwater: A Mixed Blessing." In *Focus on Environmental Geology.* Edited by R. W. Tank. New York: Oxford University Press, 1976.

CHAPTER 9

Waste Disposal and Treatment

Contents

INTRODUCTION

As populations have increased and industrialization has occurred, human demand for and usage of natural resources for domestic, industrial, and agricultural purposes has steadily increased. Waste, whether solid or liquid, is simply the resource after we have used it. The variety of waste is endless but includes everything from the worn-out family automobile or television set to the water that goes down the sink drain.

When populations were low and spread out over the land, and people were more self-sufficient in terms of what they produced and consumed, waste disposal was not a significant concern or problem. We long ago, however, surpassed the era when the purifying ability of natural water systems cleaned the then minimal amounts of waste material that was reaching streams, lakes, estuaries, and coastal waters. Most people do not live on large parcels of land that can easily absorb the waste of a single family. Instead people have become more and more

concentrated in towns and cities, and as a result their waste has become concentrated. The sheer volume of waste water and solid waste being produced daily in the United States has become a major nationwide problem. The failure to deal adequately with the issues has led to widespread degradation of water quality and destructive alteration of the land surface area.

In rapidly developing areas, or in regions with large existing populations, vital land and water resources can be endangered or seriously damaged if waste disposal and land use decisions are made that fail to account for the existence of certain pollution hazards (Hines, 1974). These hazards are basically a function of several interacting factors:

1. the type, location, and emission characteristics of pollution sources

2. the proximity and sensitivity of land

and water resources to these pollution sources

3. the existence of one or several critical physical conditions that may affect the generation, transport, and distribution of pollutants in the environment

An understanding of these factors and their interactions is a prerequisite for evaluating and controlling waste disposal and pollution problems.

TYPES, SOURCES, AND EFFECTS OF POLLUTANTS

The types of contaminants being produced and discharged today are getting increasingly complex (see table 9.1). Hundreds of different kinds of artificially induced materials can now be found scattered over the landscape and in polluted water systems, both fresh and saline. Agricultural chemicals such as phosphorous from fertilizers, synthetic herbicides and pesticides, domestic wastes with their nutrients and microorganisms, and a myriad of industrial wastes including various heavy or trace metals, acids, and organic and inorganic compounds are among the most common pollutants. Each of these has some effect on biological systems, many of which are not clearly understood.

Organic Wastes

Organic materials are derived from domestic sewage, livestock wastes, and a variety of industrial processes including pulp and paper production, petroleum refining, and fruit, vegetable, and meat processing. The organic matter itself is not poisonous but its effects are more subtle. As the organic material is broken down in water, for example, the oxygen in water is removed. The greater the amount of organic matter, the greater will be the demand on available oxygen. This demand for oxygen is usually called biological or **biochemical oxygen demand** (BOD); it is a useful indicator of organic pollution. The BOD is important to the life of a lake or stream because the resultant oxygen level is one of the most critical factors affecting what life forms can be supported. Vertebrates (such as fish), invertebrates, and bacteria, in that order, require progressively less oxygen to survive. In other words, as the oxygen levels decline from excess organic matter, the vertebrates will be the first to die off or disappear, followed by the invertebrates, and finally by **aerobic bacteria**.

Nutrients, principally nitrates and phosphates, are fertilizers for plants in both terrestrial and aquatic environments. Once in a body of water, whether from domestic sewage, detergent, industrial waste, or agricultural runoff, their effect is the same. These nutrients, usually phosphorus, stimulate aquatic plant life such as algae, commonly leading to algal blooms (see figure 9.1). The ultimate result is poorer water quality through the depletion of oxygen during subsequent decomposition of the algae. Domestic waste water appears to be the principal nitrate and phosphate contributor to water supplies. Household detergents commonly constitute about two-thirds of the phosphorus coming from municipal sources. Both livestock feedlots and increased fertilization can also contribute to the nutrient overload in some water bodies.

Because phosphorus is concentrated in organic matter and on small soil particles near the ground surface, it is susceptible to

Table 9.1. *Major Sources and Types of Pollutants*

Major types of pollutant emissions:

a. Noxious or toxic chemicals, acids, caustics, pesticides
b. Toxic metallic substances
c. Soluble and particulate organic substances
d. Nutrients, particularly compounds of nitrogen and phosphorus

e. Pathogenic microbes
f. Mineralized water
g. Airborne noxious or toxic gases, and particulate matter
h. Suspended sediment and turbidity.
i. Debris (paper, rags, cans, trash)
j. Grease and oil

Source of Pollutants	Type of Pollutant	Possible Additional Sources of Information
1. Agricultural wastes and irrigation return flows[1].	a, c, d, f	U.S. Department of Agriculture; California Department of Agriculture.
2. Animal wastes	c–e	U.S. Department of Agriculture; California Department of Agriculture.
3. Dredge spoils	a–d, h	U.S. Army Corps of Engineers, Bay Conservation and Development Commission.
4. Heavy construction or landscape alteration[1].	h, i	County planning and public works departments
5. Incinerators; open burning	b, g	Bay Area Air Pollution Control District
6. Industrial stack gases	b, g	Bay Area Air Pollution Control District
7. Junkyards	b, i, j	County planning departments
8. Mining and mine wastes	a, b, f, h	California Division of Mines and Geology
9. Motor vehicles	b, g, j	California Division of Highways; Bay Area Air Pollution Control District.
10. Pesticide spraying	a	U.S. Department of Agriculture; California Department of Agriculture.
11. Septic tanks[1]	a, c–e, j	County public health or environmental engineering departments.
12. Solid-waste disposal sites[1]	a–j	California Department of Public Health
13. Storm-water runoff[1]	a–e, h–j	County public works departments.
14. Toxic-chemical storage areas	a, b	California Department of Industrial Relations; California Regional Water Quality Control Board, San Francisco Bay Region.
15. Municipal and industrial wastewater-treatment plants and outfalls[1]	a–j	California Regional Water Quality Control Board, San Francisco Bay Region.
16. Wastewater-injection wells	a, b, f	California Regional Water Quality Control Board, San Francisco Bay Region.
17. Watercraft	e, i, j	U.S. Coast Guard

[1] Most important source of pollutants.

SOURCE: W. G. Hines, "Evaluating Pollution Potential of Land-Based Waste Disposal, Santa Clara County, California," U.S. Geological Survey Water Resources Investigation 31–73.

removal with sediment during surface erosion and runoff. Land use conservation practices in agricultural areas, such as contour plowing, cover crops, use of crop residues, strip croppings, terracing, and the like can greatly eliminate these erosional losses.

Nitrogen in the form of nitrate is water soluble and moves with water through the soil profile. Excess nitrates in water can constitute a health hazard. There is clear evidence that infants on diets involving water with a high nitrate content have developed

Figure 9.1. *Aquatic Plant Overgrowth from Excess of Nutrients, Salinas River, California. This river drains one of the most intensively farmed areas of California. Fertilizers draining into the river lead to excessive plant growth. The shiny area on the right side of the river is the only patch of clear water.* PHOTO BY: *Gary Griggs.*

methemoglobinemia, which reduces the oxygen-carrying capacity of the blood and may lead to suffocation (Chanlett, 1973).

Disease Organisms

A number of bacteria, viruses, various protozoans, and other microorganisms are detrimental to water quality and hazardous to public health, either because they are pathogenic (disease carrying) or because they alter water quality in some way. Certain microorganisms cause undesirable tastes and odors, others affect color, some may lead to undesirable growth and eventually clog filters and pumps. Coliform bacteria, present in the intestinal tract of all warm-blooded animals, are not pathogenic, but their presence in sewage is commonly used as an indicator or index of domestic sewage contamination. Cholera, typhoid, paratyphoid, amoebic dysentery, and hepatitis are all water-borne diseases transmitted by ingestion and can be prevalent where drinking water supplies are contaminated (see table 9.2). This is more common in undeveloped areas, but can occur anywhere when flooding occurs and mixes waste water and drinking water supplies. These diseases can also occur where water bodies such as rivers, lakes, and even coastal waters, which are used for recreation, are also used for waste disposal. Intertidal discharge of primary treated sewage in the midst of a recreational area in northern Monterey Bay on the California coast has led to polluted beaches and is believed responsible for numerous cases of hepatitis contacted by surfers. Much of the water in San Francisco Bay does not meet the coliform bacteria standards established for bathing water due to the discharge of inadequately treated domestic waste water. The discharge of sewage into the Bay of Naples on the coast of Italy has led to several recent cholera epidemics. Unfortunately, problems of bacterial or viral contamination of surface waters are still widespread and will continue into the future until adequate waste water treatment and disposal facilities are constructed.

Table 9.2. *Water-Borne Diseases Transmitted by Ingestion. Grouped by Types of Etiological Agent and Ranked by Likelihood of Transmission*

Disease	Agent	Comment
	Bacterial Agents	
1. Cholera	*Vibrio cholerae*	Initial wave of epidemic cholera is water-borne. Secondary cases and endemic cases are by contact, food, and flies.
2. Typhoid fever	*Salmonella typhi*	Principal vehicles are water and food. Case distribution of water-borne outbreaks has a defined pattern in time and place.
3. Bacillary dysentery (Shigellosis)	*Shigella dysenteriae* *Shigella flexneri* *Shigella boydii* *Shigella sonnei*	Fecal-oral transmission with water one transmitter. Direct contact, milk, food, and flies are other transmitters. Ample water for cleanliness facilitates prevention.
4. Paratyphoid fever	*Salmonella paratyphi* *Salmonella schottmulleri* *Salmonella hirschfeldi*	Few outbreaks are water-borne. Other fecal-oral short circuits dominate. Ample water facilitates cleanliness.
5. Tularemia	*Pasteurella tularensis*	Overwhelmingly by handling infected animals and arthropod bites. Drinking contaminated raw water infects people.
	Protozoan Agent	
1. Amebic dysentery (amebiasis)	*Entamoeba histolytica*	Epidemics, which are rare, are mainly water-borne. Endemic cases are by personal contact, food, and possibly flies.
	Viral Agent	
1. Infectious hepatitis	A filterable virus, not isolated.	Epidemics are due to transmission by water, milk, and food, including oysters and clams.
	Helminthic Agent	
1. Guinea worm disease (dracintiasis)	The roundworm, *Dracunculus medinensis*; gravid female, 1 m long, migrates to skin.	Unknown in North America. Cycle is worm larva through human skin to water to the crustacea, such as cyclops, to human ingestion of water with cyclops in infective form.

ADAPTED FROM: E. T. Chanlett, *Environmental Protection* (New York: McGraw-Hill Book Company, 1973), table 3–7.

Chemical Constituents

A vast variety of chemical constituents, primarily from industrial operations, but also included in domestic waste water and agricultural runoff, affects water quality in various ways. Acute or chronic toxicity, objectionable reactions such as foam production, or discoloration, uncertain physiological responses, and unknown effects on organisms of even trace amounts of certain elements or compounds are all reasons for concern. The heavy metals and trace elements such as lead, mercury, copper, arsenic, chromium, zinc, cadmium, and silver, many of which are utilized in various galvanizing and plating processes, can act as metabolic or respiratory blocks in many organisms. The result is inhibition or destruction of enzymes that are essential to life processes. Many of these and other pollutants, which may be absorbed by the bottom mud, can be released whenever bottom sediments are disturbed. The pollutants are then picked up by **benthic organisms** and enter the food chain. Recent concern with the possible buildup of individual toxic elements in natural systems has grown out of several large-scale tragic incidents of mercury poisoning in New Mexico and in Japan.

The inhabitants of Minimata Bay, Japan, suffered an epidemic of neurological disorders, which were eventually related to mercury poisoning from fish and shellfish. The poisoning was due to methyl mercury chloride, which was formed in the waste sludge of a plant that used mercuric oxide for acetaldehyde production. The solid wastes were discharged to Minimata Bay where the organo-mercury compound was accumulated by fish and shellfish. Each of the victims of the disease had eaten fish or shellfish from Minimata Bay. By the end of 1960, 111 cases were reported, and as of August 1965, 41 deaths had occurred (Klein and Goldberg, 1970.)

A first step in eliminating discharges of such potentially toxic substances is to examine aquatic environments such as rivers and lakes, and to delineate areas of unnatural concentrations and then determine their sources. For example, a detailed study of the distribution of mercury, copper, and lead in the surface sediment of San Francisco Bay, where the effects of automobiles and industrial or domestic waste water discharge have become obvious, has recently been completed (McCulloch et al., 1971; Peterson et al., 1972). The Federal Water Pollution Control Administration, now part of the Environmental Protection Agency (EPA), has also been heavily involved in such monitoring and recently released a comprehensive five-year study of trace metal concentrations in the rivers and lakes of the United States (see table 9.3).

Acid mine drainage is a severe problem to stream and water quality in the states of Pennsylvania, Kentucky, West Virginia, and Illinois, where coal mine operations cover vast regions (see chapter 10). The possible large-scale development of coal fields in the western states, however, would be subject to the same problems. The weathering of the pyrite (iron sulfide) included in the coal-bearing rocks contributes iron and sulfur to the surface water, produces sulfuric acid, which lowers stream **pH**, and reduces oxygen availability. These are all detrimental to water quality and biological systems. Although strict regulations are beginning to affect present operations, the older strip mine areas and shafts will no doubt discharge acid runoff for years to come.

Many industrial operations, primarily those in metal production, also contribute acid discharge to surface water, thus altering the pH and upsetting biological systems.

The vast quantities of chlorinated hydrocarbons and other pesticides and herbicides applied to agricultural land form a distinct variety of contaminants. Many of these have been produced artificially and do not break down readily under normal conditions. After application they are commonly leached from the soil, either through precipi-

Table 9.3. *Violation of Water Quality Criteria for Public Water Supplies in Rivers and Lakes of the United States (1962–1967)*

A. *Violations of Water Quality Criteria by River Basin*

Basin	Total No.	Cd	As	Fe	Mn	Pb	Cr
Northeast	6	1	1	2	0	0	2
North Atlantic	7	0	6	0	0	1	0
Southeast	12	0	2	10	0	0	0
Tennessee R.	2	0	1	1	0	0	0
Ohio River	81	2	15	0	58	6	0
Lake Erie	20	2	2	1	12	3	0
Upper Mississippi	11	0	3	1	1	6	0
Western Great Lakes	3	0	1	0	0	2	0
Missouri River	6	0	3	1	1	1	0
Southwest — Lower Mississippi	10	0	1	3	1	4	1
Colorado River	5	0	1	0	0	3	1
Western Gulf	6	0	0	6	0	0	0
Pacific Northwest	9	1	7	0	0	1	0
California	0	0	0	0	0	0	0
Great Basin	0	0	0	0	0	0	0
Alaska	1	0	0	0	1	0	0

Metal columns: Cd, As, Fe, Mn, Pb, Cr (grouped under "Metal")

B. *Number of Violations, Limits, and Mean of Values in Excess of Limits for Principal Trace Metals.*

Metal	Limit $\mu g/1$*	No. of Violations	Mean of Those Values Which Exceeded the Limits, $\mu g/1$
Cadmium	10	6	39
Arsenic	50	41	91
Iron	300	25	>676
Manganese	50	74	>586
Lead	50	27	71
Chromium^{+6}	50	4	94
Zinc	5000	0	—
Copper	1000	0	—
Silver	50	0	—
Barium	1000	0	—

*As set by the Water Quality Criteria for Public Water Supplies.

Source: *J. F. Kopp and R. C. Kroner, A Five-Year Summary of Trace Metals in Rivers and Lakes of the United States (Oct. 1, 1962 — Sept. 30, 1967)* (Cincinnati, Ohio: U.S. Dep't. of Interior, Federal Water Pollution Control Administration, p. 12).

tation or irrigation, and enter the nearby surface waters. Their uptake, biological magnification, and subsequent chronic effects on organisms at the top of the food chain, for example birds of prey and fish-eating marine birds, have been extensively studied in recent years. The Salinas River Valley, one of the most fertile and intensively farmed agricultural areas in the country, has been subjected to heavy pesticide application. Marine organisms from the adjacent coastal area into which the river drains were discovered to contain the highest concentration of DDT of marine animals found anywhere in the world (see figure 9.2). Legislation at the state level has now virtually eliminated the usage of DDT, Aldrin, Diel-

drin, and some of the other more persistent chlorinated hydrocarbons.

Thermal Discharges

The release of warm water, whether from industrial-cooling or -processing operations or from electrical power production, is continually increasing. At present 50 billion cubic meters annually, over 80 percent of all the water used by industry in the United States, are used for cooling (Wagner, 1971). Entire volumes have been written on the effects of increased water temperatures on aquatic organisms and the contrasting phrases "thermal pollution" and "thermal

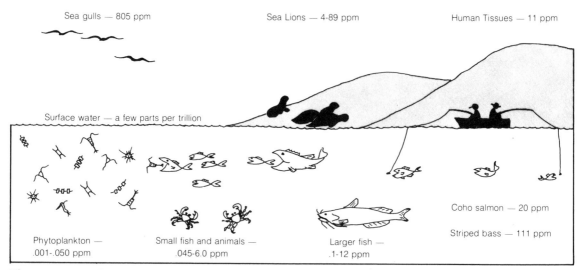

Figure 9.2. *Biological Magnification of DDT in Monterey Bay, California. Organisms collected in the late 1960s in Monterey Bay which receives heavy agricultural runoff, clearly illustrated the concentration of DDT moving up the food chain. PPM refers to DDT concentration in parts per million.* DATA FROM: *Hopkins Marine Station, Pacific Grove, Calif.*

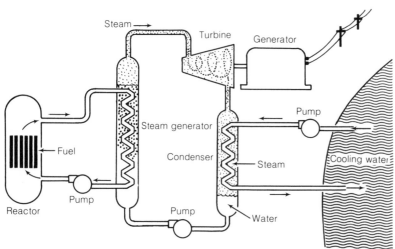

Figure 9.3. *Flow of Cooling Water in a Nuclear Power Plant. The source of cooling water may be a river, lake, bay, or the ocean.*

enrichment" have been used by opposing viewpoints to describe the same phenomena. In electric power production, water from rivers, lakes, estuaries, or the ocean is used to condense steam. After fulfilling this purpose it is returned to its source, usually 5 to 11 degrees C (10 to 20 degrees F) above normal (see figure 9.3). Because conventional power plants are at best only about 40 percent efficient (nuclear power plants are even

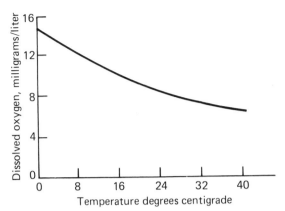

Figure 9.4. *Amount of Dissolved Oxygen in Water Related to Water Temperature. The quantity of oxygen dissolved in water is closely related to its temperature. The higher the water's temperature, the lower the oxygen content of the water. Thermal pollution can lower oxygen content below the point necessary to sustain many animals.*

less efficient), great quantities of fuel are essentially being utilized to heat water. As temperature in streams, lakes, or any water body increases, the ability of the water to hold oxygen decreases (see figure 9.4). The amount of dissolved oxygen present is a major factor determining the ecosystem composition in aquatic environments. In addition to reducing the water's ability to hold oxygen, increased temperature changes the rates of various metabolic processes, feeding, reproduction, spawning, and other biological activity. This eventually will lead to the disappearance or loss of certain species and their replacement by either warmer water organisms or more temperature resistant forms. In addition to these subtle effects, thermal shock can kill organisms outright. Fish kills of tens or even hundreds of thousands have occurred in rivers due to heated discharges from electrical generating plants (see table 9.4).

These are some rather obvious effects of thermal discharges or elevated water temperatures in natural water systems. The severity of the effects, however, is dependent upon the volume and temperature of the discharge, in contrast to the temperature and size of the water body that is being utilized as a heat sink. Although coastal waters in mid-latitude regions are far more able to cope with large quantities of heated water than are more confined inland streams and lakes, they can also suffer significant changes. In subtropical areas like Florida, where normal ocean temperatures are already in the eighties on the Fahrenheit scale, thermal discharges from power plants have elevated the water temperature above the thermal death point, resulting in massive destruction of aquatic life.

Radioactive Wastes

The mention of atomic energy probably invokes more and varied responses from people than any other energy topic. On the one hand, it seems to offer an almost unlimited supply of "clean" energy, while, on the other hand, serious questions seem to remain unanswered or unresolved. The basic uncertainties or problems are those related to radiation, either from the production of radioactive wastes, which are created throughout all stages of atomic power production, or from an accidental release due to some type of power plant failure.

There are several important points to keep in mind in the following discussion. First, natural radioactivity exists everywhere, in the atmosphere and in crustal rocks. Radiation therefore is nothing new

Table 9.4. *Reported Fish Kills Caused by Heated Waste Water Discharges from Electric Power Generating Plants*[1]

Date	State	Stream or Lake	Nearest Town or County	Degree of Severity	Number of Fish Killed
Aug. 6–8, 1962	Pennsylvania	Raystown Branch, Juniata River	Saxton	Heavy	3,441
Aug. 11, 1962	Missouri	Discharge canal to Montrose Lake	Ladue	Heavy	[2]
Sept. 7, 1963	Illinois	Rock River	Rockford	Light	
May 28, 1964	Texas	Unnamed stream	Victoria	Light	
Aug. 19, 1965	Pennsylvania	Schuykill River	Reading	Moderate	1,000
Aug. 20, 1965	Ohio	Greater Miami River	Montgomery County		11,250
Jan. 19–22, 1966	Ohio	Ohio River	Toronto	Light	200
Sept. 2, 1966	Pennsylvania	Schuykill River	Philadelphia	Heavy	50,000
Jan. 1, 1967	Ohio	Sandusky River	Sandusky		300,384
Jan. 17, 1967	Ohio	Sandusky Bay	Erie County		78,755
Jan. 1, 1968	Nebraska	Lake Hastings	Hastings	Moderate	5,000
Jan. 2, 1968	Ohio	Sandusky River	Sandusky County		250,585
Mar. 1, 1968	Utah	Price River	Castlegate	Heavy	150
July 1968	Massachusetts	Cape Cod Canal	Sandwich	Moderate	
Aug. 22, 1968	Massachusetts	Cape Cod Canal	Sandwich	Moderate	
Dec. 16, 1968	West Virginia	Ohio River	New Cumberland	Light	9,500
Dec. 24, 1968	Ohio	Sandusky River	Sandusky County		3,000
June 30, 1969	Florida	Biscayne Bay	Miami		

[1] The Federal Water Pollution Control Administration has maintained a national inventory of pollution-caused fish kills since 1962. This inventory is compiled from voluntary reports submitted by cooperating State and local water pollution control and conservation agencies. This tabulation should not be considered as completed because of the voluntary nature of this inventory and the inability of the cooperating agencies to know of all incidents of fish mortality.

[2] Several thousand.

SOURCE: "Environmental Effects of Producing Electrical Power," Hearings before the Joint Committee on Atomic Energy of the Congress of the United States (Washington, D.C.: U.S. Government Printing Office, 1969), p. 376.

or different. What is new is the human-induced increase in radiation that people are exposed to, whether from bomb testing, atomic energy or power production, or X-rays. It now seems well established that (1) no increase in radiation exposure is beneficial (except with certain types of specialized medical treatments which may also involve many complications), and (2) there is apparently no minimum or threshold dosage of radiation below which

no biological damage results. Therefore, any increase in background radiation is potentially harmful and has to be considered along with any benefits resulting from activities which lead to radiation exposure or release.

Any process or procedure involving atomic energy produces radiation. Initially, the mining, milling, and fuel fabrication of naturally occurring radioisotopes, such as uranium, exposes and concentrates a certain amount of radioactivity. Irradiation during nuclear plant operation and fuel processing produces radioactive fission products that are extremely dangerous. Reactor operation also exposes various materials to radiation, such as fuel rods, cooling water pipes, and other nonfuel elements, which creates additional radioisotopes. Specialized uses of radiation in medicine, industrial work, and scientific research also produce radioactive waste, which must be either disposed of or stored.

The distinction between disposal and storage is an important one to make clear. Disposal, whether to the air, to a body of water, or to the earth, is an irreversible action. There is no way to retrieve the materials. Storage, however, implies containment and control. Although significant accidental leakage into the ground has occurred in the case of radioactive waste "storage" at Hanford, Washington, the goal is complete containment.

Depending upon the levels of radioactivity, wastes have been divided into low level, intermediate level, and high level. Present treatment and disposal practices involving these wastes can be summarized in three basic approaches: (1) dilute and disperse, (2) delay and decay, and (3) concentrate and contain.

Dilute and Disperse

Large volumes of liquid and gaseous low level wastes are disposed of in this fashion. During normal nuclear power plant operation, low level wastes, such as activation products produced in the cooling water system, are routinely released to the water body being utilized for cooling. Unfortunately, standards for permissible levels of low level releases do not take into account the tre-

mendous ability of organisms to concentrate certain elements. Oysters, for example, living in the effluent canal of a small atomic power plant at Humbolt Bay in northern California have concentrated radioactive zinc-65 12,000 times above normal. The effects of these isotopes on marine ecosystems, their concentration within the food chain, the **synergism** with other environmental variables, and the ultimate impact on people are not well understood.

Low level solid wastes cannot be disposed of in the same manner. Common practice with these materials involves placement in metal drums, commonly filled with cement, and then shallow burial in trenches or ocean dumping. The volume of such wastes, produced by both the activities of industry and the Atomic Energy Commission (superseded by the Energy Research and Development Administration [ERDA]), has continued to increase. Metal drums filled with radioactive wastes have in the past been washed ashore and been recovered on beaches. This obviously indicates either lack of understanding about the consequences of the disposal practice or simply gross negligence. Wastes have not always been dumped at the proper depths or far enough from shore. Extreme pressure at depth may also crush the barrels and release the contents to the marine environment before decay has occurred.

Delay and Decay

Intermediate level wastes cannot be routinely discharged to the environment, so they are treated in several ways. Certain materials, such as those with short **half-lives**, may be stored at the site until sufficient decay has occurred so that they can be discharged without exceeding radia-

Figure 9.5. *Shallow Burial of Solid Nuclear Wastes at Oak Ridge National Laboratory.* PHOTO COURTESY: *Energy Research and Development Administration.*

tion limits. With wastes of higher activity, burial in concrete-lined wells, or storage and subsequent discharge into open trenches is practiced at Hanford, Washington (see figure 9.5). As the radioisotopes penetrate downward they tend to bind to clays and soil particles. If their movement is not carefully monitored, or if the groundwater regime and subsurface conditions are not completely understood, aquifer contamination can become a serious threat.

Concentrate and Contain

Although high level liquid wastes only amount to about 1 percent of the total volume of radioactive wastes produced in the United States, they contain over 99 percent of the radiation. These liquids are produced primarily at nuclear fuel reprocessing plants and are extremely hazardous due to the presence of strontium-90, cesium-137, and plutonium-239. Because of their long half-lives, these radioisotopes need hundreds of years to decay and are, therefore, long-term hazards. These isotopes are also biologically

very active. Strontium-90 and cesium-137 behave chemically like calcium and potassium and are easily incorporated into the tissues and organs of plants and animals. Plutonium, even in extremely low concentrations, is the most toxic known agent producing lung cancer, which makes it also an extremely dangerous material.

The Atomic Energy Commission (AEC) originally established the concept that radioisotopes should be isolated from the biological environment and allowed to decay for at least twenty half-lives. Plutonium-239 has a half-life of about 25,000 years. In other words, this material should be safely stored and carefully isolated from the biosphere for about 500,000 years!

Up to the present the high level radioactive wastes produced in this country have been stored in concrete-encased steel tanks (see figure 9.6). The greatest quantity of wastes, approximately 320 million liters (85 million gallons), is contained at three federal repositories: the Hanford Reservation near Richland, Washington, the National Reactor Test Site near Arco, Idaho, and the Savannah

Figure 9.6. *High Level Radioactive Waste Storage at Hanford, Washington. These are the tanks that are still being utilized to contain all the high level radioactive waste produced in the United States. They are the same tanks that have leaked thousands of gallons.* PHOTO COURTESY: *Energy Research and Development Administration.*

River Reservation near Barnwell, South Carolina (Micklin, 1974). Tank storage of liquid wastes, according to the AEC, "has proved both safe and practical." However, there have been serious problems at Hanford where liquid wastes have been confined the longest. Sixteen leaks, which have totaled nearly 1,350,000 liters (350,000 gallons), have been found over the past sixteen years. The largest and most recent accident occurred in 1973 when 450,000 liters (115,000 gallons) leaked from a twenty-nine-year-old 2 million liter tank. Although the Hanford facility had been warned in 1968 of the serious situation that existed in regard to the condition of its tanks, particularly the older ones, and some improvements had been made, the existence of a "leak" of this magnitude as recently as 1973 indicates that the problem has not been adequately dealt with. Tank design is being altered to double-walled tanks and liquid wastes are slowly being solidified for safer storage.

To safely store these high level wastes and also those that are being continually produced in increasing volumes for 500,000 years, a permanent solution needs to be found. There are some critical requirements that must be met by any safe and permanent disposal or storage site. Initially, the site must be completely isolated, now and in the future, from the biological environment. There must not be any opportunity for groundwater to be contaminated by radioactive material. Any disposal site must be located in a geologically stable area, without threat from earthquakes, subsidence, volcanic activity, erosion, or other processes that would threaten containment any time in the next half-million years. The storage area must be continually protected from sabotage, airplane crashes, bombs, and other human threats. A reliable method of transporting the wastes to the site and safely storing them on arrival must exist. Meeting these criteria at present seems difficult enough, but imagine guaranteeing geological and political stability so that the site will remain safe for 500,000 years; the thought is incomprehensible. If one even stops to think about the geologic events of the past 500,000 years — vast continental glaciations, lowering and raising sea level, major climatic changes, and total landscape transformation — the predicament becomes clearer. The problem of a permanent and safe storage or disposal system is one of the most difficult humanity may ever have to face. But to move forward with any additional atomic energy development, or even to maintain our current level of involvement, a solution must be found.

The two major storage or disposal methods being considered at present are deep well injection or disposal and solidification and storage in salt mines.

Deep Well Injection Method. Two techniques involving deep well injection have been suggested, both of which require pumping. With one method radioactive wastes would be mixed with a special cement and injected under high pressure into relatively impermeable shales. Injection under pressure would cause fracturing of the shale, allowing the cement slurry to flow into the fractures where it would harden, holding the waste immobile. Some intermediate level wastes are now disposed of by this method. The second technique would involve pumping radioactive liquids into permeable formations at great depth where they would presumably remain permanently (Zeller, 1973).

The first method, and in part the second, would require high pressure pumping of radioactive wastes with the associated accident risks. Cement mixtures capable of holding nuclear wastes completely immobile under prolonged high temperature and high radiation conditions have not been developed. Dehydration of the shales under the high temperatures could cause leaching, thereby carrying the wastes from the immediate area of the cement. The radioactive wastes could escape into water- or petroleum-bearing formations by moving along joints or fractures. Diffusion or migration of the liquids would be accelerated by high temperatures. There is also the possibility of earthquake generation as occurred in Denver, due to the lubrication of fault planes. Disposal in areas far removed from oil, coal, or other mineral deposits would also be necessary to ensure accidental penetration would not occur in future drilling or extraction operations (Zeller, 1973).

Exploratory drilling for deep well disposal in the basalt flows around Hanford, Washington, and investigations of the metamorphic and sedimentary rocks beneath the AEC's Savannah River Plant in South Carolina have recently been carried out. At the South Carolina site, the disposal shafts would pass through a thick aquifer that is a prolific source of fresh water. A clay layer exists between the aquifer and the deeper metamorphic and sedimentary rocks that would be used for disposal. The hope is that this clay would form an impermeable boundary and prevent the migration of radioactive ions into the fresh water above it. The plan, known as Operation Bedrock, although offering inexpensive on-site disposal, seems to involve some dangerous uncertainties. Exploration at the Hanford site is apparently still underway.

Salt Mine Storage. Storage in abandoned salt mines, which underlie vast areas of Kansas, Oklahoma, Texas, New Mexico, and New York, has probably received the most thorough investigation of any storage or disposal method. Initially salt mines seemed to offer a number of advantages: (1) the areas are stable geologically; (2) the existence of the salt deposits indicates a general lack of circulating groundwater; (3) salt behaves plastically under stress and would tend to heal any fractures or cracks formed from tectonic stress or other causes; and (4) salt is a fairly good conductor of heat so thermal energy would be partly dissipated.

The Atomic Energy Commission, when in existence, proposed solidification of the radioactive wastes in ceramic or glass cannisters that would then be placed in metal containers for storage. These cannisters would disintegrate within a period of several years, however, leaving the salt to hold the wastes. Questions have been raised about the effect of long-term heat buildup on the salt and about heat transfer and subsequent volume change of the salt due to the radioactive heat.

The AEC in 1970 announced that an abandoned salt mine near Lyons, Kansas, had been selected as a demonstration facility for radioactive waste emplacement. Inference was also that central Kansas would most likely become the national radioactive waste depository by the mid-1970s. However, this proposal met fierce resistance from local residents, the governor, the state's congressional representatives, and the state geological survey. The survey director pointed out that the rock and salt underlying the Lyons area, owing to the presence of abandoned drill holes, shafts, and mines, was like Swiss cheese, and a distinct possibility existed that wastes buried in the salt might reach the groundwater (Micklin, 1974). Support was given to this argument by the unexplained loss of 660,000 liters (175,000 gallons) of water in an adjacent mine during some experiments with hydraulic fracturing. The uncertainties raised by this loss of containment, and public opposition as well, led the AEC to abandon the plan for Lyons, Kansas.

Other Possibilities. Several other solutions to the high level radioactive waste disposal problem have been suggested, but for various reasons have not been given serious consideration.

The thick ice caps covering much of Greenland and Antarctica have been proposed as disposal sites. Such disposal would involve the transport of containers of solidified waste across the ocean to the center of the land mass at either site. The hot containers would be set on the ice and would melt their way downward until they came to rest at the rock-ice interface beneath the ice cap. Ice would refreeze behind the descending containers and seal them off from the surface. Apparently little research has been carried out on this proposal, however. Transport safety, economic costs, possible migration of the hot fluids after container decomposition, in addition to the international treaty preventing the deposition of nuclear waste in Antarctica are immediate concerns.

Deep sea trenches or **subduction zones** have also been suggested as disposal sites for radioactive waste in addition to other hazardous materials. Three characteristics of trenches have been discussed in considering these areas as waste disposal sites: (1) the relatively high sedimentation rates in trenches, (2) the possible liquefaction of these sediments during major earthquakes, and (3) the subduction of the sediments, or their incorporation into the crust during underthrusting (Silver, 1972). Do these factors, either individually or collectively, indicate that the criteria listed previously for safe radioactive waste disposal can be met? Although sedimentation rates in trenches are relatively high, it would still require 500 to 1000 years to bury a layer of material 1 meter thick. The highest rates of underthrusting or **subduction** are about 10 centimeters per year. At this rate, and assuming the waste being dumped would be restricted to a narrow zone only 1 kilometer wide right at the trench floor plate boundary, subduction of this strip of material would take about 10,000 years. However, uncertainties in the behavior of sediments, whether in fact they liquify or not, or whether they are simply added to the base of the continental slope, as they seem to be in places, adds additional questions. For these reasons, trenches seem to be no more appropriate for toxic waste disposal than any other deep sea floor location.

While the AEC was spending billions of dollars to develop military and commercial applications of atomic energy, the amount spent on waste disposal research was minimal. The nuclear industry is still growing, but a permanent and acceptable solution to high level waste disposal has yet to be found and proven. One of the AEC's own scientific advisory committees has characterized some of the agency's waste disposal practices as "expedience designed to make the best of poor locations." This extremely critical aspect of atomic energy utilization and development clearly is a problematic one that needs to be considered seriously before more liability is accumulated.

Because of the difficulties encountered in establishing a permanent waste respository in salt, the AEC returned its attention to the concept of interim storage. In 1972 the agency announced that it would use the technique of retrievable surface storage to manage commercial high level wastes. This method, it is contended, will provide safe management for at least 100 years and perhaps for several centuries. Over this period, suitable permanent disposal means can be established (Micklin, 1974). From past experience and the increasing volumes of radioactive waste, the uncertainties, difficulties, and human error involved in encapsulating, shipping, and eventually storing the wastes still pose serious future problems.

WASTE WATER TREATMENT AND DISPOSAL

The domestic and industrial waste water problem is a serious one. Continued discharge of untreated or inadequately treated wastes into streams, lakes, or coastal waters is harmful to aquatic ecosystems as well as to human health.

About three-fourths of the population of the United States is served by domestic sewers. The rest of the people discharge their waste water into septic tanks or directly into the ground or water. Of those wastes in domestic sewers, one-tenth is discharged untreated into bodies of water and about one-third is discharged after only primary treatment, or settling. Although billions of dollars have been spent on sewers, treatment plants, and related facilities, there are still hundreds of communities dumping untreated or inadequately treated waste water into waterways. Thousands of industrial plants discharge effluents into sewage plants that are unequipped to process many of the industrial pollutants.

The usage of water continues to expand along with the volume and complexity of pollutants, while the fresh water supply remains essentially constant. There are a number of methods of treating waste water that depend upon the nature and strength of the waste, the water quality level desired or demanded, and also the economic constraints or considerations.

Septic Tanks

Septic tanks, which are usually single family waste disposal systems, were originally used and intended for farms and widely separated residences in rural areas. During the post-World War II building boom, however, septic tanks were employed in many subdivi-

sions as areas beyond city sewage disposal facilities were rapidly developed. Although perhaps 25 percent of the nation's population depends directly upon soil absorption for their waste disposal, septic tanks are unsuited for use in many areas. The density of housing, and therefore, septic tanks in suburban developments, the use of small parcels of land as both a source of water and a disposal site for waste water, and the limitations of the soil itself are some of the factors that have begun to limit the use of septic tanks. Septic systems have remained essentially unchanged through the years and in many cases they do not operate effectively. Where appropriate conditions exist, however, they are still a sanitary solution to a single family's waste water problems.

A complete septic system consists of a septic tank and a drain or **leach field** (see figure 9.7). The septic tank is simply a

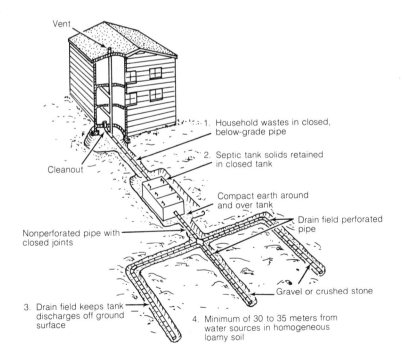

Vent

1. Household wastes in closed, below-grade pipe

2. Septic tank solids retained in closed tank

Cleanout

Compact earth around and over tank

Nonperforated pipe with closed joints

Drain field perforated pipe

3. Drain field keeps tank discharges off ground surface

4. Minimum of 30 to 35 meters from water sources in homogeneous loamy soil

Gravel or crushed stone

Figure 9.7. *Typical Septic Tank System for a Single Family Dwelling.* REDRAWN FROM: *E. T. Chanlett, Environmental Protection (New York: McGraw-Hill Book Company, 1973), p. 141.*

watertight container with an inlet from the waste water source (a residence) and an outlet that leads to the drain field. The outlet pipe is perforated and can drain either into a deep ditch, or perhaps into several deep pits, backfilled with gravel. Within the tank, anaerobic bacteria decompose the solids and allow the liquid waste to slowly flow through a baffle and then through the outlet into the drain field.

The rate at which the effluent is absorbed is critical to the operation of the disposal system. If the soil is impermeable or only slightly permeable, the liquid effluent will not be absorbed rapidly enough. Eventually the water may back up into the house or may rise in the soil and collect at the ground surface, both undesirable for obvious reasons. If the subsurface material, on the other hand, is very permeable, such as coarse, clean sand or gravel, the effluent may move quickly away from the area. In this case, little natural filtration occurs, and eventually the waste may contaminate a well, stream, or the groundwater system (see figure 9.8; Cain and Beaty, 1973).

Soil permeability is a function of the grain size and sorting of soil particles. Where any doubt exists about permeability, a percolation test (which determines the rate at which the water level will drop in a satu-rated test hole) should be required before approval is given for installation of a septic tank system. Such a test should preferably be required during winter or spring months when the water table is highest. Percolation rates change with time, however, due to mechanical breakdown of soil particles, clogging of pores by organic matter from the effluent, and soil saturation. Septic tank systems may eventually fail or become troublesome and a health hazard during winter months when the soil is saturated. Failures can be due to faulty or leaky tanks, insufficient drain field, high seasonal water table, or poor permeability. Some older systems have been built by simply laying a perforated drain pipe in a ditch perhaps one or two feet deep and covering this with soil. Failure or soil clogging and saturation in such a system is only a matter of time.

No septic tank system should be placed in close proximity to any water system. The contamination of groundwater by septic tank effluent has been a concern for good reason. The increased suburban usage of septic tanks and the widespread use of nondegradable household detergents has led to some interesting effluent-tracing work. A number of studies in midwestern and eastern communities, where each parcel has its own well and septic tank, have shown that 20 to 30

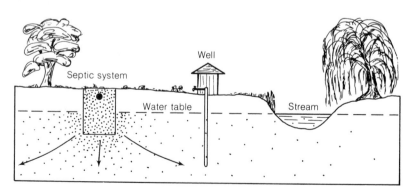

Figure 9.8. *Contamination of a Well and Stream by an Improperly Placed Septic Tank. If a septic tank is placed too close to a stream or well, if the water table is very high, or if the subsurface material has very high permeability, contamination can occur.*

percent of the well water samples contained measurable quantities of detergents, and some were unsafe for human usage from a bacteriological standpoint (Cain and Beaty, 1973). Minimum parcel sizes (1 acre) should be established for new septic tank systems and large parcels (2 to 3 acres minimum) should be required for installation of a septic tank and well or water system on the same parcel. Although each area will have its own suitabilities and limitations, it is questionable that smaller parcels can consistently provide both a source of fresh water and a sink for waste water, especially when many such parcels are adjacent to one another.

Thus the use of septic tanks on small lots in crowded subdivisions with inadequate disposal area, the installation of faulty or inadequate tanks and drain fields, and the placement of such systems in poorly suited soils are all problems with which both geologists and planners must deal. Areas inadequate for septic tank disposal systems, those with rock outcrops, steep slopes, impermeable or poorly drained soils, or with seasonally high water tables should be delineated as unsuitable on planning maps by experienced geologists or sanitary engineers. Marginal areas should also be determined, along with known locations of streams, wells, springs, and other water systems that must be avoided.

Municipal Waste Water Treatment

All the purification processes of natural water systems can be compressed in time and space within a series of sewage treatment processes. Three principal processes are normally involved (see table 9.5; Chanlett, 1973), and the completeness of these is a function of the treatment level, the type of system used, the efficiency of the plant, and the strength of the waste water. Prior to processing, several pretreatment steps are usually involved, primarily to protect the pumps and other equipment in the treatment plant. Bars or screens, and grinders, either remove or pulverize large material, grit chambers settle sand and grit, and grease and scum are skimmed from the surface.

Settling

This process simply allows the waste water to sit in or pass slowly (quickly during high flow periods) through a sedimentation or settling tank, also called a clarifier. Where material that settles and floats (sludge) is removed. This sludge is sent to a sludge digester that, operating at a constant temperature, slowly breaks down or digests the organic matter. The digested sludge is usually disposed of on land, some is routinely dumped in the ocean, and a small amount is dried, packaged, and sold as a fertilizer or soil conditioner.

Bio-Oxidation

Although oxidation occurs in natural water systems, within a treatment plant it is accelerated and controlled. The waste water, having been through a settling tank, is brought into contact with microorganisms and oxygen, which break down the organic material and oxidize it. The two most commonly utilized bio-oxidation processes are the use of the high-rate "trickling filter," and the creation of activated sludge (see figure 9.9). A trickling filter consists of a circular concrete tank filled with coarse gravel that is covered with a microbiological film. The effluent is slowly sprayed over the rocks and as the water trickles down, breakdown, as-

Table 9.5. *Sewage Treatment Processes and Their Effects*

Treatment Action	Process	Sewage Constituent Effected	Cumulative Percentage Removed	Sequential Step Required
Sedimentation	Primary settling	Settleable solids BOD Bacterial count	35–65 25–40 50–60	Sludge digestion, dewatering, and final disposal
Bio-oxidation: each process listed here has been preceded by primary settling	High-rate "trickling filters"	Settleable solids BOD Bacterial count	70–90 65–95 70–95	Secondary settling and digestion, dewatering, and final disposal of the additional sludge
	Activated sludge	Settleable solids BOD Bacterial counts	80–95 85–95 90–95	Secondary settling, thickening, digestion, dewatering, and final disposal
	Intermittent sand filters	Settleable solids BOD Bacterial counts	90–95 85–95 95 and over	None
Disinfection	Chlorination of settled sewage	Settleable solids BOD Bacterial count	35–65 in primary settling 25–40 in primary settling Plus 90–95	None

SOURCE: E. T. Chanlett, *Environmental Protection* (New York: McGraw-Hill Book Company, 1973), p. 151.

similation, and oxidation occur. With activated sludge, large aeration tanks are employed that also bring waste water, oxygen, and microorganisms into contact. In both cases, the organic matter or microorganisms that flow out of this process are settled out and pumped into a sludge digester. Two other oxidation methods, those of intermittent sand filters and oxidation ponds are also utilized (see figure 9.9). The first is simple and efficient but requires large areas of sand beds; the second requires little care or investment, only land areas somewhat removed from people where the waste water can be ponded and exposed to sunlight and microorganisms.

Chlorination

This final process is essentially one of disinfection, but it provides other benefits as well. Where swimming, surfing, fishing, or other water usages are involved, the disinfection (usually incomplete) of the treated effluent by either chlorine, ozone, or some other chlorine compound is usually demanded. Chlorine, if enough is present, can also sig-

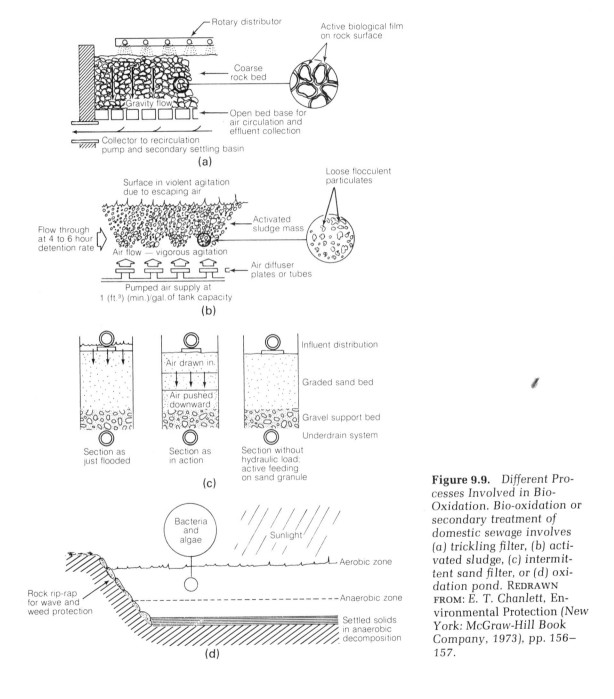

Figure 9.9. *Different Processes Involved in Bio-Oxidation. Bio-oxidation or secondary treatment of domestic sewage involves (a) trickling filter, (b) activated sludge, (c) intermittent sand filter, or (d) oxidation pond.* REDRAWN FROM: *E. T. Chanlett, Environmental Protection (New York: McGraw-Hill Book Company, 1973), pp. 156–157.*

nificantly reduce the BOD by direct oxidation of organic compounds.

Treatment Levels

The terms **primary** and **secondary treatment** are heard more often than sedimentation or bio-oxidation. Primary treatment involves only the first, and in some cases the third, steps of this system, settling and chlorination (see figure 9.10). The effectiveness of this level of treatment varies somewhat, however, as a function of settling time and the "strength" of the waste water. The longer the settling time, the more complete will be the removal of solids and bacteria and the lower will be the biochemical oxygen demand (see table 9.5). Due to the large amount of unsettled material remaining after primary treatment, even chlorination cannot adequately disinfect effluent in many cases. Many coastal cities in the past have gotten by with either no treatment or primary treatment followed by shallow water or even beach discharge. Public health hazards have, however, led to demands for better treatment and disposal systems. Along the California coast, over 3 million cubic meters of primary treated waste water is discharged each day.

Many communities have gone to secondary treatment, which subjects the effluent from primary processing to either the activated sludge or the trickling filter method (see figure 9.9). An efficient secondary plant will reduce degradable organic waste by 90 percent, and, with chlorination, 95 to 99 percent of the pathogenic microorganisms can be eliminated. Cities or communities on rivers, lakes, bays, or the ocean have commonly relied on disinfection, dilution, and dispersal by the receiving water body. As mentioned earlier, increasing loads of this

sort on some natural water systems have led to our existing water pollution problems. San Francisco Bay provides a good example. A few years ago, the treated municipal effluents of 5,256,000 people entered the receiving waters of the bay directly. Each day over 1.8 million meters^3 of municipal effluents, 1.4 million meters^3 of industrial effluents, and 13.7 million meters^3 of recirculated bay and river cooling waters were discharged to receiving waters (Kaiser Engineering, 1969). It is not surprising then that much of the bay is in its present polluted condition. Swimming and shellfish collecting are prohibited in many areas. Better levels of treatment are now being slowly required by both state water quality control boards and the EPA at the federal level.

Advanced Water Treatment

Conventional secondary treatment does not remove phosphorus or nitrogen from waste water, but turns the organic forms of these nutrients into mineral forms that are more usable by algae and other plants (Bylinsky, 1971). Some pathogenic organisms and many other dissolved chemical pollutants also remain after secondary treatment. The demand then for a higher quality effluent in some places and also the need to produce a safe, clear, and clean domestic water supply have led to more advanced treatment processes, often discussed or included in the phrase "tertiary" treatment. A typical domestic water treatment plant employs various processes (see figure 9.10) including addition of coagulants, flocculation, settling, filtration through sand and/or charcoal beds, chlorination, and often other specific additives or processes for specific removals (i.e., iron, manganese, and so forth). Water

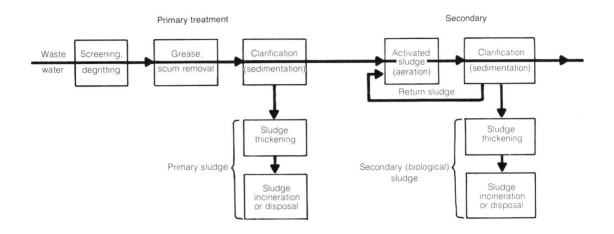

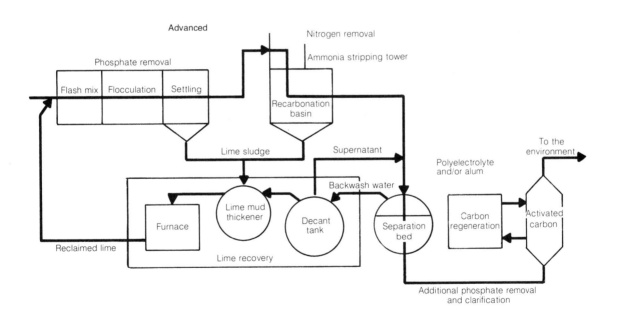

Figure 9.10. *Sequence of Processes Involved in (1) Primary, (2) Secondary, and (3) Advanced Waste Water Treatment of Domestic Sewage. Most communities in the United States utilize only primary treatment, and advanced treatment is only utilized in a few experimental sewage plants.*

that has been through this tertiary treatment usually becomes drinking water. It makes no economic or practical sense to treat water to this level for subsequent discharge. Tertiary treatment is designed to produce high quality domestic water.

At a model waste water treatment plant at Lake Tahoe, California, several additional processes are utilized. One is the removal of phosphate, and the other the conversion of nitrogen to the ammonia form followed by ammonia "stripping." Because of the desires to protect this lake, even after nutrient removal, the effluent is pumped out of the basin to a recreational lake, rather than back into Lake Tahoe.

Land Disposal and Reuse

The application of sewage or treated effluent to the land surface directly has been used for years around many large European cities (Edinburgh, Paris, Berlin) and is also practiced by over 130 communities in the American Southwest. The Golden Gate Park of San Francisco has been watered by reclaimed sewage of local origin for years. For a short time some Las Vegas, Nevada, golf courses have maintained greens from thoroughly processed waste water. At Whittier Narrows in southern California, waste water is reclaimed for groundwater recharge at the rate of 240 million liters per day (63 million gallons per day) in the Los Angeles Basin. Further south, near San Diego, the Santee project reclaims waste water for recreational use by a series of processes: activated sludge, thirty-day retention in oxidation ponds, water spreading at the surface followed by infiltration and subsurface filtration, and finally chlorination. The processed water flows into recreational lakes used for fishing, boating, and swimming (see figure 9.11). That we can still point out virtually all the existing projects of this sort indicates just how little effort has been expended in this direction.

Another useful approach to land disposal and reuse of waste water, first used by a food processor (Seabrook Farms) in New Jersey

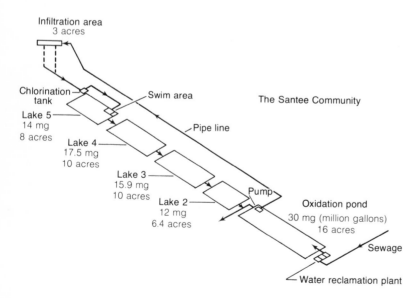

Infiltration area
3 acres

Chlorination tank

Swim area

The Santee Community

Lake 5
14 mg
8 acres

Pipe line

Lake 4
17.5 mg
10 acres

Lake 3
15.9 mg
10 acres

Pump

Lake 2
12 mg
6.4 acres

Oxidation pond
30 mg (million gallons)
16 acres

Sewage

Water reclamation plant

Figure 9.11. *Waste Water Reclamation Facilities at Santee, California. After secondary treatment and percolation through a series of ponds, treated waste water is used for recreation.* Redrawn from: *"Santee Recreation Project, Santee, California". Final report, Federal Water Pollution Control Administration, Cincinnati, Ohio, (1967).*

and later developed and refined by a research team at the Pennsylvania State University, uses the soil and its plants and animals as a living filter. Effluent from a treatment plant is sprayed over fields, pastures, or forests at rates commensurate with soil absorption. Plant roots and soil microorganisms are efficient at removing phophates and nitrates, the two most problematic nutrients in aquatic systems. Experimentally grown crops have even benefited from the fertilizing effects of the nutrients; hay yield increased 300 percent and its nitrogen content increased by 30 percent from waste water irrigation (Wagner, 1971).

The reclamation or reuse of treated waste water is a complex issue with numerous deterrents, some legal, some economic (usually the overriding considerations in most cases), some involving ignorance and tradition, but none are technological (McGauhey, 1968). We seem to be making progress in dealing with each issue as we see our present situation more clearly. We simply have to begin to understand, and this is most evident in places like Southern California, that we cannot afford to import water from 800 kilometers away, use it once, and flush it into the ocean. Serious consideration has to be given to reusing and reclaiming all our waste water. This involves publicizing successful existing efforts and understanding the necessity and beneficial effects of water reuse. By changing our economic thinking, we can actually consider and weigh the indirect costs of our present system against the potential future benefits of reclamation.

The principal choices for usage of reclaimed water are industry, recreation, groundwater recharge, and to a lesser degree, agriculture. Although direct reuse for drinking and domestic water does occur in a few places, this will probably not be common in the near future because of people's traditional feelings about "waste water." It has been said, however, that the only difference in cities drawing water from a lake, and those from a stream, is that the first are drinking their own waste water, whereas the second are drinking someone else's. It is something to think about.

SOLID WASTE DISPOSAL

Although solid wastes are generated by agriculture, industry, mining, businesses, and homes, we are concerned here primarily with the urban problem of garbage and rubbish disposal. The urban population of the United States continues to grow and now produces about 630,000 metric tons of solid waste each day. These wastes are collected, transported minimal distances commensurate with public acceptance, and then dumped (Schneider, 1970). The greatest expenditure ($14 per metric ton) in the past has been for collection, with considerably less ($4 per metric ton) having been spent on disposal. As long as the refuse has been collected and the disposal site is not a health hazard or is not esthetically too unpleasant, the operation is usually labeled successful. The effects of solid waste disposal sites on the water quality, both ground and surface, have not always been given consideration.

Types of Solid Wastes

Residential, commercial, and industrial activity generate a wide variety of wastes that have different effects on water quality. Although general categories might include garbage, rubbish, construction wastes, abandoned vehicles, and other kinds of material, it is more useful to deal with the individual components or constituents (see figure 9.12). In a general way, these can be broken down into food or organic wastes, paper products, metals, glass, plastics, cloth, brick, rock, and dirt, leather and rubber, yard wastes and wood (see figure 9.13), with organic wastes and paper products constituting almost 75 percent of the total weight.

Some of these materials, such as brick, rock, and glass are essentially inert and pose no pollution problems. Organic matter, however, needs to be decomposed and, therefore, requires a significant amount of oxygen. With decomposition, methane and carbon dioxide may be given off; the carbon dioxide increases the hardness and acidity of water, which may lead to solution and leaching. Ashes are generally rich in nitrates, phosphates, and other elements that can be leached out by percolating water. The nature of the solid wastes, therefore, and the concentrations of various soluble elements or compounds are of primary importance in determining the potential threat to the hydrologic environment.

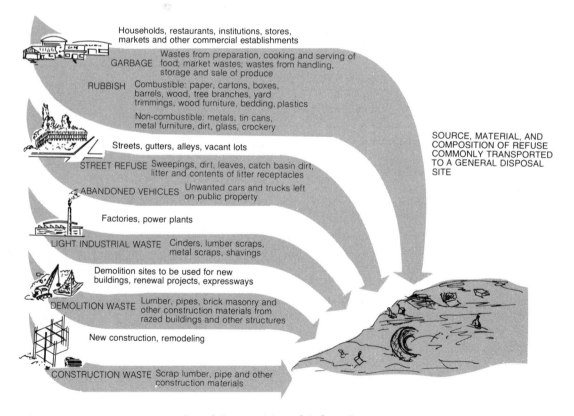

Households, restaurants, institutions, stores, markets and other commercial establishments

GARBAGE Wastes from preparation, cooking and serving of food; market wastes; wastes from handling, storage and sale of produce

RUBBISH Combustible: paper, cartons, boxes, barrels, wood, tree branches, yard trimmings, wood furniture, bedding, plastics

Non-combustible: metals, tin cans, metal furniture, dirt, glass, crockery

Streets, gutters, alleys, vacant lots

STREET REFUSE Sweepings, dirt, leaves, catch basin dirt; litter and contents of litter receptacles

ABANDONED VEHICLES Unwanted cars and trucks left on public property

Factories, power plants

LIGHT INDUSTRIAL WASTE Cinders, lumber scraps, metal scraps, shavings

Demolition sites to be used for new buildings, renewal projects, expressways

DEMOLITION WASTE Lumber, pipes, brick masonry and other construction materials from razed buildings and other structures

New construction, remodeling

CONSTRUCTION WASTE Scrap lumber, pipe and other construction materials

SOURCE, MATERIAL, AND COMPOSITION OF REFUSE COMMONLY TRANSPORTED TO A GENERAL DISPOSAL SITE

Figure 9.12. *Source, Material, and Composition of Refuse Commonly Disposed of at a Solid Waste Disposal Site.* REDRAWN FROM: *F. F. Davis, "Urban Ore,"* California Geology 25 (1972): 99.

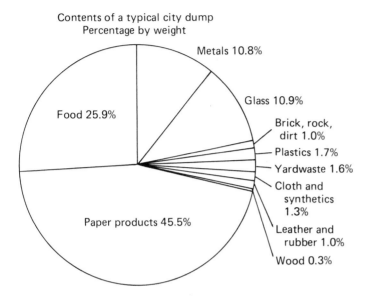

Contents of a typical city dump
Percentage by weight

Metals 10.8%

Glass 10.9%

Food 25.9%

Brick, rock, dirt 1.0%

Plastics 1.7%

Yardwaste 1.6%

Cloth and synthetics 1.3%

Paper products 45.5%

Leather and rubber 1.0%

Wood 0.3%

Figure 9.13. *Contents of a Typical City Dump: Percentage by Weight.* RE-DRAWN FROM: *F. F. Davis, "Urban Ore," California Geology 25 (1972): 101.*

Methods of Disposal

The type of waste disposal method used in any particular area depends upon local conditions, such as land availability, and also, in part, upon public attitude, which may be expressed through local government and planning. Nearly all urban wastes today are disposed of through one of three methods: (1) open dumps, (2) sanitary land fills, and (3) incineration.

Open Dumps

Approximately three-quarters of the solid wastes that are collected daily in the United States are disposed of at open dump sites (see figure 9.14). Practices at individual sites may vary considerably: the refuse may be partly burned, it may be compacted, or it may simply be piled up as high as equipment permits. Little effort is extended toward reducing the effect or impact of the waste on the surrounding hydrologic environment. Open dumps may be located in low areas or depressions, which eventually are filled and ultimately may be covered over with earth and revegetated or reclaimed. Due to the lack of compaction at some sites, reclamation may be difficult, and the sites may simply be abandoned. Too many sites have been used as open dumps simply on the basis of their availability and access, with no attention paid to problems of leaching of the wastes resulting in groundwater or surface water contamination.

Sanitary Landfills

Sanitary landfill, also known as a cut-and-cover process, is a relatively new method and is at present one of the most acceptable ways of disposing of waste. This method is still only utilized for about 13 percent of the nation's solid wastes, however. These operations may take place in depressions, such as canyons, ravines, or old quarries, or may

Figure 9.14. *Typical Open Dump Site. Many dump sites in the United States simply allow trash and garbage to pile up day after day with no cover to eliminate pests and odors.* Photo by: *Gary Griggs.*

even be carried out on a flat area or often along bays and estuaries (*see* figure 9.15). Waste material is dumped, compacted each day into layers or cells, and then covered with a layer of soil (*see* figure 9.16). The ratio of compacted refuse to soil usually is about 4:1. The soil covering serves to eliminate odors, unsightly appearance, and also the pest problems that plague open dump sites. In addition, the refuse is protected by the soil covering from direct precipitation, and consequently surface runoff of polluting substances or liquids is decreased. Sanitary landfills, despite their daily soil coverings, may still be unsanitary and produce water contamination problems if located in hydrologically unsuitable areas. The type of management and operation of individual landfill sites may produce conditions ranging from

nearly open dumps on the one hand to very clean operations on the other.

Incineration

The combustion or burning of waste material is an attractive solution in urban areas where there is a shortage of open land for dumps or landfills. About 8 percent of solid wastes collected in the United States are disposed of in this way. Although the resulting ash is usually only about one-fourth the volume of the original waste, it still must be disposed of periodically. Transportation costs are reduced with incineration because the combustion is usually carried out within the city itself. The release of oxides of sulfur and nitrogen with such a system, however, can produce serious air pollution problems,

Figure 9.15. *Sanitary Landfill, San Francisco Bay, California. Many of the communities around the margins of San Francisco Bay have used the bay for waste disposal, thereby decreasing its size and also eliminating highly productive estuarine environments.* PHOTO COURTESY: *U.S. Army Corps of Engineers.*

(a) Trench method

(b) Ramp method

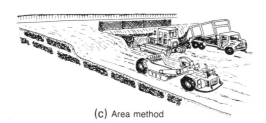

(c) Area method

Figure 9.16. *Operation of a Sanitary Landfill. Three different methods of sanitary landfill are shown. (a) Trench method. A trench is excavated where waste is to be dumped and the soil from each new trench is used to cover the waste in the trench just filled. (b) Ramp method. The solid waste is spread out and compacted on a slope. Soil scraped from the base of the slope can be used to cover the waste. (c) Area method. Waste is compacted and covered at the end of each day's operation with soil hauled by a scraper.* REDRAWN FROM: *U.S. Public Health Service, Sanitary Landfill Facts, publication 1792 (Washington, D.C.: U.S. Government Printing Office, 1968).*

especially within a city. Increasing concern for air quality in the past few years has led to the closing of thousands of municipal incinerators as well as numerous small backyard installations. On the positive side, some pilot incineration plants are now separating and reclaiming metals prior to burning and are also utilizing the heat generated by combustion to produce steam and to generate electricity (see Energy Recovery, p. 361).

Compaction

Compaction is a process utilizing tremendous pressures (up to 210 kilograms per square centimeter or 3000 pounds per square inch) to compress waste into small units. One system squeezes over 2.0 metric tons of trash into a cube 1.25 meters square. Although there is little data available, compaction has the potential for resolving some of the odor and pest problems associated with landfills, and, at the same time, reducing the land area needed for disposal.

Hydrologic Considerations of Solid Waste Disposal

The dumping of solid wastes or refuse in any open area leads to the possibility of contamination of either surface water or groundwater or both. The type and volume of waste material and the manner in which it is disposed are of principal importance. High concentrations of organic matter will demand large amounts of oxygen for decomposition; under anaerobic conditions carbon dioxide is produced and with water forms a weak acid. This acid acts on metals and other soluble ions, which then are incorporated into the **leachate** or the fluid that

has passed through the refuse and accumulated various contaminants.

The amount of precipitation or the rate at which water flows through a disposal site also has a major effect on the degree to which wastes are leached and on potential pollution problems. In arid climates, where rainfall is low and runoff limited, the amount of leaching and possible contamination of water is minimal. In humid climates, however, far more water passes through the refuse and potential pollution problems are more severe. The greater the volume of water that passes through the waste, the greater the degree of leaching, and the greater the concentration of organic and inorganic constituents in the leachate.

Within a landfill, the permeability and infiltration capacity of the soil used as cover determine the amount of surface water or precipitation that passes through the refuse and garbage. The water with its dissolved constituents percolates downward to the soil and rock underlying the disposal site. Of critical importance here is (1) the level of the water table, (2) the rate and direction of groundwater flow, and (3) the permeability of the underlying material. When and if the leachate reaches the groundwater system it will become part of the system and move with it (see figure 9.17). Any undesirable or potentially harmful substances leached from the refuse then are added to the groundwater. The result may range from increased hardness of the water (or increased concentration of calcium and magnesium) to total contamination with bacteria, salts, metals, nutrients, and other undesirable substances.

The nature of the underlying rock or sediment, chiefly its grain size and permeability, is of equal importance. Permeable soil, sand, gravel, or well-fractured bedrock obviously permit rapid movement of percolating

Figure 9.17. *Generalized Transport of Contaminants to Surface and Groundwater from a Dump Site. Water percolating through solid waste leaches out contaminants that can pollute nearby water supplies.* RE-DRAWN FROM: *W. J. Schneider, "Hydrogeologic Implications of Solid Waste Disposal", Geological Survey Circular 601-F (1970).*

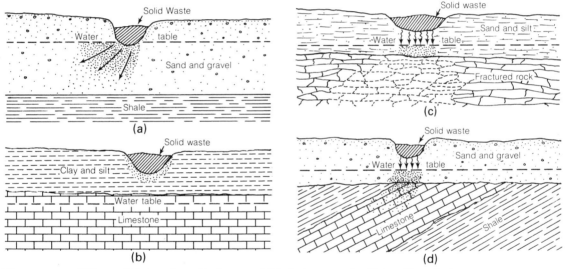

Figure 9.18. *Effects of Subsurface Geology and Hydrology on the Fate of Leachate from a Solid Waste Disposal Site. (a) A high water table and permeable rock can lead to rapid transport of leachate and groundwater contamination. (b) An impermeable rock type underlying a solid waste disposal site will contain any leachate and prevent contamination. (c) An aquifer consisting of fractured rock may also be contaminated by leachate. (d) If the water table is high beneath a dump site, contaminants can enter the permeable layers, which may also contain groundwater.* ADAPTED FROM: *W. J. Schneider, "Hydrogeologic Implications of Solid Waste Disposal"* Geological Survey Circular 601-F (1970).

water. Although some filtration may occur, the chemical contamination moves downward relatively quickly under the effect of gravity to the water table (*see* figure 9.18). Clays and more impermeable bedrock or soils retard the movement of leachate and restrict it to the immediate area around the disposal site. Under these conditions, pollution is frequently limited to the shallow local groundwater and contamination of deeper aquifers is negligible (Schneider, 1970). It is easy to see that where the water table is deep, the potential for contamination is reduced because the leachate must pass through more soil or rock before reaching groundwater. Where the water table is high or shallow, or in the extreme where the waste is actually in contact with the groundwater, leaching and contamination will be far more severe.

Once in the groundwater system the dissolved contaminants or pollutants will flow until they are absorbed, filtered out, or leave the aquifer. They may be removed at a well or may be discharged at the surface as a spring or stream base flow. Groundwater flow through fine-grained materials is very slow because of their low permeability. The large surface area of clays and their chemical activity also lead to significantly decreased concentration of dissolved ions and other substances even after transport over short distances. Groundwater flow through coarse-grained sands and gravels, in contrast, is relatively rapid and little is filtered from the water. Studies at a disposal site in Illinois showed that passage of leachate through 1 to 2 meters of silt-rich clay lowered the total dissolved solids (TDS) concentration by the same amount of flow through 200 meters of sand.

The permeability of the underlying soil or rock has to be given serious consideration in the selection of any solid waste disposal site. The direction of groundwater flow, the usage of ground and surface waters in surrounding areas, and the discharge of subsurface flow to streams is of critical importance. Highly permeable soils or rocks, areas with high water tables, and those where ground and surface water resources are intensely used must be delineated and avoided. These kinds of hydrologic considerations are straightforward and necessary in the earliest stages of solid waste disposal planning. Consequences of failing to make these kinds of determinations or studies are now obvious from past experiences.

Wasteload Reduction

Reducing the consumption of materials and products lowers the generation of wastes. Reducing consumptions means educating consumers to change their fundamental habits and initiating product controls by legislative enactment. The Environmental Protection Agency has evaluated various methods of source reduction and has found that taxes or charges on products, deposits, product design regulation, and outright bans on goods that require use of scarce raw materials are the most feasible methods at the present time (EPA, 1974).

The control of nonrefillable beverage containers has received considerable attention because it is relatively easy to implement at a state or local level. Two types of strategies are possible: a mandatory deposit for all beverage containers and a ban on production and sale of nonrefillable containers. Various jurisdictions have passed mandatory deposit legislation including the states of Oregon and Vermont and the cities of Bowie, Maryland, and Ann Arbor, Michigan.

The Environmental Protection Agency, in 1973, did a study of potential results of a national mandatory deposit based on the Oregon model. The study group assumed the $0.05 deposit would be repaid to consumers when empty containers were returned to the retail store. The retailer would be required to accept any empty container of the kind, size, and brand sold by that retail outlet. Retailers, in turn, could return empty containers to the distributor, who would also be required to pay the refund. Projected benefits would be as follows:

1. reduction of litter from beverage containers by 60 percent (Oregon's litter was reduced 75 percent)

2. a total reduction in beverage container consumption from 8.8 to 2.8 million tons

3. significant energy savings (container production is energy intensive)

4. air pollution reduction

5. water pollution reduction

6. reduced production of mine waste (EPA, 1974)

With this type of legislation, an attempt is being made to return to a period twenty years ago when almost all soft drinks and beers were sold in returnable deposit bottles. At that time the average deposit bottles made twenty-five trips from manufacturer to consumer and back. The replacement of deposit bottles by nonreturnable containers has not only produced unsightly litter and pollution problems at disposal sites, but has also caused a needless drain on natural resources.

Resource Recovery

Recycling

No one ever really consumes anything. Whether it is an automobile, a television set, or the daily newspaper, the object is used for a time and then is usually dumped somewhere. Recycling is becoming an increasingly important alternative to land disposal that can serve to reduce the amount of waste

accumulation and at the same time allow reuse of materials and reduce the drain of our depleted natural resources.

Neighborhood recycling or "ecology centers" accept a certain segment of the waste stream — usually newspapers, glass, aluminum, tin, and bi-metal cans — for return to industrial processors. Recycling is desirable from the standpoint of resource conservation, but it does have certain economic limitations. Except for those centers operated by major industry, most recycling centers require some subsidy in the form of donated labor, facilities, or equipment to remain in operation. Proposals to require house separation of recyclable wastes (which was actually being done years ago), and separate or "piggyback" collection by refuse companies, may make household recycling more viable in the future.

Materials Recovery

The disposal of automobiles, due to the value of the materials involved, presents an obvious opportunity to recover reusable resources. Until recently, the value of scrap steel and salvageable parts was great enough to encourage the automobile owner to sell an old car to a wrecking yard. However, because of escalating labor and shipping costs, it now often does not pay to recycle the automobile. Consequently, many junk cars are abandoned, resulting in eyesores and the loss of several tons of valuable iron ore and coal that are used needlessly to produce new cars.

Two recent events are reversing this trend, however. First is the growth of automobile-shredding plants that, in contrast to conventional wrecking processes, produce a high value scrap for which there is considerable demand (see figure 9.19). Sec-

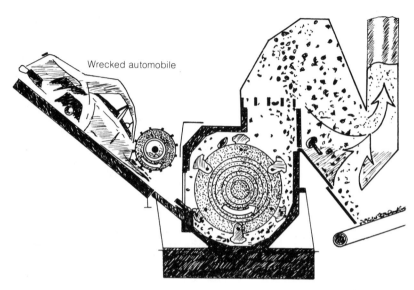

Wrecked automobile

Figure 9.19. *Cutaway Model of a Hammer Mill Auto Shredder. The large automobile shredder has undoubtedly been the greatest single factor in moving large numbers of junk vehicles back into the economic mainstream in the form of recycled steel and other materials. With a daily capacity of 800 to 1000 cars, a single shredder can greatly reduce a region's junk auto inventory within only a few years.* REDRAWN FROM: *National Center for Resource Recovery, "Junk Autos — a National Resource," NCRR Bulletin (Fall, 1973): 11.*

ondly, many communities have undertaken programs to remove abandoned vehicles by contracting with salvage companies through local law enforcement agencies. Where the car registration can be identified, the former owner is charged for removal of the vehicle. These fees, along with any returns from salvageable scrap, can then be put into a fund to perpetuate the program. In addition, California and several other states have offered financial assistance from vehicle registration fees to local communities for expenses incurred in the removal of abandoned autos.

Large-scale resource recovery requires the use of highly mechanized processes to retrieve materials and/or energy from refuse. It also provides the opportunity to significantly reduce environmental degradation and hazards to human health, as well as to provide for optimal use and conservation of resources.

Recovery potentials fall into two basic

groups of processes. The first is "materials recovery," or mechanical recovery, which involves removal of the glass and metal portion of the waste stream for use as a relatively pure raw material. The process begins with shredding of the heterogeneous refuse for size reduction, following which, the waste is channeled into a high velocity air column to blow off the light material (primarily organic) and drop the heavier material (primarily inorganic). The heavy inorganic portion is then channeled through a series of separation processes for metal and glass sorting. The particular systems used vary, but the remaining processes can include secondary shredders, air classifiers, screens, magnetic separators, froth flotation, and other sorting devices (see figure 9.20). The final glass or metallic material is then saleable to manufacturers (NCRR, 1975).

In the second series of processes, the organic portion of the refuse is converted into some type of secondary product, as for example, compost fertilizer, fiberboard building material, or energy utilizable fuels (see figure 9.21). Paper waste, primarily newsprint and corrugated cardboard, is recoverable both as a material and as an organic product. Its primary potential, however, is recovery as fuel in the organic portion of the waste stream, due in large part to contamination problems where paper is mixed with other refuse in disposal and collection. These contaminants are difficult to remove and are not acceptable in high speed paper-making processes (Abert et al., 1974).

Energy Recovery

The single greatest pressure to implement recovery systems may be the current energy shortage. As fossil fuel prices increase, many communities and utilities are exploring the use of the combustible organic portion of waste — commonly called refuse-derived fuel (RDF) — as a fuel supplement. In many cases RDF is not only a lower cost fuel option, but also is a fuel with significantly less sulfur, thereby reducing air pollutant emissions.

One method of obtaining RDF is to add unprocessed organic refuse directly to power-generating equipment. The Union Electric Company in St. Louis, Missouri, has conducted an experimental program in which organic refuse, after shredding and air classification, was added with pulverized coal to its boilers for electricity generation. Based on this successful experience, Union Electric is designing a $70 million system to be operational in mid-1977 that will process 8000 tons of solid waste per day from the metropolitan St. Louis area (NCRR, 1975).

In addition to direct use of processed waste, another method of energy recovery is the conversion to other fuels through a process called **pyrolysis**. In this process wastes are decomposed at high temperatures in a nearly airless environment. There are no pollutants discharged into the air and a number of by-products are recoverable, including gas, oil, tar, and various chemicals. Gas and oil are utilizable directly by utilities or industries to generate electricity or as a supplementary fuel source (Midwest Research Institute, 1973). Some communities, as for example Baltimore, Maryland, and San Diego, California, are using or currently designing pyrolysis recovery systems. The facilities designed for both Baltimore and San Diego include "front end" materials recovery of inorganic glass and metal that are not usable in the pyrolytic process (NCRR, 1975).

Anaerobic breakdown of organic matter at depth in a landfill produces methane gas. An experimental project is now underway at the Mountain View, California, sanitary landfill site to tap into this methane by drilling through the landfill and then utilizing the gas as fuel. The project would recover about 28,000 cubic meters (1 million cubic feet) of raw gas through some eighteen wells drilled into a portion of the landfill beside San Francisco Bay. After treatment to remove impurities and to increase the heating value, about 17,000 cubic meters a day of processed gas would be available. Although this is in itself a relatively small volume, if this project is successful other large landfill garbage dumps could also become economic supplemental sources of energy.

Outlook for Resource Recovery

Before total resource recovery becomes the prevalent method for future disposal of

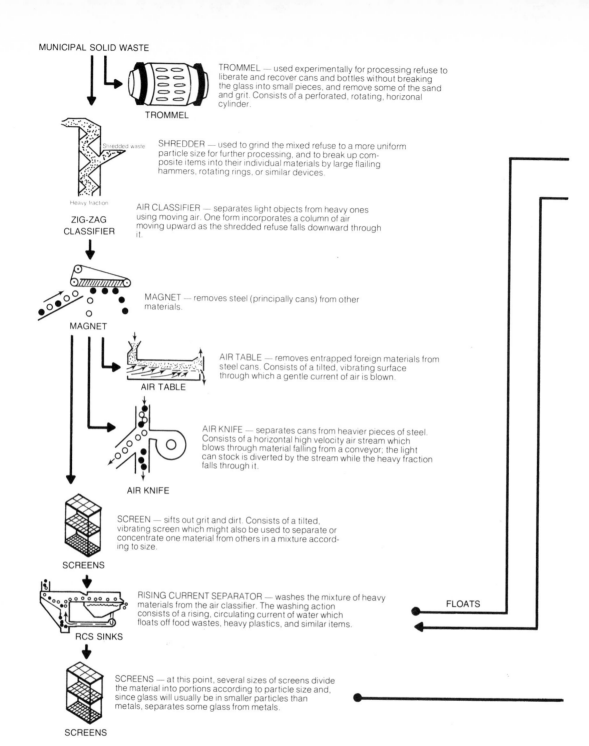

MUNICIPAL SOLID WASTE

TROMMEL

TROMMEL — used experimentally for processing refuse to liberate and recover cans and bottles without breaking the glass into small pieces, and remove some of the sand and grit. Consists of a perforated, rotating, horizonal cylinder.

Shredded waste

SHREDDER — used to grind the mixed refuse to a more uniform particle size for further processing, and to break up composite items into their individual materials by large flailing hammers, rotating rings, or similar devices.

Heavy fraction

ZIG-ZAG CLASSIFIER

AIR CLASSIFIER — separates light objects from heavy ones using moving air. One form incorporates a column of air moving upward as the shredded refuse falls downward through it.

MAGNET

MAGNET — removes steel (principally cans) from other materials.

AIR TABLE

AIR TABLE — removes entrapped foreign materials from steel cans. Consists of a tilted, vibrating surface through which a gentle current of air is blown.

AIR KNIFE

AIR KNIFE — separates cans from heavier pieces of steel. Consists of a horizontal high velocity air stream which blows through material falling from a conveyor; the light can stock is diverted by the stream while the heavy fraction falls through it.

SCREENS

SCREEN — sifts out grit and dirt. Consists of a tilted, vibrating screen which might also be used to separate or concentrate one material from others in a mixture according to size.

RCS SINKS

RISING CURRENT SEPARATOR — washes the mixture of heavy materials from the air classifier. The washing action consists of a rising, circulating current of water which floats off food wastes, heavy plastics, and similar items.

FLOATS

SCREENS

SCREENS — at this point, several sizes of screens divide the material into portions according to particle size and, since glass will usually be in smaller particles than metals, separates some glass from metals.

362

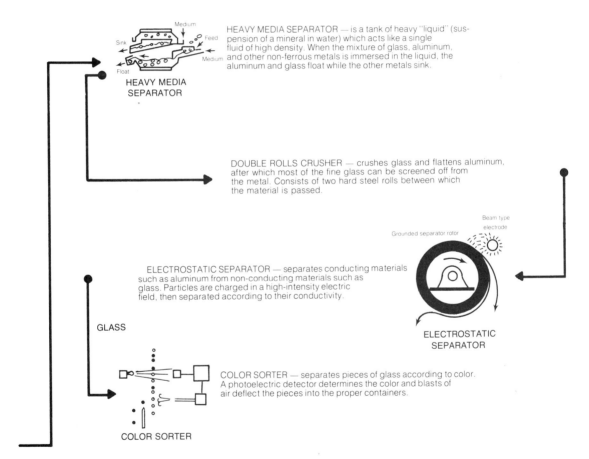

SECONDARY SHREDDER — reduces large piece sizes for second pass through the rising current separator. Without secondary size reduction, cans and large pieces of metal may float out with the organic material, or be lost in subsequent processing steps.

SECONDARY SHREDDER

HEAVY MEDIA SEPARATOR — is a tank of heavy "liquid" (suspension of a mineral in water) which acts like a single fluid of high density. When the mixture of glass, aluminum, and other non-ferrous metals is immersed in the liquid, the aluminum and glass float while the other metals sink.

HEAVY MEDIA SEPARATOR

DOUBLE ROLLS CRUSHER — crushes glass and flattens aluminum, after which most of the fine glass can be screened off from the metal. Consists of two hard steel rolls between which the material is passed.

ELECTROSTATIC SEPARATOR — separates conducting materials such as aluminum from non-conducting materials such as glass. Particles are charged in a high-intensity electric field, then separated according to their conductivity.

ELECTROSTATIC SEPARATOR

GLASS

COLOR SORTER — separates pieces of glass according to color. A photoelectric detector determines the color and blasts of air deflect the pieces into the proper containers.

COLOR SORTER

Figure 9.20. *Typical Materials Recovery Facility.* REDRAWN FROM: *National Center for Resource Recovery, "NCRR Material Test Program — A Recovery Facility Preview," NCRR Bulletin (Fall, 1973):* 6–7.

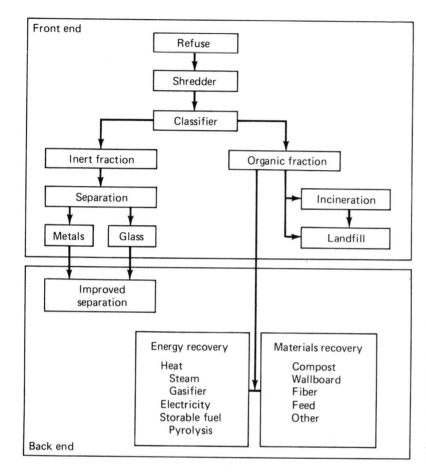

Figure 9.21. *A Total Resource Recovery System. Front end refers to materials recovery (see figure 9.20) and includes that portion of the refuse stream that is available for mechanical extraction and reuse as a relatively raw material. Back end refers to organic materials that, because of their physical characteristics, can only be recovered as a derived product through conversion.* REDRAWN FROM: *G. Abert James, H. Alter, and J. F. Bernheisel, "The Economics of Resource Recovery from Municipal Solid Waste," Science 183 (March 15, 1974): 1053. Copyright © 1974 by the American Association for the Advancement of Science.*

municipal solid waste, major economic, technological, and institutional difficulties must be overcome. Although the technology is present today, refinements of existing recovery systems will take place over the next few years. A major stumbling block in many areas is that solid waste disposal in dumps and sanitary landfills is less expensive than resource recovery. The cost comparisons normally used, however, do not account for leachate pollution, odor and pest problems, and irretrievably lost resources — major difficulties of land disposal that, until very re-

cently, officials have refused to recognize. Recent developments do give definite hope that substantial progress will be made in implementation of resource recovery systems in the near future.

REFERENCES

References Cited in the Text

Abert, James G.; Alter, H.; and Bernheisel, J. F. "The Economics of Resource Recovery from

Municipal Solid Waste." *Science* 183(March 15, 1974): 1052–58.

Bylinsky, G. "The Limited War on Water Pollution." In *Man's Impact on Environment.* Edited by T. R. Detwyler. New York: McGraw-Hill Book Company, 1971.

Cain, J. M., and Beatty, M. T. "Disposal of Septic Tank Effluent in Soils." *Journal of Soil and Water Conservation* 20(1973): 101–05.

Chanlett, E. T. *Environmental Protection* New York: McGraw-Hill Book Company, 1973.

Environmental Protection Agency. *Second Report to Congress-Resource Recovery and Source Reduction: Publication SW-122.* Washington, D.C.: U.S. Government Printing Office, 1974.

Hines, W. G. *Evaluating Pollution Potential of Landbased Waste Disposal: Santa Clara County, California.* U.S. Geological Survey Water Resources Investigation 31–73, 1974.

Kaiser Engineering, Inc. *San Francisco Bay Delta Water Quality Control Program: Final Report to the State of California,* 1969.

Klein, D. H., and Goldberg, E. D. "Mercury in the Marine Environment." *Environmental Science and Technology* 4(1970): 765–68.

McCulloch, D. S. et al. *Distribution of Mercury in Surface Sediments in San Francisco Bay Estuary, California.* U.S. Geological Survey Basic Data Contribution 14, 1971.

McGauhey, P. H. *Engineering Management of Water Resources.* New York: McGraw-Hill Book Company, 1968.

Micklin, P. P. "Environmental Hazards of Nuclear Wastes." *Science and Public Affairs* 30(1974): 36–42.

Midwest Research Institute, *Resource Recovery: The State of Technology.* Washington, D.C.: U.S. Government Printing Office, Feb. 1973.

National Center for Resource Recovery. "Resource Recovery Systems: A Review." *NCRR Bulletin* 5,2(Spring 1975): 26–33.

Peterson, D. H. et al. *Distribution of Lead and Copper in Surface Sediments in the San Francisco Bay Estuary, California.* U.S. Geological Survey Basic Data Contribution 36, 1972.

Schneider, W. J. *Hydrogeologic Implications of Solid-Waste Disposal: U.S. Geological Survey Circular 601-F,* 1970.

Silver, E. A. "Subduction Zones: Not Relevant to Present-Day Problems of Waste Disposal." *Nature* 239(1972): 330–31.

Wagner, R. H. *Environment and Man.* New York: W. W. Norton and Company, 1971.

Zeller, E. J. "The Disposal of Nuclear Waste." *California Geology* 26(1973): 79–87.

Other Useful References

Caffrey, P.; David, M.; and Ham, R. K. "Evaluation of Environmental Impact of Landfills." *American Society of Civil Engineers, Journal of the Environmental Engineering Division* 101(1975): EE1

Franks, Alvin L. "Geology for Individual Sewage Disposal Systems." *California Geology* 25(1972): 9–14.

National Center for Resource Recovery. *NCRR Bulletin.* Published quarterly.

CHAPTER 10

Resources and Planning

Contents

INTRODUCTION

MOST mineral or resource recovery today is either carried out through surface mining or through underground operations, which may involve mining, or drilling and pumping, such as in the extraction of oil and gas. With the removal, transport, and processing of any resource there are a number of wide ranging environmental impacts that can occur, depending upon the resource and the extraction technique or method. In the past these side effects, such as the acid runoff associated with the strip mining of coal, were scarcely given any consideration and were simply accepted as part of the operation. Fortunately this has begun to change.

If new recovery operations are preceded by thorough studies and careful planning, the potential adverse impacts can be recognized at the onset. Legal controls and site rehabilitation plans can be employed to minimize the problems and eventually to return the site to something approaching a natural environment or a useful area.

The increasing per capita usage of mineral and energy resources is leading not only to expanded resource extraction operations with their associated impacts, but also to dwindling reserves. Proper land use planning is necessary to ensure protection of areas containing valuable mineral or energy resources so that other land uses do not conflict with eventual resource recovery.

METHODS AND ENVIRONMENTAL EFFECTS OF RESOURCE EXTRACTION

Subsurface Mining

Mining beneath the ground surface removes that surface support supplied by the extracted material. Where a vein of ore or a coal seam is close to the surface, subsidence or collapse may eventually occur regardless of the mining method and the nature of the roof support. The subsidence of the ground surface as a result of underground mining, especially coal mining, has led to extensive damage to buildings, streets, and utilities in developed areas adjacent to mines as well as to loss of life (see chapter 5). Subsidence also alters drainage patterns. This allows surface water to infiltrate mine workings, which creates large underground impoundments of water. In the case of coal mines, the water usually becomes acid. Its return to the surface, whether by pumping or subsurface flow, can generate serious pollution problems (see figure 10.1).

In addition to the problems of disposal of mine waters, the waste rock that is removed must be brought to the surface and disposed of or dumped. The ground surface is altered and water percolating or leaching through these mine tailings may produce water quality problems downslope. The magnitude of the problem is directly related to the nature of the material involved — whether coal with its associated iron sulfide, mercury, or some other metal — and the volume that must be disposed of.

Figure 10.1. *Acid Mine Drainage. Highly contaminated water often collects in an abandoned strip mine area.* PHOTO COURTESY: *U.S. Bureau of Mines.*

Surface Mining

The relatively simple sounding process of surface mining has produced large-scale environmental destruction in virtually every state in the nation. By 1965 surface mining had disturbed 13,000 square kilometers (3.2 million acres) of land in the United States (see figure 10.2). About 95 percent of this area was related to the extraction of only seven commodities. Coal constitutes the major portion, 41 percent of the total land area mined; sand and gravel make up about 26 percent; and stone, gold, clay, phosphate, and iron together constitute about 28 percent (see figure 10.3; U.S. Dep't. of Interior, 1967).

Surface mining offers distinct advantages over underground mining: safer working conditions, more complete recovery of deposits, recovery of deposits not normally reachable for physical reasons, and lower costs. In some situations, however, the amount of overburden that must be removed to reach a given amount of product places an economic limit on the surface operation; in this instance, underground mining may be a solution.

A number of different methods may be

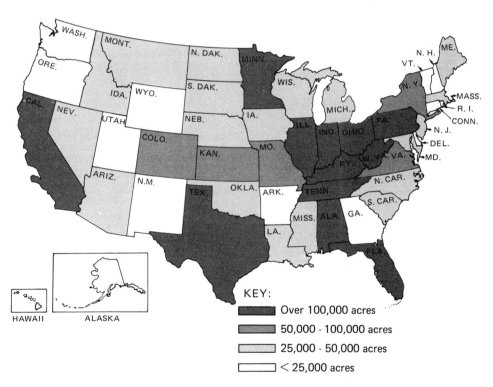

KEY:

- Over 100,000 acres
- 50,000 - 100,000 acres
- 25,000 - 50,000 acres
- < 25,000 acres

Figure 10.2. *Acreage by State Disturbed by Surface Mining as of 1965.* REDRAWN FROM: *U.S. Department of Interior, Impact of Surface Mining on the Environment (Washington, D.C.: U.S. Government Printing Office, 1967).*

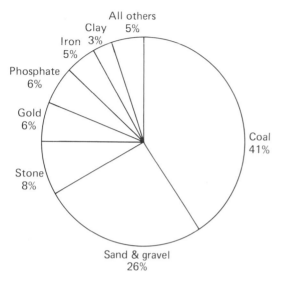

All others 5%
Clay 3%
Iron 5%
Phosphate 6%
Gold 6%
Stone 8%
Coal 41%
Sand & gravel 26%

Total, 3.2 million acres

Figure 10.3. *Disturbance of United States Land According to the Various Commodities Being Mined as of 1965.* REDRAWN FROM: *U.S. Department of Interior, Impact of Surface Mining on the Environment (Washington, D.C.: U.S. Government Printing Office, 1967).*

utilized for surface mining. The open pit method creates an increasingly larger hole from which material is removed. Open pit mining is utilized for limestone, sandstone, sand and gravel, marble, granite, and often iron, copper, and other metals where a larger quantity of ore or material is present. The amount of overburden removed is usually relatively small compared to the quantity of resource present. Many of the well-known mines of the world, including the iron ore mines in the Mesabi Range, Minnesota, and the copper mines at Bingham Canyon, Utah, and Santa Rita, New Mexico, are large open pit operations.

Strip mining may be carried out in either flat or hilly terrain and is most often utilized for coal. The overburden is removed along a strip or cut to uncover a bed or seam of coal. After the coal is removed from the cut the overburden from a second parallel strip is removed and placed in the cut previously mined (see figure 10.4). This system of cut-

Figure 10.4. *Typical Strip Mine Operation.* PHOTO COURTESY: *Tennessee Valley Authority.*

ting and filling parallel strips can totally overturn many square kilometers of land surface. Unless leveled, the result is a continuous series of ridges resembling a giant washboard (see figure 10.5).

Dredging is usually carried out along water courses and involves either a suction system or some mechanized device such as a dragline or mechanical buckets. Suction dredges utilize vacuum pressure to remove material from underwater areas. The dragline method uses a bucket suspended from an arm to scrape material from either land or underwater areas. Sand, gravel, gold, and tin in **placer deposits** have all been extracted using these techniques. Marked changes in the hydraulics of streams leading to substantial changes in erosion and sedimentation patterns may result from these types of operations. Gold mining, for example, along

streams in the foothills of the Sierras has resulted in piling of river bottom gravel over many square kilometers of land surface (see figure 10.6).

The mining methods utilized, the proportion of mineral to waste rock, the magnitude of the operation and nature of the materials involved, and the terrain and climate of the area are all factors affecting environmental impact. Where the terrain is steep and seasonal precipitation is high, for instance, the effects of land surface disruption and a mining operation on runoff, sedimentation, and slope stability will be very pronounced (see figure 10.7).

Regardless of the method employed, surface mining involves:

1. the preparation of the site, which entails the removal of vegetation, if present,

Figure 10.5. Landscape Alteration By Strip Mining. PHOTO COURTESY: Tennessee Valley Authority.

Figure 10.6. *Gravel Piles from Placer Mining. Many stream beds, in addition to the surrounding landscape, were permanently altered in the search for gold during the California gold rush.* PHOTO BY: *Gary Griggs.*

Figure 10.7. *La Grange Placer Mine During the Gold Rush Era in California. This destructive technique was widely used 100 years ago but is no longer practiced.* PHOTO COURTESY: *U.S. Forest Service.*

and the development of access roads and areas for plant construction

2. the removal and disposal of overburden or waste material

3. the excavation and removal of the ore or material

A final step, land surface reclamation, has not often been considered as an essential part of the process in the past.

The most obvious impact of surface mining is the complete alteration of the land surface. Vegetation and topsoil are totally removed, a massive excavation or hole is created, and overburden is dumped, scattered, or piled over the landscape. The result is a drastic reshaping of the land and an alteration of the normal surface and subsurface drainage patterns. Many square kilometers of land may be literally overturned to depths of 20 to 30 meters (60 to 90 feet) or more. With topsoil removed, vegetation is not readily established on spoil or overburden banks, or on the exposed rock in

a quarry. In addition, the spoil piles are usually highly erodible and with even moderate precipitation can quickly add sediment to a stream course along with harmful chemical constituents (see figure 10.8). Undesirable materials may be associated with the mineral itself or may be produced or discharged by on-site processing of the mineral. In addition to the acid drainage associated with coal mining, some other potentially harmful chemicals are the arsenic compounds associated with silver deposits and cyanide used to extract gold. Many heavy metals are highly soluble and are frequently found in fluvial systems near mining operations. Recent recognition that mercury in surface sediments in aquatic environments may enter the food chain through biological activity makes it important to evaluate the existing reservoirs of mercury contained in sediments. Although mining and smelting operations as well as industrial activities contribute to this reservoir (see figure 10.9), industrial waste water discharges are probably of far greater significance in most areas.

Figure 10.8. *Stream Destruction from Placer Mining. This stream valley, now filled with 125 to 150 feet of fill, is located downstream from the La Grange placer mine shown in figure 10.7.* PHOTO COURTESY: *U.S. Forest Service.*

Figure 10.9. *Gullying at Kennett, California. This old photograph shows the effects of vegetation destruction from the fumes of a smelter.* PHOTO COURTESY: *U.S. Forest Service.*

The large-scale mining of the iron ore, taconite, in Minnesota, and its processing along the shores of Lake Superior has led to lake discharge of waste material (Mitchell, 1975). Every day around the clock, a pasty slurry of 60,000 tons of solids suspended in 1.8 million tons of water pours from the Reserve Mining Company into the lake at Silver Bay. While the heavier particles move toward the bottom of the lake, the lighter ones are carried out into the water. Asbestiform fibers, associated with the taconite, enter the water, circulate through parts of the lake, and are subsequently picked up by public water supply intakes, primarily at Duluth, Minnesota. In just twenty years, the mining company has cast its effluent over 5200 square kilometers (2000 square miles) of the world's greatest fresh water body and into the water supplies of over one-quarter million people. Although the discharge contains copper, nickel, lead, zinc, and chromium, these are insignificant compared to the trillions of tiny asbestiform fibers. Many scientists and doctors believe that the fibers, which are in many of the surrounding water systems, can cause cancer in humans. Because of the uncertainties and the time before the effects may be observed, the court cases against the company have dragged on for months. In August of 1976, however, a federal judge ordered the lake discharge halted.

Even areas removed from mining sites can be damaged by stream and water impoundment, pollution from sediment and harmful chemical constituents, isolation of areas or drainages by artificial barriers, loss of natural beauty, and creation of rubbish dumps and abandoned plants and equipment. All these add to the negative impact of surface mining.

As richer ores or mineral resources are exploited, deposits of lower concentration or marginal economic value at present will probably become economical. In many instances this will mean removal of more overburden or processing greater volumes of material to extract the same amount of resources. Mining operations and their potential impact will no doubt grow considerably in size and the land will come under even greater stress. Effective environmental planning and safeguards must precede these future efforts.

PLANNING FOR MINERAL EXTRACTION

Preserving Sites for Future Extraction

Urban growth provides both a need and a threat to full utilization of mineral resources in accordance with sound environmental practices. Because the optimum location of many minerals, such as construction materials (sand and gravel, limestone for cement, and so on), is normally close to urban or urbanizing areas, conflicts often develop between the extraction operation and neighboring residential or commercial land uses. Complaints normally heard center on the noise, dust, vibrations, truck traffic, water quality, and unsightliness of mining or quarry operations and abandoned pits.

Planning for the optimum use of mineral resources, and for the most suitable use of the site after excavation in terms of the surrounding land use, should begin with a comprehensive geologic study of the region. Such studies cannot only locate mineral deposits, but they can also provide a community with information on slope stability, flooding, zones of high water table, and

other hazards that can serve to guide development. Geologic maps of mineral deposits enable a planner to designate areas as "mineral preserve zones" on the community's general plan. Deposits in areas subject to development pressures should not undergo urban encroachment until after orderly extraction of the minerals has been accomplished. Open space buffer zones may have to be created to prevent conflicts between future mining or resource recovery sites and residential areas. Deposits in remote rural areas can serve a variety of open space needs for a number of years until extraction takes place.

Inclusion of a mineral preserve area in a jurisdiction's general plan serves a number of purposes: (1) It conserves the natural resource. (2) It alerts owners and prospective buyers of neighboring properties to proposed mining operations. (3) It provides an opportunity for planners to evaluate the effects of prospective mining activities on streams, groundwater basins, biologic communities, and other environmental resources. (4) It permits the development of long-range plans for use of sites when the deposit has been exhausted.

Regulating Mining Activities

Various types of land use regulations are needed to implement and refine policies stated in the general plan. One type of regulation is the establishment of a specific zoning district (see p. 405, chapter 11). When applied to a particular property such a regulation indicates the uses that are not consistent with the long-range goals of the general plan and therefore are not permitted. Usually, mineral extraction zones in a zoning ordinance will set general operational standards for noise, air and water pollution, setbacks from adjacent land uses, truck access and landscaping.

Another method of regulating mining operations, similar to the zone district, is by conditional use permit. This procedure allows a prospective operator to file a request to extract a particular mineral resource. After the application is reviewed, a public hearing is held before the planning commission to determine whether a permit should be issued, and if so, the conditions under which the operation should be conducted. Conditions of operation (regulation of noise, dust, air and water degradation, and the like) created in this manner are tailored to each site.

When evaluating an application for mineral extraction, planners must consider the following critical environmental factors: creation of impervious surface layers, degree of surface compaction, extent and implications of devegetation, increase in surface runoff, increase in erosion or landslide hazard, siltation of nearby streams, increase in air pollutants, removal of underground or surface water, stockpiling and disposal of overburden, disposal of chemical by-products, and site rehabilitation. When these factors are sufficiently negative and cannot be adequately mitigated by operating conditions placed on the permit, the applicant should not be allowed to operate.

In addition to these factors, the following operational methods should be encouraged in all mineral extraction operations (Young, 1968):

1. *Phasing* — The phasing of operations so that only certain areas of the total site are cleared, mined, and rehabilitated will reduce adverse environmental and aesthetic effects.

2. *Construction of berms* — Berms, or low embankments, provide a more effective noise and visual barrier than mere vegetation screening. Berms can also provide a storage area for overburden (inside the berm) and topsoil (top and sides of berm).

3. *Preservation of topsoil* — During excavation topsoil can be placed on berms and used for landscape screening. After extraction operations are completed, topsoil can then be used for permanent landscaping of the rehabilitated site.

4. *Preservation of natural vegetation* — Natural vegetation preserved on the perimeter of a mining operation, and in unexcavated areas in a phased operation, helps prevent soil erosion and off-site siltation, controls dust and noise, facilitates water percolation, and helps control runoff.

5. *Desilting basins and erosion control* — Siltation basins strategically placed, and other erosion control measures, help to reduce erosion, siltation, and flooding from extraction sites.

6. *Faithful performance bond* — A bond put up by the mining operators for compliance with a mining permit would provide insurance for the governmental agency and neighboring property owners that the operation will be completed as planned.

Rehabilitation

Experience has shown that if land reclamation is integrated into both the planning and operation of a mining process, it can be carried out more effectively and economically than as a separate operation. Extraction and reclamation should progress as a single operation with the pattern of excavation prearranged to facilitate the creation of useful lands. A rehabilitation plan should be compiled in the initial planning stages of the extraction operation. The plan should indicate the topography of the site before and during excavation to minimize earth moving necessary after mining activities are finished. It should also indicate where topsoil and overburden are to be stockpiled, suggest design of landforms, landscaping materials, and methods and timetable for completion of the rehabilitation operation. The overall goal of the plan should be the elimination of any continuing environmental or aesthetic problem, and the preparation of the site for a use that is of overall benefit to the community. Possible uses of spent quarry sites are illustrated in figure 10.10.

Since 1955 the National Sand and Gravel Association has taken a leadership role in establishing standards for improved operational practices as well as actively promoting land reclamation. A number of commendable projects have resulted. For example, near Grand Rapids, Michigan, a 200-acre gravel pit was transformed into a community park with swimming, fishing, picnicking, and other desirable recreational facilities. At Sansbar Estates, Michigan, a water-oriented recreation community was developed from a sand and gravel mining operation. Many other sites have been restored to practical public and private uses, including a sewage treatment plant in Dallas, Texas; a state park in Nebraska containing a chain of nine lakes used for swimming and fishing; and a commercial park in Columbus, Ohio (Bauer, 1965).

However, because of the lack of proper environmental safeguards, legislation, or controls in many jurisdictions, about two-thirds of the total acreage affected by surface mining in the United States (8700 square kilometers or 2 million acres) has not been reclaimed by either natural processes or human effort. The amount of land area affected by surface mining continues to increase. Reclamation and revegetation plans must be required and carried out as a condition of operation.

MINERAL RESOURCES

The previous discussions of resource extraction methods and their environmental effects and planning for extraction were general for the most part. Because a number of major nonfuel mineral resources and particularly energy resources have very distinct problems and side effects associated with their removal, processing, and transport, these are discussed in the following sections in more detail.

The nonfuel minerals included are all recovered by surface mining or quarry operations; the energy resources are extracted by surface and subsurface techniques.

Sand and Gravel

The use of sand and gravel follows a growth curve similar to population in a particular region. Extracted sand and gravel are used for virtually all forms of urban construction. A critical characteristic of this industry is its high bulk and low unit value, which in effect means that transportation expense is a major portion of the cost per unit volume of the material. Generally 55 kilometers is the maximum distance for hauling sand and

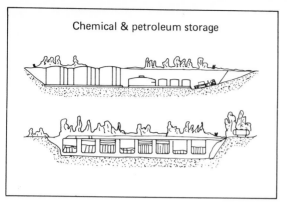

Chemical & petroleum storage

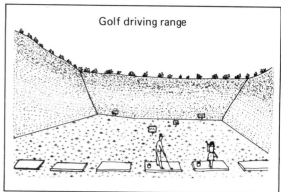

Golf driving range

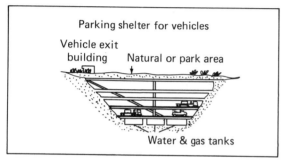

Parking shelter for vehicles

Vehicle exit building   Natural or park area

Water & gas tanks

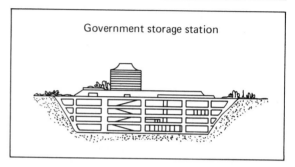

Government storage station

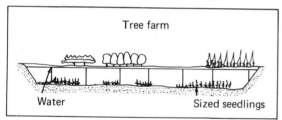

Tree farm

Water   Sized seedlings

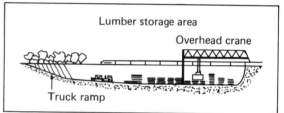

Lumber storage area

Overhead crane

Truck ramp

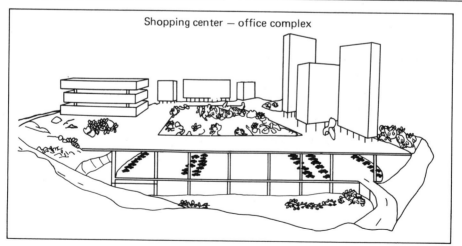

Shopping center — office complex

Figure 10.10. *Rehabilitation Plans for Spent Quarries. Many very worthwhile uses can be made of exhausted quarries, especially where they lie within close proximity to urban areas.* REDRAWN FROM: *E. Young, "Urban Planning for Sand and Gravel Needs," Mineral Information Service 21 (1968): 147–50.*

gravel. Because of this constraint, deposits located near urbanizing areas should be those of the most immediate interest to both the construction industry and planners. However, the paradox here is that the wave of building and road construction, which has escalated the demand for sand and gravel products, is at the same time causing urban conflicts which make deposits less extractable (see figure 10.11). In the Denver, Colorado, metropolitan area for example, a 1960 survey estimated 925 million tons of sand and gravel lay within 24 kilometers of downtown Denver. However 58 percent of these deposits were designated as inaccessible because they had been built upon or were too close to newly constructed residences to permit mining (American Society of Plan-

ning Officials, 1961). This conflict must be addressed by the city and regional planner to allow the orderly extraction of sand and/or gravel resources and to avoid undue escalation of construction costs.

The quarrying of sand and gravel has become the second largest mining business in the United States. Much of this material is excavated from stream channels because they are often a convenient source of clean high quality aggregate. Many streams in the southwestern United States are dry for much of the year and are, therefore, easily mined. The mining of a stream course for any resource has a potential impact on the aesthetics of the area, the hydraulics of the stream system itself, and on engineering structures along the stream. Considerations need to be

Figure 10.11. *Resource Extraction in an Urban Area. This sand and gravel quarry in the bed of Santiago Creek, California, is in the midst of an urban area.* Photo courtesy: *California Div. of Mines and Geology.*

made of these impacts before allowing new mining enterprises to begin or allowing on-going operations to expand.

The net effects of most urban activities or expansion on streams in the Southwest are those of increasing flood peaks, decreasing low flow discharges, and increasing sediment loads during construction periods. Although the mining of sand and gravel in stream channels can be carried out so as to cause minimal changes in water and sediment discharge, several other effects are common (Bull and Scott, 1974). Removal of aggregate from a stream bed initially causes local lowering of the channel. The tendency is then for downcutting to occur upstream from the area to reestablish a uniform gradient. Removal of material usually significantly exceeds the rate at which sand and gravel are being added from the watershed. The lowering of the stream bed and removal of material increases the potential for undermining or undercutting such structures as pilings or foundations of bridges. A bridge in Tucson, Arizona, for example, had 7.6 meters of sand and gravel above the base of the piers when it was built in 1965. By 1973, 4.0 meters of the material had been removed, leaving the bridge much more susceptible to structural damage during periods of peak stream discharge.

Considerable information is needed by decision makers to plan and control the multiple uses of urban streams. The following questions need to be answered in regard to stream bed mining (Bull and Scott, 1974):

1.   Are sand and gravel being removed from the stream bed at a more rapid rate than they are being replenished?

2.   During high stream flow, what is the effect of an aggregate pit in an active channel on stream bed scour at different distances upstream and downstream?

3.   What are the effects of stream bed mining on groundwater recharge?

4.   What are the magnitudes and time lags of the effects of mining in active channels at distances of several kilometers from mining areas?

Coarse-grained **aggregates** may be quarried wherever relatively clean sand and gravel formations are exposed in sufficient volume and at appropriate locations to make exploitation economical. There are environmental side effects even in these instances, however.

One problem in the production of aggregate, whether sand, gravel, or decomposed granite, is the need to wash the material to remove the fine particles (silt and clay). This requires a source of water in significant volume as well as washing and settling facilities. Many past operations simply discharged the wash water into the closest waterway, substantially increasing turbidity, leading to siltation with the resultant loss of fish-spawning areas and other aquatic habitats. Somewhat more conscientious operators constructed wash basins next to stream courses. However, the failure of inadequately constructed basins during excessive precipitation and runoff can lead to the same results (see figure 10.12). When processing aggregate, these kinds of discharges have to be terminated, if stream quality is to be maintained or improved, and other methods, such as properly engineered settling ponds, must be utilized.

Often the very characteristics that make a deposit good aggregate material for quarrying (that it is relatively clean, coarse-

Figure 10.12. *Effects of Wash Basin Failure. The siltation of this stream bottom resulted from the failure of poorly constructed upstream wash basins associated with a gravel quarry operation, Santa Cruz Mountains, California.* PHOTO BY: John Gilchrist.

grained, weakly consolidated, and of significant thickness or areal extent) also make it a good aquifer. The removal of this resource for construction sand, therefore, reduces the extent of the aquifer and its water storage capacity. This obviously conflicting use of a resource has to be resolved in accordance with regional needs. On the one hand, a permanent groundwater reservoir may potentially serve many people; on the other hand, a quarry operation provides sand for construction, some employment, but may add little else to the county or municipality in which it operates beyond property tax.

Most mineral extraction industries pay nothing to the county or municipality for operation in terms of taxes on raw materials removed. A small tax on each ton of sand or gravel or coal removed could be utilized to return the land to something resembling natural conditions after the completion of the quarry operation. The requirement to have quarry or mining operators post bonds to guarantee reclamation of the terrain and ground surface after completion is another solution being successfully employed.

Limestone and Shells

Lime comprises approximately 45 percent of the raw material in cement. Cement can be used in many diverse manners and constitutes a primary construction material. Manufacture of cement occurs primarily in proximity to urban areas, and transportation costs necessitate that extraction sites also be nearby. Lime is derived either from shell deposits in coastal areas, exploited by suction methods, or from limestone, recovered by open pit methods. Extraction of limestone involves environmental problems similar to those of other surface mining operations. Major shell mining operations occur in Texas, Florida, and the San Francisco Bay in California. Few investigations have been made to determine whether shell mining is environmentally harmful; however, dredging and washing operations do create turbidity, which biologists believe can be detrimental to marine life.

Stone

Stone is divided into two categories: (1) crushed rock and broken stone and (2) **dimension stone**. Crushed and broken stone is used for projects that do not have rigid size specifications for the material. This category includes aggregate base rock used for pavement, ballast, and fill; riprap used for ground stabilization for retaining walls, jetties, breakwaters, and seawalls; refractory stone used for brick and furnace linings; and agricultural stone used as fertilizer, soil conditioner, or for poultry grit. Dimension stone is a more specialized industry, requiring rock shaped to specific size requirements. Dimension stone must be free from cracks and weakening joints and have uniform texture together with an attractive color. The rock must be free of minerals that oxidize upon weathering, causing deterioration or surface staining. Both types of stone are extracted by open pit methods, with environmental effects similar to those discussed under sand and gravel, but usually relative to the tonnage removed and the size of the quarry.

ENERGY RESOURCES

Coal

The removal of coal from beneath the ground surface generates the greatest overall problems of any type of surface mining operation in the United States, either past or present. Thousands of square kilometers of the Appalachian region, including much of the states of Pennsylvania, West Virginia, Kentucky, and Illinois, are underlain by coal and have been devastated as this fuel product has been extracted over the years. Large areas in Arizona and New Mexico and in the Montana-Wyoming-North Dakota region are also underlain by coal, which is beginning to be exploited.

Pyrite, or iron sulfide, is a common constituent of coal, which, when exposed to the atmosphere and to water, oxidizes to form sulfuric acid in addition to other iron and sulfur compounds. Acid- and iron-rich streams and groundwater result, which most aquatic life cannot tolerate. The water becomes useless for almost any purpose as well as rusty colored and unsightly. The Monongahela River, which drains parts of Pennsylvania and West Virginia, annually discharges the equivalent of 180,000 metric tons of sulfuric acid into the Ohio River. It takes about 270 kilometers of river to neutralize this acid (Flawn, 1970).

The high acidity of spoil banks inhibits the growth of vegetation until complete oxidation has occurred and the acids have all been leached out. In the meantime, however, the barren piles of spoil are subject to rapid erosion. The results are drainages choked with sediment and a scarred and gullied landscape. It has been estimated that the yield of sediment from strip-mined areas is as much as 1000 times larger than sediment production from the same area under natural conditions (Strahler and Strahler, 1973). In one strip-mined area in Kentucky, spoil banks produced about 9500 tons of sediment per square kilometer in a four-year period, while the yield from undisturbed forested areas nearby was only about 9 tons per square kilometer.

All these disastrous effects of coal strip mining can be avoided with proper site planning and reclamation. Initially the topsoil can be stockpiled as mining proceeds. After termination of the operation, the spoil banks can be leveled and the topsoil redistributed over the ground surface (see figure 10.13). Experiments with revegetating strip-mined areas show that the application of lime serves to raise the pH of the soil and allow for growth of certain plant species (Magnuson and Kimball, 1968). As certain plants become established, sufficient soil restoration of minerals and organic matter allows for the successional growth of other plants. The leveling of the ground surface and development of a vegetative cover re-duces leaching of iron and sulfur in addition to retarding surface erosion.

These reclamation measures all cost money, but in amounts dependent upon initial planning and the resultant terrain. In the past, the costs were not included in mining costs because little or no restoration was carried out. Laws requiring land surface reclamation now exist in many areas, and the cost of leveling or grading spoil banks, replacing the topsoil, and planting cover crops have to be borne by someone. There is no technological reason why every area that has been strip mined cannot be reclaimed and transformed into a natural, aesthetically pleasing landscape. In fact, a natural landscape can sometimes be improved by the introduction of varied landforms — particularly on more level lands in the Midwest.

Available data indicate a range of 1 to 2¢ per mined ton of coal (about $50 per acre) for grading alone, to 40 or 50¢ per mined ton of coal (about $2500 per acre) for complete reclamation and reforestation (Brooks, 1966). Although these costs are still low in relation to the direct and indirect costs of increased erosion, sedimentation, and lowered water

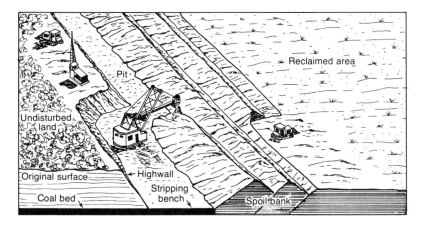

Figure 10.13. *Reclamation of Strip-Mined Area Concurrent with Coal Removal.*

quality, with coal selling at $4.50 per ton complete reclamation would lead to a price increase of about 10 percent or more (Flawn, 1970). An increase of this magnitude would eventually have to be borne by the consumer, as is any price increase, but could easily have a major impact on the competitive fuel industry.

Petroleum

Oil and gas together provide about three-fourths of the energy needs of the United States; however, a continually increasing portion of the oil is supplied by foreign imports. In addition, as land reserves are depleted, an increasing amount of oil exploration and recovery is now taking place in the offshore area. The very existence of petroleum in near-surface sedimentary rocks and

its extraction and transport have and will probably continue to produce significant environmental impact.

Exploration and drilling for oil on land involves a certain amount of land surface alteration. Roads are cut, sites for drilling rigs leveled, and drilling mud and oil may, in fact, reach streams or surface water. Perhaps the area of greatest concern today for the United States is the north slope of Alaska. Problems here include the passage of large vehicles, and the placement of equipment and structures including the 1250-kilometer (800 mile) trans-Alaska pipeline on permafrost. The effect of the structures on permafrost and the effect of changes in the permafrost on the structures are of some uncertainty, but of considerable potential significance (see Permafrost in chapter 5).

The most severe problems with petroleum involve the spillage of oil, especially

Figure 10.14. *Areas Throughout the World's Coastal Regions Where Offshore Drilling Is Occurring.*

into the marine environment. This has occurred through both offshore drilling, particularly in the Santa Barbara Channel and in the Gulf of Mexico, and more significantly, through accidental or intentional discharge of petroleum by oil tankers.

Offshore Drilling

Operations for oil exploration, drilling, or recovery are occurring today on virtually every offshore area in the world (see figure 10.14). High winds and sea conditions, icebergs, earthquakes, deep water, steep submarine slopes, and unstable bottom conditions are the major hazards that confront offshore drilling (see figure 10.15). In addition, the high pressures under which most oil occurs, the still developing technology of offshore oil drilling and recovery, and the fierce economic competition in the energy field produce added complications for these operations.

In the Santa Barbara Channel, for instance, oil deposits occur beneath relatively deep water. The sea floor is criss-crossed with faults and numerous earthquakes have occurred in the area in the past. The oil-bearing strata are so close to the surface that oil has been seeping out naturally for years.

Figure 10.15. *Semisubmersible Drilling Platform in the North Sea During Moderately Rough Weather. Wind and wave action are some of the many difficulties that offshore drilling platforms must withstand.* Photo courtesy: *British Petroleum.*

Measurements indicate a natural seepage rate of about 50 to 70 barrels per day (a barrel contains 42 gallons or about 160 liters). For comparison, during the month and a half of major leakage of the Santa Barbara oil spill in 1969, oil was being released at a rate about 1300 times as rapidly as natural seepage. Considering the environmental constraints of the Santa Barbara Channel with the intensive use of the ocean water and beaches in the area, the consequences of a major oil spill or well blowout are clearly disastrous (see Steinhart and Steinhart, 1972).

The Santa Barbara Channel is not unique, however, as the offshore operations along the Gulf Coast have led to far greater oil spills and disasters. One oil company was recently taken to court and indicted on charges of failure to install storm chokes on 147 wells on an offshore platform. These are safety devices that can seal wells off to avoid spillage in the event of a mishap, but they take time to install. As a result one well ended up gushing 600 to 1000 barrels of oil per day into the sea for a period of several weeks before being brought under control.

Many offshore failures seem to be a direct result of cost- and time-saving considerations, which simply were given more importance than required safety precautions. The overall or long-range environmental and economic impacts, however, are going to override the short-term economic gains from such practices. For instance, although many thousands or hundreds of thousands of dollars in time and material may have been "saved" by installing only shallow casing in the wells of Platform A in the Santa Barbara Channel, the direct cost to the oil companies to clean up the spill was over $12 million. The increased costs of advertising to compensate for the oil company's image after the spill will probably never be known. Environmental costs are difficult or in some cases impossible to evaluate monetarily but are obviously significant to the local residents. The environmental costs are either borne by or passed on to the public at large as the oil companies continue to increase their profits.

The existing regulations and controls need to be enforced and variances for arbitrary reasons can no longer be routinely granted by the government. The environmental and geologic hazards in areas of offshore leases have to be considered at the outset and taken into account in platform and pipeline engineering, placement, and operations on a continuing basis.

Oil Tankers

By far the greatest source of oil contamination of the ocean is accidental and intentional discharge by tankers. With increased oil demands in northern Europe and the United States, and the development of large oil fields in the Middle East and also South America, transport of oil at sea has grown to represent 55 percent of all commodities moved across the ocean. Tanker size has grown from 16,000 tons during World War II to 475,000 tons in the mid-1970s, and plans are being considered for even larger ships (see figure 10.16). A ship of 300,000 tons is about 400 meters long, draws 24 to 26 meters of water, and can carry 3,750,000 barrels of oil. This total volume is about fifty times the amount released during the Santa Barbara spill.

Growth in tanker size has been motivated principally by economic arguments. By reducing operating costs and increasing cargo capacity, shipowners foresaw great savings

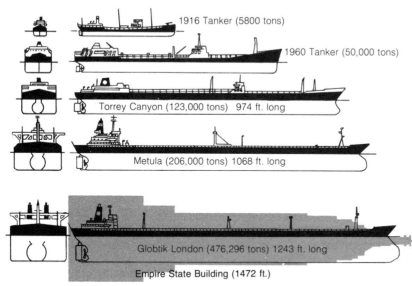

1916 Tanker (5800 tons)

1960 Tanker (50,000 tons)

Torrey Canyon (123,000 tons)   974 ft. long

Metula (206,000 tons) 1068 ft. long

Globtik London (476,296 tons) 1243 ft. long

Empire State Building (1472 ft.)

Figure 10.16. *Comparative Growth in the Size of Oil Tankers. Economic considerations have led to the construction of larger and larger tankers with increasing potential for environmental disasters. The 206,000-ton Metula went onto rocks at the tip of South America on August 9, 1974, and spilled 400,000 barrels of oil.* REDRAWN FROM: *N. Mostert, "The Age of the Oilberg," Audubon* 77, 3 (1975) 18–45.

in the new ships. The rise in conventional tanker accidents and costly oil pollution incidents, however, have increased insurance rates for large tankers. Profit margins have declined. Structural problems and tank explosions in some of the early super tankers led to modifications, which raised the investment as well. Due to the deep drafts of most super tankers there are still severe limitations as to the ports that they can enter to unload. Super tankers also have limited maneuverability because of their large size and single engine design. In face of obstacles they cannot respond with instant action. For instance, it takes up to 6.5 kilometers (4.0 miles) and twenty-five minutes to stop a 250,000-ton tanker cruising at sixteen knots. Although the new ships are generally built with improved oil pollution control equipment, their sheer size increases the possibility of disasters involving truly massive oil releases at sea or along the coast. The names and locations of major tanker collisions con-

tinues to grow: the Torrey Canyon off Land's End, England; the San Francisco Bay spill involving two sister ships; the Ocean Eagle off of San Juan, Puerto Rico; and the list goes on.

Late at night on August 9, 1974, the 206,000-ton super tanker *Metula* went on the rocks at the eastern end of the Straits of Magellan, opening up a 78-meter gash in her 320-meter-long hull. There, in waters where few persons have supposed such ships might venture, the disaster finally occurred that had been predicted but desperately feared: the wrecking of one of the giant new super tankers with a full cargo of crude oil. Some 400,000 barrels of oil (63 million liters) were spilled, a little over one-quarter of the total cargo, making this the second largest oil spill ever to occur. The fact that it happened where it did, remote from general public awareness, meant an obscurity for the *Metula* that she did not deserve and that would not have happened elsewhere.

The location of a spill, therefore, the nature or type of the oil, the volume released, the sea and weather conditions, the marine flora and fauna in the immediate area, and the cleanup equipment and personnel available are all important in determining the overall impact of any particular oil spill.

In addition to accidents, oily ballast water is routinely discharged to the sea by many tankers. After a ship unloads its cargo of oil, it still contains hundreds of barrels that coat the insides of the tanks and pipes. To make the return trip the tanker takes on ballast water for stabilization. Although the oil in the tanks, now mixed with the ballast water, can easily be concentrated in a single tank and can be safely unloaded at the port of destination, the lack of storage tanks, problems of tank ownership, and tariff problems dealing with oil off-loading often lead to discharge of the ballast water and oil at sea, usually not far off shore. These types of discharges are nearly impossible to detect or prevent, but occur daily. The off-loading of this oily ballast water must become routine procedure, but the appropriate storage facilities need to be constructed and international off-loading problems simplified.

Effects of Oil in the Sea

The fate of oil that enters the marine environment is not totally understood, but research and repeated oil spills have begun to make some processes apparent. Because oil contains some volatile fractions (or components that will evaporate), within a few days perhaps 25 percent of the oil may evaporate. Through bacterial breakdown and photooxidation more of the oil is gradually broken down, and after about three months at sea, only about 15 percent of the original petro-

Figure 10.17. *A Coastal Bird Rendered Helpless by Oiled Feathers Following the San Francisco Bay Oil Spill.* PHOTO BY: *Gary Griggs.*

leum remains as an asphaltic material (Wagner, 1974). However, when spills occur close to shore, as is commonly the case, the oil washes onto land with little time for breakdown or evaporation.

The deaths of many marine birds due to oiled feathers (see figure 10.17), the destruction of certain intertidal flora, fauna, and shellfish beds, prevention of commercial fishing, and the general aesthetic damage to the coast, harbors, and boats have been the major impacts of past oil spills. Along the coast of northern Europe, for example, especially along the English Channel, seabirds have repeatedly been killed by the thousands as a result of continued oil spills. Even so, it is still generally believed by many, and proclaimed by the oil companies, that following the evaporation of the more toxic and volatile components, the tarry residue remaining is not particularly harmful to marine organisms and that no evidence exists for long-term damage.

A spill of fuel oil in Buzzards Bay near the Woods Hole Oceanographic Institution in 1969 led to an intensive long-term study of the fate of the oil (Blumer et al., 1971). Much of the oil sank to the bottom as a result of weighting by sediment and contaminated the sea floor. There was immediate destruction of about 95 percent of the benthic organisms in the area within the first few days. The lighter fractions of the oil began to evaporate and the news media reported the area as beautiful as ever. The study, however, continued, utilizing sensitive gas **chromatography** to determine the presence of oil in sediments and surviving bottom organisms. Eight months later oil could still be detected in marine animals. There had not been the "swift and complete bacterial breakdown" usually reported after other spills. Shellfish still maintained traces of hydrocarbons after up to six months in running seawater. The oil residues from continued study are now believed to alter behavior, physiology, and sexual reproduction in shellfish and possibly other marine organisms. The implications of long-term biological effects are becoming clearer, especially in the light of increased oil tanker traffic and potential spillage. We can no longer assume just because the oil seems to be out of sight that it is no longer affecting the biological environment and is not having other long-term impacts.

Oil Shale and Tar Sands

Oil shales are sedimentary rocks that contain hydrocarbons in solid form disseminated throughout the rock. Petroleum also occurs in certain sands known as tar sands. The basic difference between these oil sources is that the bituminous matter in oil shale is solid, whereas it is a very viscous liquid in tar sands. This liquid will flow if heated by steam, so petroleum can be recovered by conventional methods without great difficulty. The hydrocarbons in oil shale, however, do not begin to flow except in special retorts or containers at temperatures of 800 to 900 degrees F. With this treatment, ten to twenty-five gallons of oil or more can be produced from each ton of shale.

In the United States, the most extensive oil shale deposits underlie about 44,000 square kilometers of Wyoming, Utah, and Colorado (see figure 10.18). The world's largest and best-known tar sands are in northern Alberta, Canada. These sands are being commercially exploited at present. Approximately 45,000 barrels of oil are being recovered from about 90,000 tons of sand daily. Estimates on oil shale reserves vary between 1000 and 2000 billion barrels of oil; these values are comparable to the estimates of ultimate world capacity of liquid petroleum (1800 to 2500 billion barrels). In other words, oil shales represent a major untapped potential petroleum source. Under present conditions, however, some estimates indicate only about 10 percent of the oil may be recoverable.

The recent "energy crisis," increasing costs and problems of oil imports, and the overall limits of liquid petroleum reserves have led to research and investigations of the potentials of oil shale and problems associated with its exploitation. Because the hydrocarbons cannot be pumped from shale, the rock must be mined and then treated. Although some oil shale lies at the surface, the deposits also extend to depths of 600 to 700 meters. The mining of oil shale by underground room-and-pillar methods has

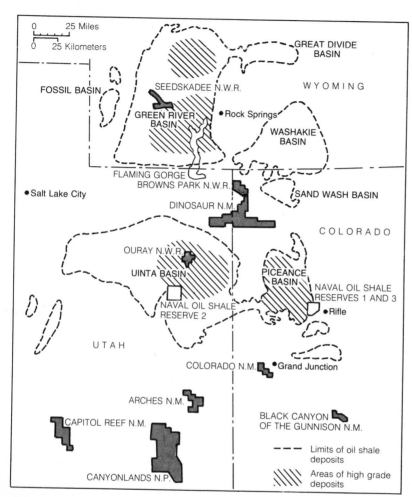

Figure 10.18. *Location of Known Oil Shale Deposits in the Western United States and Their Relation to National Parks, Monuments and Wildlife Refuges.* ADAPTED FROM: *D. C. Duncan and V. E. Swanson, "Organic-Rich Shale of the United States and World Land Areas", Geological Survey Circular 523 (1965).*

been demonstrated to be practical. Massive strip mining operations could be utilized for surface or near-surface material, which would involve removal and disposal of overburden and removal of the shale itself. Depending upon the mining method utilized and the quantity of overburden, 1.4 to 1.8 tons of shale and waste will be produced for each barrel of oil. An industry producing a million barrels of oil per day will have to dispose of 1.4 to 1.8 million tons of waste each day. This is enough material to cover a square kilometer to a depth of 2 meters. The effects of this material on the land surface, problems of erosion, stabilization, revegetation, and the expense and difficulties of land reclamation all need to be evaluated prior to embarking on large-scale oil shale development. Fortunately there are many large, arid arroyos and canyons where the oil shale occurs that offer large-scale disposal sites for spent shale and overburden. Research is needed to establish methods and procedures for placing and controlling the spent shale into the disposal area to prevent dusting, leaching, and erosion.

Geothermal Energy Development

Geothermal energy, or that heat that exists within the earth, is now receiving greater attention as an alternate energy source. In addition to the natural geothermal gradient within the earth, or the progressive increase in temperature with depth, areas of volcanic activity and often plate boundaries provide concentrations of heat that are potential sources of power. Volcanoes, hot springs, and geysers may exist at the earth's surface in such areas. Where crystalline rocks are sufficiently fractured or otherwise permeable, and circulating fluids are present, wells can be drilled to tap the steam and/or hot water.

Although electricity has been produced from geothermal energy for over seventy years, development of this energy source has been slow. At present, energy from geothermal sources is being utilized in Italy, El Salvador, Mexico, New Zealand, Japan, the Soviet Union, Iceland, and California. The Geysers site in California is still the only operating power plant of its type in the United States. It has been in production since 1960 and has been gradually increased in size so that it now provides about half the power for the city of San Francisco. During its fifteen years of operation and increasing production, no significant decline in the resource has been detected. Additional exploratory drilling has identified large geothermal fields in the Salton Sea-Imperial Valley area of southeastern California. Other potential sites in the western states are being investigated as well.

Normal production of geothermal power involves drilling to tap the heat, piping the steam or hot water through a heat exchanger, and then passing the steam under pressure through a turbine, which generates electricity. Exhaust steam may be condensed to water in cooling towers. The water being extracted, however, is actually a brine and presents a number of potential problems. The presence of salts may raise the salinity of waste water to perhaps six times that of seawater. Toxic trace elements or metals may be contained in harmful concentrations as well. These salts and trace elements can corrode or precipitate out in pipes, leading to costly repair and maintenance problems and, if discharged, can destroy or contaminate local surface water supplies. Subsidence and possible resultant damage to wells or surface facilities can result from fluid withdrawal. By utilizing waste heat to distill the mineralized water and by reinjecting the waste brines, fresh water can be produced, some salts can be reclaimed as a byproduct, and surface subsidence can be decreased or eliminated.

Geothermal drilling must be carried out carefully and with adequate casing to avoid blowouts and wastage of this energy source. One well at the Geysers site in California was apparently sheared off at depth, possibly by a deep-seated landslide, and has been releasing steam ever since.

Air pollution can also be a significant problem with geothermal energy development. Approximately five tons of hydrogen sulfide gas per day is discharged at the Geysers. Sulfur oxide production in some proposed geothermal power plants may be comparable to that produced by an equivalent sized fossil fuel plant. Certainly the composition of steam or hot water is a variable that needs to be considered as is the sulfur content of the oil being burned at a fossil fuel plant. Geothermal power production does have important potential in selected areas but it is not without environmental impact. The degradation of air and water quality, potential subsidence, adjacent land usage, all need to be considered in the development of this resource.

Uranium Mining Wastes

The mining of uranium produces huge piles of sandy tailings, which contain radium-226, with a half-life of 80,000 years. Radium is absorbed by bones, as is strontium, and therefore is a potential health hazard. Millions of tons of these radioactive tailings lie scattered in piles throughout the Colorado River Basin. As dust from the piles blows into rivers, and ultimately into Lake Powell and Lake Mead, the radium-226 levels of these reservoirs have risen to twice the maximum permissible levels for human consumption in some locations. Nearly a million tons of tailings containing radium are gradually eroding into the Animas River near Durango, Colorado. Crops grown on irrigated water from the river downstream from mine **tailings** contain twice the radium as crops growing upstream. It appears that the base of the food chain here, including hay and alfalfa for cattle, is already being contaminated and concentration of radium up the food chain is likely (Wagner, 1974).

The stabilization of these potentially hazardous uranium mine tailings by grading and planting, and the protection of surface drainage from contamination by eroded material, is a logical solution being em-ployed. Although some work of this type is being carried out, it is apparently not keeping pace with the problem.

Near Grand Junction, Colorado, about 180,000 tons of the sandy uranium tailings were collected by local residents and contractors and used as fill for house foundations, for mortar in masonry walls and cement slabs, as a soil conditioner, and even in children's sand boxes. With the decay of radium-226, a radioactive gas, radon, is given off. This gas has been established as the prime cause of lung cancer in uranium mine workers. In open areas where air circulates, the gas will be carried off, and because of its short half-life, will not become a problem. However, due to the radium contained in walls and slabs of homes, the gas may accumulate to dangerous levels in basements and rooms. In some instances the Colorado houses have had to be abandoned because of excess levels of radioactivity. In 1966 the Colorado State Health Department ordered a halt to the use of tailings for home construction and related uses.

There are still, however, tremendous accumulations of this material exposed to wind and water erosion throughout the state of Colorado. The pathways of this material through the environment and the ultimate site of deposition or residence, and the effects, are unknown. Covering the tailings with soil and then establishing permanent vegetation may be the best long-term solution to this problem by alleviating it at the source.

Increased radiation levels have recently been detected in Florida homes built over reclaimed phosphate mines. Radiation is coming from radon released by the decay of the trace amounts of uranium found in phosphate rock. Large areas of Florida, which produces approximately 90 percent of the nation's phosphate fertilizer, may be affected. The highest level in several hundred homes would double the risk of lung cancer for persons living continuously in the buildings for ten years according to the Environmental Protection Agency.

NECESSITY FOR RESOURCE CONSERVATION

Although proper planning can minimize or reduce the environmental impact of resource extraction and utilization, there are many clear reasons why every conceivable effort should be made to conserve both material and energy resources.

Perhaps the most obvious reason is that most of the world's resources are very lim-

ited in availability. The supplies of high-grade fossil fuels — coal, petroleum, and natural gas — for example, are finite and will probably be consumed within a few hundred years; oil and gas will be gone much sooner and coal will probably last somewhat longer (see figure 10.19). Hydroelectric, tidal, and wind energy are unlimited in the sense that they are always available, but the number of suitable sites limits their development. Geothermal and solar energy are still in early stages of development and their future is uncertain at present but certainly seems optimistic. Atomic power has limitations in terms of economics and environmental impact.

The world's supply of ferrous, nonferrous, and precious metals is very limited. In addition, the United States is highly dependent on foreign sources for many of its basic

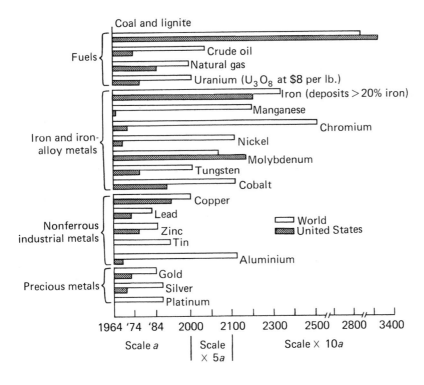

Figure 10.19. *Lifetimes of Estimated Recoverable Reserves of Fuel and Nonfuel Mineral Resources. Reserves are deposits of a high enough grade to be mined with available techniques. Increasing population and consumption rates and unknown deposits are not considered.* REDRAWN FROM: Population, Resources, Environment, Second Edition, by Paul R. Ehrlich and Anne H. Ehrlich. W. H. Freeman and Company. Copyright © 1972. Originally modified from Cloud, Realities of Mineral Distribution, (1968).

raw materials, for instance, copper and nickel. At present, the United States constitutes less than 6 percent of the world's population, but it consumes about 30 percent of the world's raw materials. As more and more nations begin to industrialize, and worldwide standards of living are raised, needs for critical raw materials will rise. The resources are, however, still finite.

The obvious limitations to the world's supply of mineral and energy resources, the overconsumption of these by the United States, and our dependence on often politically unstable foreign countries are strong reasons for the conservation and reuse of all possible resources.

The environmental effects of essentially any type of resource extraction and utilization discussed in this chapter are another basic reason for implementation of conservation practices. As demands increase, ores and fuels of lower quality and concentration will be exploited. This, in many cases, will mean larger scale mining and processing operations to produce the same amount of resource. The environmental impact of these expanded operations must be given major consideration. Conservation and recycling of minerals, substitute materials, use of renewable energy sources, and extraction methods that adequately consider environmental safeguards and site rehabilitation are critical for future planning.

REFERENCES

References Cited in the Text

American Society of Planning Officials. *Land Use Control of the Surface Extraction of Minerals.* Planning Advisory Service Information Report no. 153, December 1961.

Bauer, A. M. *Simultaneous Excavation and Rehabilitation of Sand and Gravel Sites.* Silver Springs, Md.: National Sand and Gravel Association, 1965.

Blumer, M. et al. "A Small Oil Spill." *Environment* 13,2(1971): 2–12.

Brooks, D. B. "Strip Mine Reclamation and Analysis." *Natural Resources Journal* 6(1966): 13–44.

Bulle, W. B., and Scott, K. M. "Impact of Mining Gravel from Urban Stream Beds in the Southwestern United States." *Geology* 2(1974): 171–74.

Department of Interior. *Impact of Surface Mining on the Environment.* Washington, D.C.: U.S. Government Printing Office, 1967.

Flawn, P. T. *Environmental Geology.* New York: Harper and Row, 1970.

Magnuson, M. O., and Kimball, R. L. *Revegetation Studies at Three Strip-Mine Sites in North Central Pennsylvania.* U.S. Bureau of Mines Report of Investigation no. 7075, 1968.

Mitchell, J. G. "Corporate Responsibility in Silver Bay." *Audubon* 77,2(1975): 46–61.

Steinhart, C. E., and Steinhart, J. S. *Blowout: A Case Study of the Santa Barbara Oil Spill.* North Scituate, Mass.: Duxbury Press, 1972.

Strahler, A. N., and Strahler, A. H. *Environmental Geosciences.* Santa Barbara, Calif.: Hamilton Publishing Company, 1973.

Wagner, R. H. *Environment and Man,* 2nd ed. New York: W. W. Norton and Co., 1974.

Young, E. "Urban Planning for Sand and Gravel Needs." *Mineral Information Service* 21(1968): 147–50.

Other Useful References

Ahearn, V. P. *Land Use Planning and the Sand*

and *Gravel Producer.* Silver Springs, Md.: National Sand and Gravel Association, 1964.

Cargo, D. N., and Mallory, B. F. *Man and His Geologic Environment.* Reading, Mass.: Addison-Wesley Publishing Company, 1974.

Davis, G. H., and Wood, L. A. *Water Demands for Expanding Energy Development.* U.S. Geological Survey Circular 703, 1974.

National Academy of Science-National Academy of Engineering. *Underground Disposal of Coal Mine Wastes: Report to the National*

Science Foundation. Washington, D.C.: NAS-NAE, 1975.

Piper, A. M. *Has the United States Enough Water?* U.S. Geological Survey Water Supply Paper 1797, 1965.

Sawyer, L. E. "The Strippers: Experiments by Coal Operators in Rehabilitation." *Landscape Architecture* 16(1966): 132–35.

Schellie, K. L., and Roger, D. A. *Site Utilization and Rehabilitation Practices for Sand and Gravel Operators.* Silver Springs, Md.: National Sand and Gravel Association, 1963.

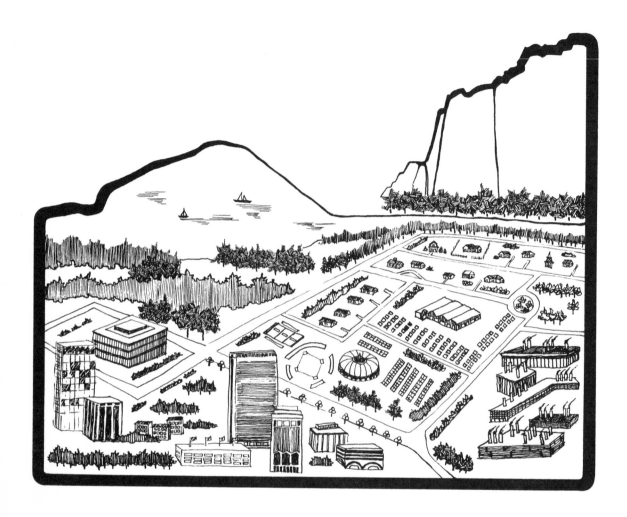

CHAPTER 11

Land Use Planning and Environmental Impact

396

Contents

INTRODUCTION

LAND use planning provides a means for managing growth to permit maximum use of a limited resource base and to retain for future decisions a maximum number of available alternatives. In this context, however, it is not enough to think only of protecting what we already have. Planners must see protection as part of a larger effort to create what we want in neighborhoods and regions in the future. In these efforts a conflict often arises between conservation and creation. In environmentally sensitive areas, natural preservation has to be an important attainable objective. Development, no matter how respectful of nature, will, however, violate natural values. In other areas, "creation" must focus less on prohibition than on sensi-

tive accommodation and balance. Geologic hazards and resources must be recognized and evaluated, and the information derived from these evaluations must be used in the planning process to make intelligent decisions about land use.

In the first five chapters, major geologic hazards — earthquakes, volcanic activity, landslides, and subsidence — were considered. In subsequent chapters, coastal erosion and flooding were discussed as additional hazards that must also be recognized in planning the use of land. To these, three allied subjects were added — waste disposal, mineral extraction, and the degradation of groundwater and surface water resources. Although these latter processes normally do not result in loss of life or property, they can be hazardous to the long-term environmental and economic health of a society. This last chapter will consider some fundamental concepts in land use planning and how effective these are in relating geology to wise land management.

THE INTEGRATION OF EARTH SCIENCE AND PLANNING

We are presently in the midst of resolving problems that will integrate geologic research and data collection with land use planning. One of the first steps, that of a partial redirection of many earth scientists into the area of "environmental geology," has been taken. A number of professional geological agencies and individual geologists are now working in areas having direct application to the planning process. However, the usefulness of the information is not yet clear to many planners, and although growing in numbers, there still exist relatively few channels whereby scientific information can be incorporated into the planning effort. Planners are laboring under similar difficulties. Until recently, very few planners had any geology training, and relatively few had any geology background at all. In recent years with the advent of environmental planning, more planners are receiving formal technical training in scientific disciplines such as geology and hydrology. At the same time, extension colleges are offering short technical courses to professional planners without this background.

Although the incorporation of geological data into land use planning in a meaningful way is relatively new, several states — notably Illinois, Colorado, and California — have taken significant steps forward and are now using geology as a planning tool on various levels. There remain, however, many areas of the country in which geologic constraints are not even considered in the planning process. Efforts to provide this input meets resistance from local budget analysts, who contend it is a luxury expenditure, or even from planners or geologists themselves, who contend there are not enough recent accurate geological data for comprehensive planning. The failure of decision makers to realize the relevance of geology in land use planning is also a significant obstacle.

In some cases, a good deal of geologic research with environmental or land use implications has been completed. However, the reports have usually only presented the geologic data rather than conclusions that might indicate that alternate land uses are necessary. Although the conclusions might be obvious from the data, these reports often receive little attention; and since the agencies involved have been cautious in the past, little if any effort has been extended to publicize such reports.

An excellent example of this precise problem in Anchorage, Alaska, was summarized by Flawn (1970). Five years before the destructive Alaskan earthquake in 1964, a USGS bulletin pointed out that the Bootlegger Cove clay, which underlies much of the Anchorage area, was an unstable material that when wet could fail under vibrational stresses such as those of an earthquake (Miller and Dobrovolny, 1959). The clay is a glacial estuarine-marine deposit that contains zones of high water content and low shear strength. The survey report stated that the low cohesive strength of the clay should be considered in designing large structures that were to have footings within it.

> They published the engineering properties of the clay. They said that shocks such as those associated with earthquakes set into motion material that is stable under most conditions. They pointed out that Anchorage is an earthquake region. They documented a number of previous slumps and slides within the Bootlegger Clay. Nowhere in this report, however, did the authors condemn the Bootlegger Clay as material that provided an unsafe foundation for structures. This report apparently passed unnoticed and unappreciated by city authorities. (Flawn, 1970)

It is obviously difficult with any geologic hazard to condemn existing structures or make decisions about them in potentially unstable areas. However, once recognized, a hazardous area can certainly be avoided or perhaps accommodated in future construction. In retrospect, how many buildings that were constructed on the Bootlegger Cove clay following the publication of this report failed during the Good Friday earthquake five years later? The USGS summarized the damage and causes of damage from the earthquake in a professional paper (Hansen, 1965). Most of the damage in Anchorage, which included the severe damage to or destruction of about thirty blocks of dwellings and commercial buildings, was traceable to failure within this same clay (see figures 11.1 and 2.32). The report again made no recommendations about future land use in the area, but clearly discussed the realities of geologic processes and foundation materials in and around Anchorage. In the rebuilding of Anchorage, one wonders how much consideration has been given to the Bootlegger Cove clay.

Typical of early reports concerning geologic hazards and land use planning that are appearing with increasing frequency are the following: those by the USGS concerning the effects of a heated pipeline on permafrost in Alaska (Lachenbruch, 1970); and indirect effects of regulating the flow of the Sacramento River (by use of a peripheral canal in the California Water Project) on the water quality of San Francisco Bay (McCulloch et al., 1970); and reports by the California Division of Mines and Geology, such as the one on the geologic and engineering aspects of San Francisco Bay fill (Goldman, 1969). Such reports need to be publicized and their conclusions and recommendations or warnings presented as clearly as possible to the planning agencies and government decision-making bodies. When public and private property is at stake, when lives can be lost, or when serious environmental degradation can occur, all agencies having relevant information need to present it so it can be utilized and informed decisions can be made.

METHODS OF LAND USE CONTROL

Land use control is one of the most significant, but controversial, activities conducted by local, state, and federal agencies. Controls are needed to ensure the best use of property and to preserve resources for future generations. However property owners often see land use restrictions as an unwarranted

Figure 11.1. *Turnagain Heights Landslide, 1964 Alaska Earthquake. Shaking during the 1964 earthquake liquefied a clay layer causing the destruction of more than seventy homes in the Turnagain Heights section of Anchorage, Alaska. This damage might have been prevented if there had been better communication between geologists and urban decision makers.* PHOTO COURTESY: *U.S. Geological Survey.*

interference in their right to use their property as they see fit.

General Plans

A comprehensive plan or "general plan" is an official document adopted by a local government for use as a policy guide for deci-sions regarding the physical development of a community. The plan is normally a policy in graphic and written form that sets forth in generalized terms basic land uses that the community intends to realize through the application of other planning tools, such as zoning and subdivision regulation. The land uses designated in a comprehensive plan usually include residential, commercial, industrial, and open space areas (see figure

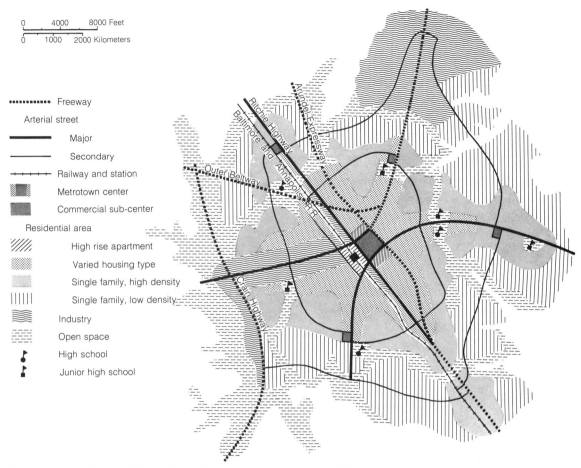

Figure 11.2. *General Plan, North Bay Area, Anne Anundel County, Maryland. This is a schematic proposal for the development of a rapidly growing area near Baltimore. Confronted with critical problems of population growth and transportation pressures, the regional planning agency developed a concept for self-contained "metrotown" community centers. The plan, centered around an existing transportation network, provides for housing, employment, and community facilities surrounded by open space reserves.* RE-DRAWN FROM: *Baltimore Regional Planning Council, "Metrotowns for the Baltimore Region," Planning Report No. 2 (1962), p.Q.*

11.2). The plan indicates in a general way how the community wishes to develop in the succeeding twenty to thirty years. The pattern projected by this plan is based on housing, open space, and other land use de-

mands, and the relationships of primary land usage and physical development. The plan should, but does not always, recognize the limitations of the physical environment on community development. Concern for the environment is a rather recent phenomenon, and many general plans do not reflect the need for the balanced use of land for natural resources and urban development.

The general plan must contain a clearly stated set of general, long-range policies that will enable a planning commission or legislative body to make decisions on development proposals brought before it each week. Current issues then can be viewed against a clear picture of what a community deems desirable for the future. Unfortunately, it is the failure of communities to actively use general plans in making everyday decisions that results in frustration of planning as a useful tool. A plan must be periodically updated to reflect changes in policy direction, but the plan that is adopted and then put on the shelf to gather dust represents an exercise in futility to both planners and the community.

How do geologic hazards and resources relate to general plans? In the formulation of the general plan, data should be gathered defining areas of geologic hazards and resources, and policies developed that delineate compatible land uses. For instance, in communities where geologic surveys show evidence of active faulting and landsliding, the general plan should designate these areas as open space reserves (agriculture or community park uses might be appropriate, for instance). Alternatively, a community might permit very low density residential development where this is compatible with these hazards. A geologic report could be required prior to construction to ensure adequate protection from suspected hazards. Similarly, floodplains and wetlands should be protected from inconsistent development by an appropriate general plan designation.

Conservation and Open Space Elements

In California, local agencies are now required to prepare conservation and open space elements to their general plans. With these elements or portions of the general plan, communities can integrate environmental constraints into the long-range planning process and thereby formulate a future land use plan that balances development with environmental protection. Regarding the conservation element, the California Legislature has decided that general plans must address "the conservation, development and utilization of natural resources including water and its hydraulic forces, soils, rivers, . . . minerals and other natural resources" (California Government Code, Section 65302[d]). Thus a community must plan for such things as mineral protection and extraction, flooding, stream pollution, erosion control, groundwater utilization, and watershed protection. However, these conservation programs may also be approached using the open space element, which according to the legislature, must provide for protection of lands devoted to open space uses — recreation, scenic beauty, agriculture, and conservation or use of natural resources (see figure 11.3). An open space general plan must later be accompanied by an open space zoning ordinance that implements the various policies stated in the plan (California Government Code, Section 65302[e]).

Seismic Safety Considerations

In 1971 California passed additional legislation, the Seismic Safety Element Bill, requiring that seismic safety considerations also be made a mandatory part of all community general plans. The state legislature specified that there be "an identification and appraisal

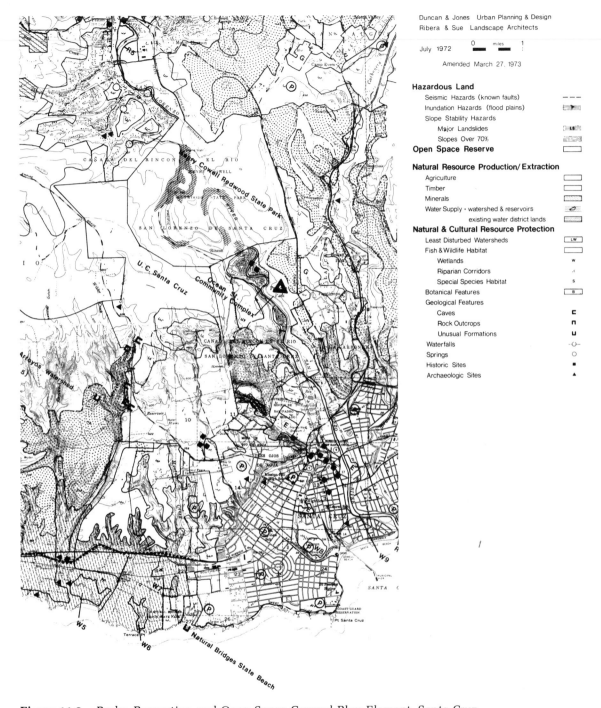

Duncan & Jones Urban Planning & Design
Ribera & Sue Landscape Architects

July 1972   0   miles   1

Amended March 27, 1973

Hazardous Land
Seismic Hazards (known faults)
Inundation Hazards (flood plains)
Slope Stability Hazards
    Major Landslides
    Slopes Over 70%
Open Space Reserve

Natural Resource Production/Extraction
Agriculture
Timber
Minerals
Water Supply · watershed & reservoirs
    existing water district lands
Natural & Cultural Resource Protection
Least Disturbed Watersheds
Fish & Wildlife Habitat
    Wetlands
    Riparian Corridors
    Special Species Habitat
Botanical Features
Geological Features
    Caves
    Rock Outcrops
    Unusual Formations
Waterfalls
Springs
Historic Sites
Archaeologic Sites

Figure 11.3. Parks, Recreation and Open Space General Plan Element, Santa Cruz County, California. This is a portion of a plan that shows some uses of natural resources and open space amenities. Open space planning should not be programmed alone, but must be integrated with the other land use, economic, and transportation considerations involved in community planning. PHOTO COURTESY: Duncan and Jones and Ribera and Sue, Inc., "Parks Recreation and Open Space Plan" (Santa Cruz, Calif.: Santa Cruz County Planning Dept., 1972).

of seismic hazards such as susceptibility to surface ruptures from faulting, to ground shaking, to ground failures, or to effects of seismically induced waves such as tsunamis and seiches [waves in lakes or reservoirs]" (California Government Code, Section 65302[f]). The effect of this bill has been to force cities and counties in California to consider seismic hazards in their planning programs. The basic objective is to reduce loss of life, injury, damage to property, and economic and social dislocations resulting from earthquakes (California Div. of Mines and Geology, 1972). Since it is nearly impossible to provide absolute safety from earthquakes, a community, when preparing this plan, must determine through its elected officials the level of risk it is willing to accept. Often officials will accept more risks in the siting of homes than they will in the siting of hospitals and schools.

As is typical, the intention of the California bill is good but several omissions are unfortunate. The bill does not specify who is qualified to evaluate technical geologic data for a seismic safety general plan, allocates no money to local planning agencies for its preparation, and provides no sanctions for communities that do not implement provisions of their plan when adopted. However, the California legislature has set a progressive example for other states in planning for geologic hazards.

It is sometimes difficult to see how all these individual general plan elements — seismic, land use, conservation, and open space, among others — interrelate in determining a community's overall general planning program. For this reason, several areas — most notably Tulare County, California, and Montgomery County, Pennsylvania — have adopted a resource management general plan that combines the basic general plan elements relating to the environment.

Because the general plan is only a guide or pattern for orderly development, it does not have the force of law. It must be implemented by ordinances such as zoning, subdivision, and grading controls that are enforceable (*see* figure 11.4). These are discussed below.

Zoning

While the general plan is a policy guide setting forth community goals and future land use, the zoning ordinance precisely delineates areas in which stipulated uses are permitted and quantifies such aspects as building height, setback from property lines, and minimum lot size (*see* figure 11.5). In other words, the zoning ordinance is a key method for implementing the general plan, and the legal tool most communities use for regulating land use. One requirement is that zoning must be reasonable rather than arbitrary and must relate to the health, safety, and general welfare of the public. Courts have consistently held that there must be a reasonable basis for classifying certain properties as different from others.

To be effective a zoning ordinance must contain enlightened planning (consistency with the general plan is a prime requirement), provisions for good administration, and recognition of environmental processes operating within a jurisdiction. (The establishment of zoning that is compatible with the geologic processes acting within an area, such as a floodplain or an active fault trace, has been discussed in previous chapters.) These are very logical and necessary restrictions, which, nevertheless, are often strongly

LAND USE PLANNING PROCESS

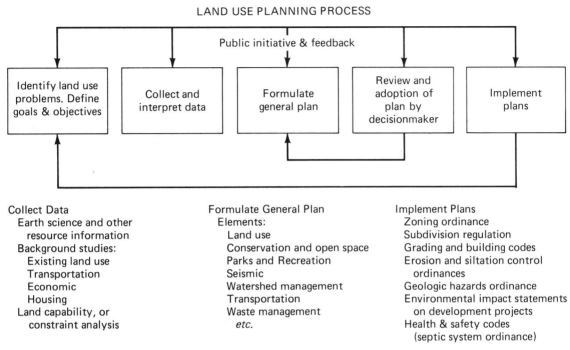

Collect Data
  Earth science and other
    resource information
  Background studies:
    Existing land use
    Transportation
    Economic
    Housing
  Land capability, or
    constraint analysis

Formulate General Plan
  Elements:
    Land use
    Conservation and open space
    Parks and Recreation
    Seismic
    Watershed management
    Transportation
    Waste management
      etc.

Implement Plans
  Zoning ordinance
  Subdivision regulation
  Grading and building codes
  Erosion and siltation control
    ordinances
  Geologic hazards ordinance
  Environmental impact statements
    on development projects
  Health & safety codes
    (septic system ordinance)

Figure 11.4. *Diagram of the Land Use Planning Process.*

opposed by property owners intent on developing their land regardless of its geological stability or setting. In 1974 the planning staff in Santa Cruz County, California, proposed floodplain zoning for a parcel of land that had been covered with 3 feet of water during a "thirty-year flood" in 1955 (see figure 11.6). The owners, who wanted to construct a large savings and loan building on the site, opposed the zoning and could not understand why the planning commission would rezone their potentially valuable parcel to "worthless floodplain." The number of such instances is overwhelming.

A zoning code should also contain provisions to protect water and mineral resources. Many jurisdictions have adopted "natural resource protection" zones, which

provide for open space uses compatible with the resource involved. In areas where mineral resources occur, for example, the zone district should provide for preservation or holding of lands for the proper extraction of minerals and protect against preemption by other uses such as industrial, business, or residential structures. Similarly, highly erodible soil, land used as watershed for surface water supply, and aquifer recharge areas should be zoned to prevent incompatible development. This type of zoning is particularly critical where natural resources are serving urban demands; however, it should not be limited to these areas.

Zoning for resource or hazard protection may be accomplished through the use of a combining zone district, which provides

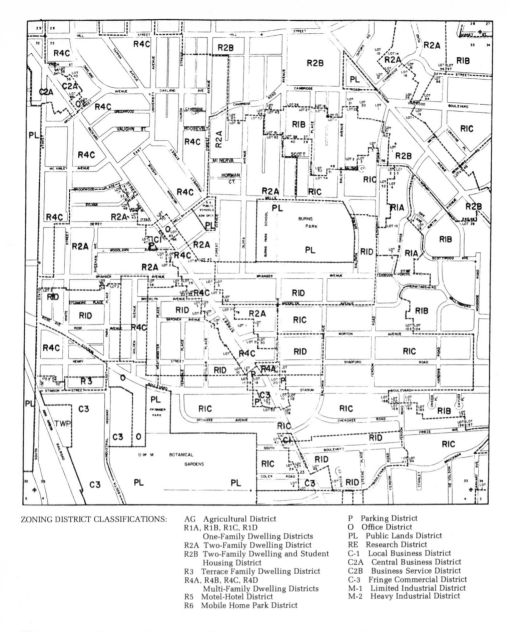

ZONING DISTRICT CLASSIFICATIONS:

AG	Agricultural District
R1A, R1B, R1C, R1D	One-Family Dwelling Districts
R2A	Two-Family Dwelling District
R2B	Two-Family Dwelling and Student Housing District
R3	Terrace Family Dwelling District
R4A, R4B, R4C, R4D	Multi-Family Dwelling Districts
R5	Motel-Hotel District
R6	Mobile Home Park District
P	Parking District
O	Office District
PL	Public Lands District
RE	Research District
C-1	Local Business District
C2A	Central Business District
C2B	Business Service District
C-3	Fringe Commercial District
M-1	Limited Industrial District
M-2	Heavy Industrial District

Figure 11.5. Zoning Map, Ann Arbor, Michigan. A zoning ordinance establishes the types and density of uses for an area. Zoning follows general designation established by the community's general plan, but delimits more specific uses along precise boundaries. The basic residential, commercial, and industrial districts may be subclassified according to density, types of use permitted, or on some other basis. As shown in this example, special districts such as parking or research may be added to meet the special needs of the area concerned. PHOTO COURTESY: William I. Goodman and Eric C. Freund, eds., Principles and Practice of Urban Planning (Washington, D.C.: International City Managers Association, 1968), p. 410.

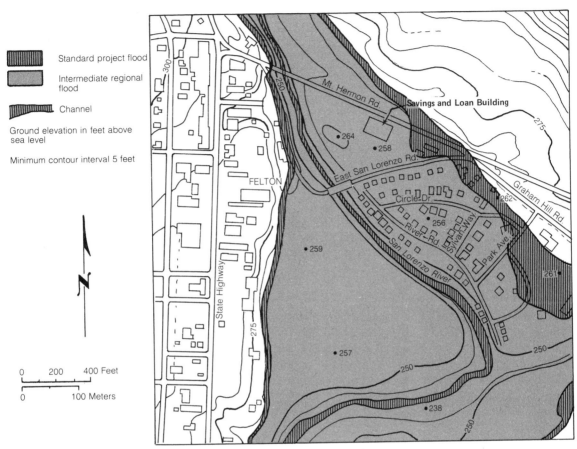

Figure 11.6. *Proposed Savings and Loan Building in a Floodplain, Santa Cruz County, California. The site of the proposed structure would be inundated by up to 3 feet of water during a major flood. The structure was not approved, and the site was zoned to a floodplain designation contrary to the wishes of the property owners. Now only uses compatible with the flood hazard are permitted.* REDRAWN FROM: *U.S. Army Corps of Engineers, Flood Plain Information — San Lorenzo River, Santa Cruz County, California (1973).*

regulations over and above the existing (primary) zone. The use designated by the primary zone — single family residences, for instance — would be allowed only if development can be accomplished without undue hazard or damage to a defined re-source. As an example, if a floodplain combining zone was added to a residential zone district, the ordinance could provide for elevation of the structures above the level of the flood hazard, or preferably, construction of the houses outside the flood-prone area.

Subdivision Controls

While a general plan and zoning ordinance provide for residential development, most communities use subdivision controls to regulate how this development is allowed to take place. Subdivision ordinances regulate street design, sewer or septic tank design, lot sizes, building setback distances, and other physical aspects of subdivision construction. However, with such ordinances, local governments can prevent excessive grading and consequent erosion and slope stability problems, ensure proper drainage to eliminate excessive runoff or flooding, and protect against other geologic hazards through the requirement of geologic investigations before development occurs.

The town of Portola Valley in California requires a detailed geologic investigation from a qualified soils engineer and/or engineering geologist prior to any construction or change in land use. The report must assess such hazards as landslides and active faults and must evaluate how geologic conditions might affect and be impacted by the development proposed. The town then enforces a provision of the California Real Estate Code that requires that a summary of this report be shown to the first purchaser of

each lot (Mader and Crowder, 1970). Similar provisions have been adopted in subdivision ordinances of other jurisdictions to ensure that development is compatible with geologic hazards. In many instances, however, this protection falls short if the local agency does not require that the report be

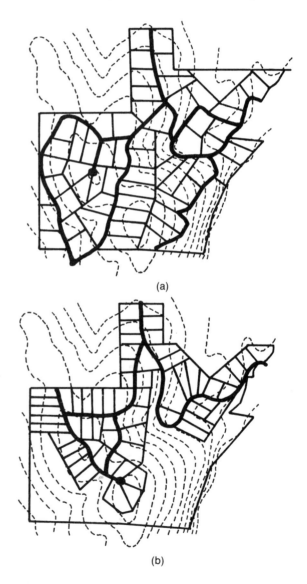

(a)

(b)

Figure 11.7. *Proposed Subdivision, Northern California. (a) shows a developer's proposed subdivision in a mountainous area of California. With 60 lots, 14,000 feet of roads, and little open space, the proposal was rejected by the planning commission; (b) shows a revised plan worked out by the developer with the assistance of the local planning department. It portrays 60 lots, 8500 feet of roads, more open space, and less disturbance of vegetation and soil. This plan was approved by the planning commission.* PHOTO COURTESY: *California Resources Agency,* Environmental Impact of Urbanization on the Foothill and Mountainous Lands of California (*Sacramento, 1971*), p. 53.

prepared and reviewed by competent professionals.

Design factors and construction practices are often important in avoiding the adverse impacts of development — particularly in steeply sloping terrain. Figure 11.7 shows a subdivision proposed with a number of undesirable features: excessive road construction, poor lot layout, and potential water quality degradation from septic tanks on steep slopes. The subdivision was redesigned (see figure 11.7), with the assistance of the local planning department, to reduce these undesirable impacts. Figure 11.8 shows two possible methods for developing a particular parcel of land. Both schemes accommodate the same number of structures; however, the cluster design can provide more open space, avoid hazardous slopes, and minimize grading and erosion from soil disturbance. Good subdivision ordinances should provide the flexibility to allow for innovative design.

Other Local Ordinances

Building and Grading Codes

Local ordinances, or statutes such as building codes, which set standards and specifications for construction have proven effec-

tive in controlling slope failures due to faulty or improper construction or engineering practices. These ordinances and codes must, however, have the backing of a strong agency with enforcing power, and must also have money appropriated for enforcement.

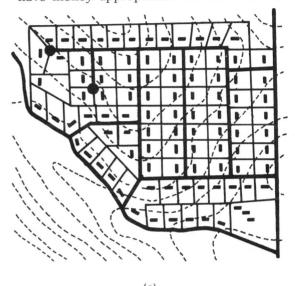

(a)

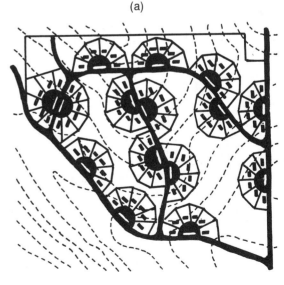

(b)

Figure 11.8. *Alternate Subdivision Plans. (a)* ▶ *Rectilinear grid layout: 94 lots, 12,000 feet of streets and utilities; (b) cluster design: 94 lots, 6000 feet of streets and utilities. Rigid adherence to the block grid design on hilly terrain can cause serious problems, including accelerated soil erosion, slope instability, destruction of scenic values, and septic system failures.* PHOTO COURTESY: *California Resources Agency,* Environmental Impact of Urbanization on the Foothill and Mountainous Lands of California *(Sacramento, 1971), p. 53.*

All too often, the most well-intentioned legislation fails, simply because either no funds or insufficient funds were available for its implementation.

The development of grading ordinances in southern California is a good example of the long but successful evolution of an effective piece of legislation governing construction practices (Scullin, 1966; and Leighton, 1966). The beginnings of uncontrolled hillside development after World War II in an area that is seismically active, that is underlain by young unstable rocks, and that commonly experiences high intensity winter rainfall soon led to enormous problems (see figure 11.9). The heavy winter rains of 1951–1952 inflicted $7.5 million in damages to the city of Los Angeles, much of this due

to failures in the areas of extensive hillside development that were improperly graded (see figure 11.10). The role of landsliding and slope failure in southern California had been underestimated by engineers and geologists. From this initial large disaster came the first grading ordinance in the city of Los Angeles and the beginnings of permits, inspections, and controls on cutting and filling and other hillside construction. The original ordinance has been modified and greatly improved so that at present an integrated system of continuing supervision by soils engineers, civil engineers, and geologists is in existence. In 1969, the ordinance was tested by severe rainstorms. Even though approximately 15 percent of all hillside homes were constructed since the adop-

Figure 11.9. *Hillside Development, Los Angeles County, California. Removal of vegetation and extensive grading for homesites, roads, and utilities can create a scarred landscape, as well as high erosion-siltation rates and increased landslide hazards.* PHOTO COURTESY: *California Div. of Mines and Geology.*

Figure 11.10. *Landslide Near Los Angeles, California. Improper grading without appropriate design and proper attention to drainage and geologic conditions contributed to this slope failure.* PHOTO COURTESY: *California Div. of Mines and Geology.*

tion of the regulations, and many homes were built in steeply sloping areas, structural damage from the storms amounted to only $5000. The example is encouraging. Much of southern California has adopted a similar system, which has now greatly reduced the occurrence of hillside failure (*see* table 11.1).

The southern California experience eventually led to the adoption of the excavation and grading section (chapter 20) of the Uniform Building Code. "The chapter sets forth rules and regulations to control excavation, grading, and earthwork construction, including fills or embankments; [and] established the administrative procedure for issuance of permits; and provides for approval of plans and inspection of grading construction" (Uniform Building Code, chapter 20). Many jurisdictions throughout the country have adopted this code, or a modification of it. Although this is, no doubt, one of the most thorough regulating statutes of its type in existence, the failure of municipalities to

adopt or enforce it places us right back at the beginning. It is encouraging to note that progress is being made, however.

Environmental Performance Standards

A new type of land use control that is beginning to emerge in a few local communities is the use of performance standards to evaluate development in sensitive environmental areas. Traditional zoning ordinances, as we have seen, establish a detailed list of permitted uses, prohibited uses, special use of conditional use procedures, and site requirements. This approach is limited because of its inability to recognize the complex interrelationships of dynamic ecological and physical systems. In most zoning ordinances there is no mechanism for environmental capabilities beyond a simple exclusion of all uses that might have some impact on a sensitive resource. The alternative to this method is to establish environmental performance standards by which a community sets

Table 11.1. *Damage Under Various Grading Codes, Los Angeles, California*

Pre-1952	1952–1962	1963 to Present
No grading code, no soils engineering, no engineering geology	Semi-adequate grading code, soils engineering required, very limited geology but no status and no responsibility	New modern grading codes; soils engineering and engineering geology required during design; soils engineering and engineering geology required during construction; Design Engineer, Soils Engineer, and Engineering Geologist all assume legal responsibility.
Approx. 10,000 sites constructed	Approx. 27,000 sites constructed	Approx. 11,000 sites constructed
Approx. $3,300,000 damage	Approx. $2,767,000 damage	Approx. $182,400 damage*
Approx. 1040 sites damaged	Approx. 350 sites damaged	Approx. 17 sites damaged
An average of $330 per site for the total number produced $\frac{\$3,300,000}{\text{Sites } 10,000}$	An average of $100 per site for the total produced $\frac{\$2,767,000}{\text{Sites } 27,000}$	An average of $7 per site for the total produced $\frac{\$80,000}{\text{Sites } 11,000}$
Predictable failure percentage: 10.4% $\frac{1040 \text{ damaged}}{10,000 \text{ total sites}}$	Predictable failure percentage: 1.3% $\frac{350 \text{ damaged}}{37,000 \text{ total sites}}$	Predictable failure percentage: .15% $\frac{17 \text{ damaged}}{11,000 \text{ total sites}}$

SOURCE: James E. Slosson, "The Role of Engineering Geology in Urban Planning," Colorado Geological Survey, Special Publication No. 1, 1969.

NOTE: It should be noted that the storms of 1952, 1957–1958, 1962, 1965, and 1969 all produced similar total losses associated with similar destructive storms.

*Over $100,000 of the $182,000 was incurred on projects where grading was in operation and no residences were involved, thus less than $80,000 occurred on sites constructed since 1963.

specific measurable levels at which key functions of the sensitive environmental system must operate, consonant with proposed changes in land use. The developer, through a licensed geologist or hydrologist, must then indicate how the proposed development will meet these standards. This approach encourages innovation and experimentation, which will result in development that is more compatible with the environment. If after development the project does not meet the stated standards, the developer is subject to corrective measures and penalties as designated by the ordinance (Thurow et al., 1975).

Streams and wetlands are resources that can be protected by performance regulations. In chapter 7 the importance of maintaining natural runoff patterns, groundwater percolation, and minimizing erosion and siltation was emphasized. DeKalb County, Georgia, recognizing these values, has proposed an ordinance controlling runoff from any development or construction project in the county. The ordinance, in part, states that storm water from developments shall not exceed runoff from the area in its natural state, calculated by using a two-year frequency storm (a storm occurring every two years on the average) in the area immediately downstream from the project. The ordinance also specifies required capacities for runoff retention basins. A similar ordinance was adopted in Leon County, Florida, requiring that any construction project be designed to include prevention of excessive runoff, erosion, and sedimentation and that plans submitted be calculated on the basis of a twenty-five-year frequency storm using rainfall-intensity curves developed by the Florida Department of Transportation.

Other resources suitable for protection by performance control regulations are groundwater aquifers, aquifer recharge areas, and steeply sloping hillsides and mountainous areas. The performance standard approach has been suggested in earlier chapters for the hazards from coastal cliff erosion (see chapter 6, pp. 226–28), and for the protection of groundwater recharge areas (see chapter 8, p. 314).

The reader interested in pursuing this subject should refer to the publication by Charles Thurow et al. (1975). The authors of this report researched a variety of environmental performance ordinances from various cities and counties throughout the United States and made a number of recommendations regarding their use by local agencies. It is an excellent review of the performance approach to land use control. The appendix should be consulted for examples of two other performance standard ordinances.

Legislation at the State Level

On a broader scale, elected representatives at the state level are responding to the recurring damage from geologic hazards and mineral extraction activities. Geologists need to keep the legislators and the public informed so that new research and discoveries can be incorporated into the legal system where necessary. The establishment of legislative committees such as the Joint Legislative Committee on Seismic Safety in California, and then the appointment of a broadly based but professional advisory group to work with these committees, seems to be the best approach to this problem. In that there is no group or industry that is going to profit directly from protective legislation involving geologic hazards, we cannot expect many paid lobbyists to be pushing in this direction. Therefore, it is important that the public be educated concerning the benefits to be derived from geologic hazards legislation and become willing participants in the legislative process. As with local ordinances, state laws are only effective when adequately administered and enforced.

It has been noted that every major seismic event in California in recent times has given substantial impetus to seismic safety legislation. One of these bills was the Field Act of 1933, which was a major piece of legislation enacted in direct response to the destruction that occurred during the 1933 Long Beach earthquake (see figure 11.11). Basically, the statute set down standards for school construction that are still being followed today. Unfortunately many schools built before the passage of the act have never been rebuilt. In the 1971 San Fernando earthquake, major damage to schools essentially was limited either to those schools built prior to the pas-

Figure 11.11. *School Damage as a Result of the 1933 Long Beach, California, Earthquake. Jefferson Jr. High School in Long Beach was almost totally destroyed by this moderate earthquake (6.3 magnitude). Although numerous other school buildings also suffered extensive damage, relatively few school casualties occurred because the earthquake took place in the late afternoon after students had been dismissed. Even so, a great public outcry caused the state legislature to pass the Field Act providing for improvements in school building codes.* PHOTO COURTESY: *U.S. Coast and Geodetic Survey.*

sage of the act or to those where poor engineering practices were allowed. This should have been an important lesson, and the eleven legislative bills enacted as a result of the 1971 earthquake and its damages indicate that perhaps it was.

Several recent summaries and critiques of geology-related legislation in California have recently appeared (Slosson and Hauge, 1973; Nichols, 1973). Below are examples of the progress and the shortcomings of such legislation in at least one state:

1. *Seismic Safety Element Bill* (1971). This bill, a direct outgrowth of the San Fernando earthquake, is discussed on pp. 402–04 above.

2. *School Siting Bill* (1971). This bill prohibits the construction or location of public school buildings over active fault traces and requires that ground shaking, ground failure, and seismically induced waves be considered in school design, location, and construction. Although the bill calls for geologic and engineering site investigations for all school construction, it does not specify what is to be included in the geologic-seismic report, nor who is to prepare the report. In light of the number of existing schools situated along the Hayward and San Andreas fault traces in California (*see* figure 2.25), it is curious that no such bill was passed years ago. One wonders what will be the fate of those schools that have already been constructed on fault traces.

3. *Dam Safety Bill* (1972). Owners of dams designated by the California Office of Emergency Services are required by this bill to prepare maps showing the areas that would be inundated should the dam fail at

varying water levels. The maps are to be submitted to local jurisdictions, which then must prepare evacuation plans and procedures. (For further discussion of seismic hazards related to dams, see chapter 2.)

4. *Hospital Safety Bill* (1972). This law requires that hospitals be designed to safety standards comparable to those for school buildings as designated by the Field Act of 1933. Geological and engineering investigations must be performed to determine foundation conditions. These investigations are subject to review by an engineering geologist on behalf of the state. A Building Appeals Board was created to consider appeals under the act and to render advice regarding the structural safety of hospitals during earthquakes.

5. *Alquist-Priolo Geologic Hazards Act* (1972). This act is perhaps the most significant piece of legislation relating land use to seismic safety. It requires the state geologist to delineate "Special Studies Zones" along all active faults. In these zones, each city or county must review every real estate development and structure for human occupancy and shall not approve the location of structures if they are likely to result in undue hazards. In most cases, a city or county would have to require a detailed geologic or engineering-geology study to determine the possible existence of seismic hazards on a particular parcel. If, for example, an active fault trace traversed a proposed site for development, this hazard would be investigated, and the city or county could then require the developer to set back any construction at least 33 meters (100 feet) from the fault trace.

Although a progressive piece of legislation from a planning-geology standpoint, the

Alquist-Priolo Act has some weaknesses. First, some if not all active fault zones are so dangerous that no development should occur along them. Direct authority to disapprove projects found from geologic investigations to be too hazardous is not provided for in the act, although disapproval of a project location is provided for. The act also indicates that local jurisdictions may establish stricter policies and criteria than those established by the state. Secondly, in 1975 the Alquist-Priolo Act was amended to exclude consideration of single family houses, not part of a larger development, built in special studies zones. The basis for the amendment was that the cost of a geologic investigation was too great a burden on a family planning to construct a single residence. Although geologic investigations are often expensive, the cost is insignificant when compared with the economic hardship and possible injury or death caused by construction of a house on a fault trace that ruptures in an earthquake.

Legislation certainly cannot solve all the problems, but for geologic hazards such as faulting and earthquakes, which do not conform to city or county boundaries, state or federal legislation is necessary. This brief review of some recent California legislation indicates that progress is being made. More and more projects are going to require professional geologic evaluations. These are being demanded by local governmental bodies and by the general public because of concern for the environmental and economic consequences of development in marginal, unstable, or otherwise hazardous areas.

Legislation at the Federal Level

As of 1976 Congress has been unable to pass federal land use planning legislation, which has been proposed during several congressional sessions. However, it seems inevitable that some type of land use planning act will eventually become law. The purpose of such an act will be to assure consistent land use policies among individual states in order to minimize misuse of resources that may have regional or national significance. Federal money will probably be provided to assist state and local governments in developing or expanding their land use planning programs.

There has been recent significant federal legislation dealing with related aspects of resource planning. The Federal Flood Disaster Protection Act, signed into law in 1973, is discussed on p. 302 chapter 7. Also, the National Environmental Quality Act (NEPA) was adopted by Congress in 1969. Because of its importance to the land use planning process, it is discussed in a separate section of this chapter.

SOURCES OF GEOLOGIC DATA

The organized formulation of land use policies and the implementation of plans to minimize geologic hazards and protect geologic resources must rely on a broad spectrum of earth science data. However, it should be recognized that the type, scope, and detail of geologic data will vary considerably in different localities, depending upon the complexity of the geology, the recent occurrence of hazardous geologic events, the type and distribution of existing and anticipated development, and the level of the planning effort (see figure 11.12). As an example, in a county where simple geologic structure and low seismic activity prevail, existing geologic mapping at a scale

of 1:125,000 may provide sufficient minimum data to develop seismic and slope stability plans. In such an area, if this data is recent and of high quality, its compilation and analysis should allow planners to prepare and implement a general plan through ordinances and land use controls. The small amount of additional data collection that might be necessary should be relatively inexpensive. In contrast, in a county of high seismicity, where complex geologic structural relationships have not been resolved, an adequate general plan cannot be prepared without detailed mapping (at a scale of 1:24,000 or larger) and detailed geologic, seismic, slope stability, and engineering studies (see figure 11.13). Costs of these studies are normally high, but vary somewhat according to the amount of additional data required, the type of disciplines involved, and the sophistication of analyses provided (Nichols and Buchanan-Banks, 1974).

Principal Sources

Principal sources of earth science information for a given area likely to be valuable for planning studies and ordinances include the following:

1. City and county departments of planning, building inspection and public works, particularly those with geologists on their staff, will normally have most of the available information. This might include geologic maps (at scales of 1:24,000 to 1:250,000); detailed geologic mapping of problem areas (at scales of 1:12,000 and larger); geologic reports on specific development projects; special purpose maps of fault zones, landslides, and areas prone to

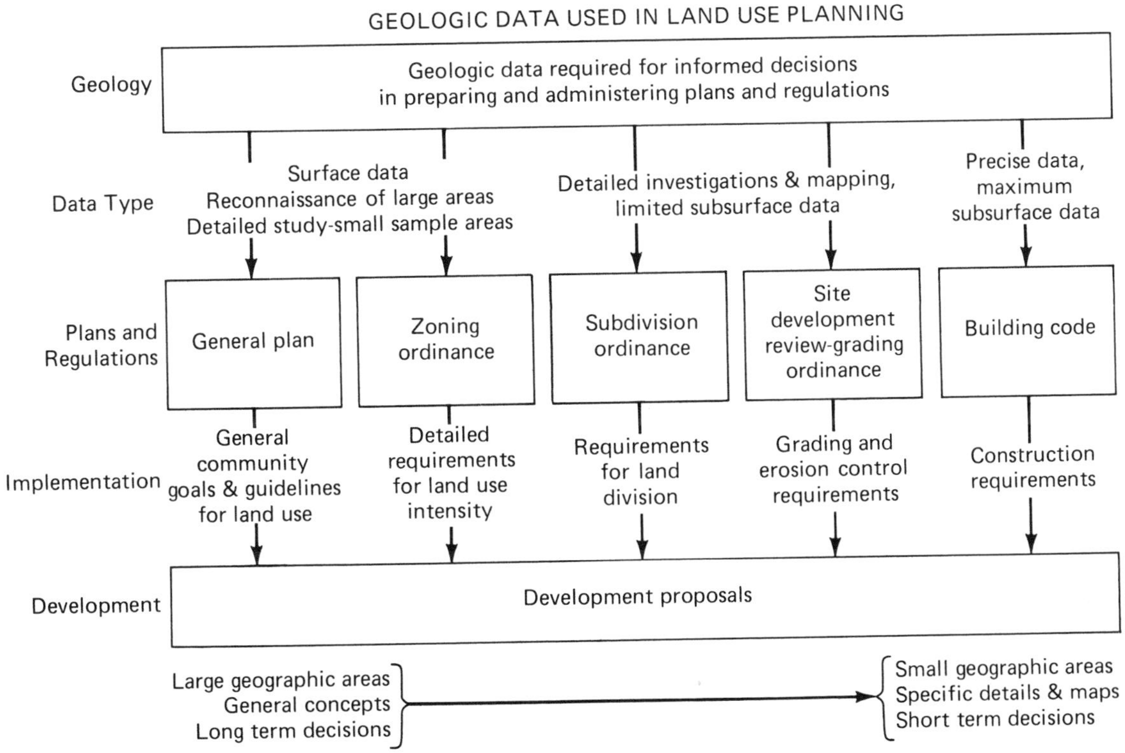

GEOLOGIC DATA USED IN LAND USE PLANNING

Figure 11.12. *Geologic Data in the Planning Process. Geologic data is important at every level of the planning process. More detailed data and mapping is needed as one progresses from general plans to precise building specifications. If reliable data and maps are unavailable for certain localities, they should be developed when preparing plans and ordinances.* REDRAWN FROM: *D. R. Nichols and L. C. Campbell eds., Environmental Planning and Geology, (Washington, D.C.: Association of Engineering Geologists Proceedings, U.S. Government Printing Office, 1969), p. 181.*

liquefaction and subsidence; and sometimes interpretative risk maps and data for specific hazards. These same departments, as well as the local flood control district, will also have maps and data on floodplains and, in most cases, other information on local streams and watersheds.

2. Geology, geography, and planning departments at local colleges and universities are often good sources of information on geology, geologic hazards, hydrology, and mineral resources.

3. Most states have a geologic survey agency with information on local geology, mineral extraction, and hydrogeology. A number of state geologic surveys have recently begun geologic mapping programs to provide planners and local agencies with in-

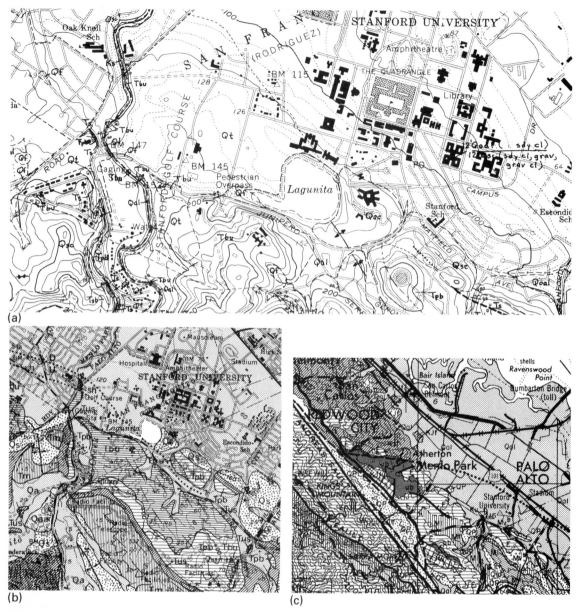

(a)

(b)   (c)

Figure 11.13. *Geologic Mapping at Varying Scales. These three geologic maps differ both in scale and amount of detail shown. Map a, at a scale of 1:24,000, is part of a geologic map of the Palo Alto 7.5 minute quadrangle, California, showing Stanford University and the western vicinity of Palo Alto. Map b shows the same vicinity at a scale of 1:62,500. Map c covers a much larger region in the same area, but at a smaller scale of 1:250,000. Successful application of earth science information to land-use planning requires that the information be available in sufficient detail to meet planning needs of the user.* PHOTO COURTESY: *W. Spangle and Associates, F. B. Leighton and Associates, Baxter McDonald and Company, "Earth Science Information in Land-Use Planning — Guidelines for Earth Scientists and Planners," Geological Survey Circular 721 (1976).*

formation and interpretative maps on various geologic hazards.

4. The U.S. Geological Survey has responsibility for conducting numerous research and mapping programs in hydrology and geology. The survey carries out seismic research, primarily in the western United States; has an earthquake information center in Boulder, Colorado; operates a worldwide seismograph instrument network; and is involved in earthquake prediction investigations. It also conducts research on volcanic activity, slope stability, groundwater and surface water hydrology, sedimentation problems, and flood inundation. Maps and interpretative reports are available for many local or regional areas and are normally detailed enough for development planning (see figures 2.52 and 2.53).

5. The U.S. Soil Conservation Service conducts a program of nationwide soil mapping, which not only provides identification maps of soil types (see figure 11.14) but also suitability interpretations for various uses that may take place on or within the particular soil type. Interpretations made in most soil surveys include evaluation of suitability for septic leach fields, shrink-swell potential, depth to seasonal high water table, and erosion potential (see table 11.2). These surveys can provide valuable data for all levels of environmental planning as well as for the design of individual projects.

6. The U.S. Army Corps of Engineers conducts geologic and hydrologic studies of streams and rivers — usually on those drainages with some potential for dam construction. Even so, these can often be a source of useful data. The corps also conducts various studies of harbors, beaches, and coastal areas that may yield data valuable in assessing coastal processes and hazards. In addition, this agency prepares floodplain information reports at the request of local communities, which delineate flood hazards in specific areas. Such reports will include maps, flood profiles, photographs, and information on expected future flooding; they can provide valuable information for local community planners interested in floodplain management (see figure 11.15).

7. Many states have agencies with regulatory control over one or more areas of geologic hazard or resource extraction. These agencies commonly collect valuable data or finance research projects. In Colorado, for example, the Division of Mines has information on mineral occurrence and extraction, while the Colorado Water Conservation Board and Division of Water Resources maintain data on groundwater basins, well locations, and areas where mining activity may have adversely affected groundwater supplies.

8. Professional organizations, such as the Association of Engineering Geologists (AEG), the American Institute of Professional Geologists (AIPG), and the American Institute of Planners (AIP), may be helpful. They organize symposia and publish journals and special booklets that provide information on a variety of subjects relating to geology and planning.

Most practicing geologists and environmental planners are aware of many of these sources, but few know all the opportunities of information that exist. Moreover, developments in recent years that add significant geologic knowledge and improve the processes for dissemination of data are

R. 24 W.

N

Scale 1:20 000

Figure 11.14. *Part of the Soil Survey Map, Scott County, Minnesota. Soil types are plotted on a series of aerial photos showing physical and cultural features of the region. Symbols AaA, Ab, AaB are alluvial floodplain soils subject to frequent flooding and drainage problems. Soil types CaA, CdB, CdB2 (Copas) are silty loams having a shallow depth to bedrock. This characteristic limits excavation for construction and use of the soil for septic filter fields. The symbol 2 indicates high erosion potential and, therefore, implies that certain precautions must be taken to control runoff and erosion on steeper slopes. Oshawa soils (Oa symbol) have a dense subsurface clay layer, creating problems of high shrink-swell potential and a high water table. Accurate soil mapping has obvious usefulness for environmental planning.* PHOTO COURTESY: *Soil Survey, Scott County, Minnesota (Washington, D.C.: U.S. Government Printing Office, U.S. Dept. of Agriculture-Soil Conservation Service, 1959), p. 16.*

420

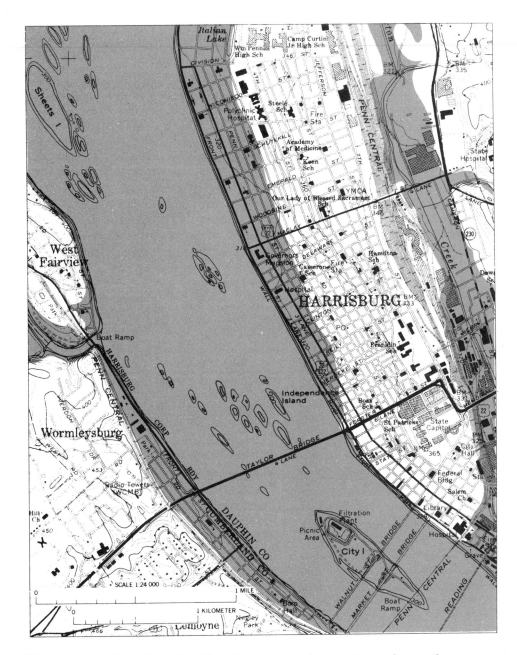

Figure 11.15. *Floodplain Map, Harrisburg, Pennsylvania. Map indicates the area that would be inundated by the 100-year flood (the flood level that, on the average, occurs once every 100 years). Although 83 percent of the Harrisburg floodplain has already been developed, this data is useful in regulating changes in land use, and planning for channel modifications, flood control reservoirs, or other structures that may reduce flood levels.* PHOTO COURTESY: *W. J. Schneider and J. E. Goddard, "Extent and Development of Urban Floodplains," Geological Survey Circular 601-J (1974).*

U. S. DEPARTMENT OF AGRICULTURE
SOIL CONSERVATION SERVICE

SOIL INTERPRETATIONS RECORD

MLRA(S) 14,15
STATE California RECORD NO. 144 AUTHOR(S) KIND OF UNIT [Series] DATE 4/75 REVISED UNIT NAME Pinto UNIT MODIFIER

CLASSIFICATION AND BRIEF SOIL DESCRIPTION

The pinto soils are very deep, moderately well drained, and formed on marine terraces from mainly sedimentary alluvium. The elevation is 20 to 1,000 feet. Native vegetation is grasses and shrubs. Mean annual air temperature is about 56°F. The surface soil is grayish brown medium acid loam 14 inches thick underlain by a light brownish gray medium acid loam about 7 inches thick and brownish yellow medium acid loam about 17 inches thick. At 38 inches is a light yellow, tan brown & reddish yellow dense clay loam substratum extending to a depth of 65 inches. Slopes range from 0 to 15%.

ESTIMATED SOIL PROPERTIES

DEPTH (IN)	USDA TEXTURE	UNIFIED	AASHO	FRACT. >3 IN. (PCT)	4	10	40	200	LIQUID LIMIT	PLAS-TICITY INDEX
A 0-21	VFSL, L	CL, CL-ML	A-6, A-4	0	95-100	90-100	85-95	50-75	20-30	5-15
21-65	SCL, CL	CL, SC	A-6	0-3	90-100	85-100	75-95	40-75	20-35	10-20

(PERCENT OF MATERIAL LESS THAN 3 IN. PASSING SIEVE: columns 4, 10, 40, 200)

DEPTH (IN)	CLAY (PCT OF <2MM)	MOIST BULK DENSITY (G/CM³)	PERMEABILITY (IN/HR)	AVAILABLE WATER CAPACITY (IN/IN)	SOIL REACTION (pH)	SALINITY (MMHOS/CM)	SHRINK-SWELL POTENTIAL
			0.6-2.0	0.14-0.17	5.6-7.3	-	MODERATE
			0.2-0.6	0.09-0.11	5.6-7.3	-	MODERATE

EROSION FACTORS K	T	WIND EROD. GROUP	ORGANIC MATTER (PCT)	CORROSIVITY STEEL	CONCRETE
.28	4		-	MODERATE	LOW
.28				HIGH	LOW

HIGH WATER TABLE / FLOODING

FLOODING FREQUENCY	DURATION	MONTHS
NONE		

HIGH WATER TABLE DEPTH (FT)	KIND	MONTHS
2,3	PERCHED	DEC-APR

BEDROCK DEPTH (IN)	HARDNESS	CEMENTED PAN DEPTH (IN)	HARDNESS	SUBSIDENCE INITIAL (IN)	TOTAL (IN)	HYD GRP	POTENTIAL FROST ACTION
60						C	

SANITARY FACILITIES

SEPTIC TANK ABSORPTION FIELDS:
0-8%: SEVERE - PERCS SLOWLY
8-15%: SEVERE - PERCS SLOWLY

SEWAGE LAGOONS:
0-2%: SLIGHT
2-7%: MODERATE - SLOPE
7+%: SEVERE - SLOPE

CONSTRUCTION MATERIAL

ROADFILL: POOR - SHRINK-SWELL, LOW STRENGTH
SAND: UNSUITED

KEYING ONLY
FILL 191
SAND 201

SCS-SOILS-5
REV. 5-76
FILE CODE SOILS-12

KEYING ONLY
RECORD NO. | CONTROL WORD | NO.
MLRA 001
STATE 011
CLASS 021
DESCR 031
PROP 041
PROP 051
PROP 061
SEPTIC 071
LAGOON 081

BUILDING SITE DEVELOPMENT

Use	Sym.	Rating
SANITARY LANDFILL (TRENCH)	TRENCH 091	0-15%: MODERATE - TOO CLAYEY — 2, 3, 4, 5
SANITARY LANDFILL (AREA)	SANARE 101	0-8%: SLIGHT / 8-15%: MODERATE - SLOPE — 2, 3, 4, 5
DAILY COVER FOR LANDFILL	COVER 111	0-8%: FAIR - TOO CLAYEY / 8-15%: FAIR - SLOPE, TOO CLAYEY — 2, 3, 4, 5
SHALLOW EXCAVATIONS	EXCAV 121	0-8%: SLIGHT / 8-15%: MODERATE - SLOPE — 2, 3, 4, 5
DWELLINGS WITHOUT BASEMENTS	DWEL 131	0-8%: MODERATE - SHRINK-SWELL, LOW STRENGTH / 8-15%: MODERATE-SLOPE, SHRINK-SWELL, LOW STRENGTH — 2, 3, 4, 5
DWELLINGS WITH BASEMENTS	DWEL 141	0-8%: MODERATE-SHRINK-SELL, LOW STRENGTH / 8-15%: MODERATE-SLOPE, SHRINK-SWELL, LOW STRENGTH — 2, 3, 4, 5
SMALL COMMERCIAL BUILDINGS	BLDGS 151	0-4%: MODERATE-SHRINK-SWELL, LOW STRENGTH / 4-8%: MODERATE-SLOPE, SHRINK-SWELL, LOW STRENGTH / 8+%: SEVERE-SLOPE — 2, 3, 4, 5
LOCAL ROADS AND STREETS	ROADS 161	0-8%: MODERATE-SHRINK-SWELL, LOW STRENGTH / 8-15%: MODERATE-SLOPE, SHRINK-SWELL, LOW STRENGTH — 2, 3, 4

FOOTNOTES

WATER MANAGEMENT

Use	Sym.	Rating
GRAVEL	GRAVEL 211	0-8%: GOOD / 8-15%: FAIR - SLOPE — 2, 3, 4, 5
TOPSOIL	SOIL 221	— 2, 3, 4, 5
POND RESERVOIR AREA	PONDRS 231	0-2%: FAVORABLE / 2-15%: SLOPE — 2, 3, 4
EMBANKMENTS DIKES AND LEVEES	DIKES 241	FAVORABLE — 2, 3, 4, 5
EXCAVATED PONDS AQUIFER FED	PONDAQ 251	NO WATER — 2, 3, 4, 5
DRAINAGE	DRAIN 261	0-2%: PERCS SLOWLY / 2-15%: SLOPE-PERCS SLOWLY — 2, 3, 4, 5
IRRIGATION	IRRIG 271	0-2%: ROOTING DEPTH / 2-15%: SLOPE-ERODES EASILY — 2, 3, 4, 5
TERRACES AND DIVERSIONS	TERRAC 281	0-2%: FAVORABLE / 2-15%: SLOPE — 2, 3, 4

FOOTNOTES

FOOTNOTES

NOTES 441

Sym.	
A	THE SUBSOIL IS SLIGHTLY CEMENTED AND AWC AND PERMEABILITY RATINGS ARE AFFECTED BY IT.
B1	ONLY TEMPORARY PERCHED WATER TABLES IN WET SEASON. EMBANKMENTS EFFECTIVELY CUT OFF PERCHED WATER SOURCES.
B2	AREAS WITH SANDY CLAY LOAM SUBSOILS - 0 +0% SLOPES. RATE MODERATE FOR EXCAVATION.
C1	AREAS WITH CLAY LOAM SUBSOILS, 0-8% SLOPES. RATE MODERATE FOR EXCAVATION.
D1	ON-SITE INVESTIGATION NEEDED BECAUSE OF VARIABILITY OF PLASTICS INDEX.
E1	SOILS ARE DRY DURING SEASON OF USE.

Table 11.3. *San Francisco Bay Region Environment and Resources Planning Study*

	A	B	C	D	E	F	G	H	I	J	K	L	M	N	O	P	Q	R
Topographic elements																		
1 Topographic map 1:125,000 scale	X	X	X	X	X	X	X	X	X	X	X	X	X	X	X	X	X	X
2 Relief model — reproducible 1:125,000 scale	X	X	X	X	X	X	X	X	X	X	X	X	X	X	X	X	X	X
3 Orthophoto mosaic 1:1,125,000 scale	X	X	X	X	X	X	X	X	X	X	X	X	X	X	X	X	X	X
4 Slope map 1:125,000 scale	X		X		X				X	X	X	X	X	X	X	X	X	X
5 Urban orthophoto map 1:7,200 scale	X	X	X	X	X	X	X	X	X	X	X	X	X	X	X	X	X	X
6 Patterns of urban growth	X	X	X	X	X	X	X	X	X	X	X	X	X	X	X	X	X	X
Geological and geophysical elements																		
1 Active faults	X	X	X	X	X	X					X	X			X	X		X
2 Slope stability	X	X	X	X	X	X							X		X	X		X
3 Rock properties and engineering behavior	X	X	X		X	X			X	X	X	X		X	X	X		X
4 Soil properties and engineering behavior	X	X	X		X	X			X	X	X			X	X	X		X
5 Seismicity and ground motion	X	X	X	X	X	X			X	X	X	X		X	X	X	X	X
6 Mineral commodity utilization	X	X					X		X	X	X			X	X		X	X
7 Open-space study	X	X	X	X	X			X	X		X	X				X	X	X

A–Hillside Development
B–Flatland Development
C–Fault Hazard Zones
D–Subsidence Potential (Lowlands)
E–Seismic Hazard
F–Water Navigation
G–Water Pollution

H–Utilization/Pollution of San Francisco Bay
I–Availability of Mineral Commodities
J–Public Safety
K–Recreation and Education
L–Environmental Aesthetics

M–Hydrologic Hazards
N–Public Facility Design, Local
O–Public Facility Design, Regional
P–Sewage Disposal
Q–Solid Waste Disposal
R–Urban Land Use; General Policy Planning

not well known, possibly due to inadequate publicity. Two of these developments are the HUD-USGS Environment and Resources Planning Study and the relatively new field of **remote sensing**.

San Francisco Bay Region Environment and Resources Planning Study

A research and demonstration study in the San Francisco Bay region jointly sponsored by the U.S. Department of Housing and Urban Development and U.S. Geological Survey investigated the physical hazards and resources of that region and the need to improve regional environmental planning and decision making. The project was a pilot study — experimental in nature — costing $5 million and requiring a three-year effort. The San Francisco area was selected for the study primarily because of numerous environmental problems caused by rapid urban development in a region having severe geologic hazards and significant natural resources. The project has established guidelines and methodologies applicable to

Hydrologic elements	A	B	C	D	E	F	G	H	I	J	K	L	M	N	O	P	Q	R
1 Public water-supply service area	X	X	X		X	X	X							X	X			X
2 The groundwater resource	X	X		X	X		X		X					X			X	X
3 Aquifer-recharge areas		X		X			X							X		X	X	X
4 Waste-water sources and pollutional loadings	X	X	X	X	X	X	X	X		X	X	X		X	X			
5 Quality of receiving waters							X	X			X	X		X	X		X	X
6 Land pollution susceptibility	X	X					X				X	X		X	X			X
7 Eutrophication of fresh water bodies		X				X	X				X	X					X	X
8 Stream channel aesthetics	X	X					X				X	X		X			X	X
9 Floodplain inundation		X		X	X		X		X		X	X	X	X	X	X	X	X
10 Coastal flooding		X							X		X	X		X	X			X
11 Storm design criteria	X	X					X				X			X	X		X	X
12 Design criteria for flood-flow facilities	X	X					X				X		X	X	X		X	X
13 Stream-borne sediment	X	X				X	X	X			X	L	M	N	O		X	X
14 Local drainage problems		X					X				X	X	X					X
15 Land subsidence	X			X							X		X	X	X		X	X
16 Physical and chemical hydrological properties of San Francisco Bay							X		X	X	X	X	X		X	X	X	
17 Planning and management considerations		X		X		X	X	X			X	X	X	X	X	X	X	X

NOTE: The table indicates data outputs (left column) and their relation to environmental planning problems (top).

SOURCE: Program design for San Francisco Bay Region Environment and Resources Planning Study, U.S. Department of Interior, Geological Survey, and U.S. Department of Housing and Urban Development, Menlo Park, Calif., 1971.

similar studies now planned or taking place throughout the country.

The technical component of the study has included consideration of geology, geophysics, hydrology, and topography. Specific data related to seismic hazards, landslides, unconsolidated sediments, and mineral utilization have been developed as part of the geological and geophysical element of the program. Emphasis in the study has been placed on relating earth science data and concepts to regional land use problems (see table 11.3). The study attempts in an innovative way to express environmental data in a form that can be readily utilized in urban planning and decision making. During the study a continuing evaluation of outputs was made through discussions with planners and other potential users of the data.

The study produced approximately eighty published maps, reports, and other data compilations that present, interpret, and evaluate various geologic and environmental information for environmental planning and management in the bay area. HUD and USGS have also published several reports presenting guidelines for effective use of environmental information in urban planning and decision making.

Remote Sensing

The term **remote sensing** is used here to describe the detection and mapping of natural hazards and resources with imagery derived from aircraft and spacecraft. The term is usually defined to include black and white aerial photography — an important tool for geologists in the study of faults, landslides, and coastal resources for the past twenty years. Recently, however, a number of different types of cameras and special sensors have been used in high altitude aircraft and spacecraft (including unmanned satellites) to greatly expand sources of hydrologic and geologic information.

Air Photo Interpretation

The examination of large-scale aerial photography can yield detailed information about surface water hydrology, seismic hazards, and slope stability. In addition, recent photographs can be compared with older ones to analyze, for example, the progressive development of landslides, coastal erosion rates, or hydrologic characteristics of river systems.

Liang and Belcher (1958) indicate three major steps in the interpretation of air photos:

1. examination of overlapping air photos with a stereoscope to get a three-dimensional perception of the landform

2. identification of ground conditions by observing certain elements appearing in the photograph

3. interpretation of photographs with

respect to specific problems by association of ground conditions with one's background and experience

The quality and reliability of any interpretation is naturally dependent upon the individual's knowledge of geology and environmental conditions of the area under study. The acquisition of such knowledge either by study of available maps and reports, or by field examination, is therefore an essential part of any photo interpretation analysis.

Applications of Other Types of Remote Sensing Imagery

Within the past few years the utilization of aerial photographs has been augmented by new techniques whereby sensing is done in several bands (in addition to the visible portion) of the electromagnetic spectrum. Electromagnetic energy travels as waves with various wave lengths. Names applied to progressively longer electromagnetic wave lengths are gamma ray, X-ray, ultraviolet, visible, infrared, microwave, radio short wave, and long wave (see figure 11.16). Using instruments that are sensitive to these varying wave lengths, investigators may gather far more information about an area than can be obtained from conventional photography, which only records images in the visible spectrum.

The types of remote sensors important for natural resource and hazards inventories include the conventional aerial camera, the multiband camera, the gamma ray spectrometer, the radiometer, and side-looking radar. The radiometer and radar scanners have the advantage of being usable at night because they do not need sunlight for their operation. The proper selection of instruments is based on the information needed

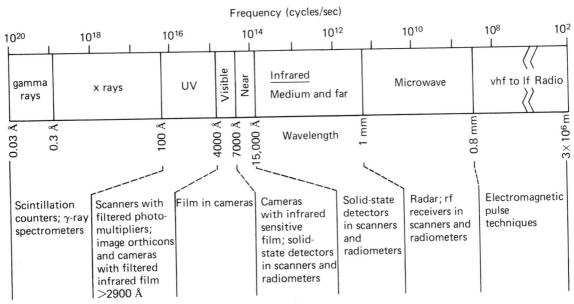

Figure 11.16. *Electromagnetic Spectrum. This diagram shows the relationship between electromagnetic energy, its respective wave lengths, and the various instrument systems that produce photographs and images (pictures by instruments other than photographic cameras) in each part of the spectrum. Each portion of the electromagnetic spectrum is capable of providing knowledge about a particular resource or management problem.* REDRAWN FROM: *American Society of Photogrammetry, "Photogrammetry," Photogrammetric Engineering, 29 (1963): 761–99.*

and the degree of interpretative resolution of the subject desired compared to its surroundings. Results of many investigations by the National Aeronautics and Space Administration (NASA) and U.S. Geological Survey indicate that in the visible and infrared spectrum many discriminations are possible by taking several simultaneous pictures, with each picture recording a different wave length band. This is called **multispectral photography** or imagery, and is usually done with a multiband camera (see figure 11.17). The resulting multiple images can assist the interpreter in recognizing features that are of interest (see figure 11.18).

Geologic and hydrologic applications of remote sensing imagery are numerous. Satellite imagery has been particularly useful for identification of structural features of regional extent because of the perspective provided by a single photo at orbital height. For example, major faults are usually composed of smaller faults and deformational features rather than a continuous single fracture. Because conventional aerial photographs of small areas will show only portions of faults, the full length and magnitude of the fault feature can evade the interpreter. In contrast, satellite imagery can provide a surprisingly identifiable linear feature that, with some

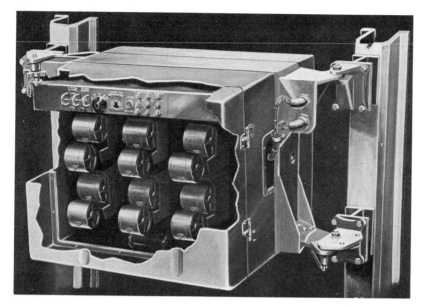

Figure 11.17. *Multiband Camera. This camera uses 12 lenses, each with a differing film-filter combination, and each therefore, designed to function best in one portion of the spectrum. A study of the distinguishing tonal or color characteristics in each of the 12 photographs enables an interpreter to obtain more positive identification of an area's resources than can be obtained from any single photograph.* PHOTO COURTESY: *U.S. Geological Survey, EROS Data Center.*

additional investigation, may be easily recognized as a fault. Photographs from the ERTS-1 (Earth Resources Technology Satellite) taken in 1972 uncovered a fault feature extending northeasterly from Lake Tahoe through Nevada, possibly to the Yellowstone National Park vicinity of Wyoming (see figure 11.19). The discovery of this fault system led to renewed exploration for mineral deposits, which in Nevada seem to be closely correlated with fault systems (Rowan, 1975). Radar imagery taken from high altitude aircraft is also useful in identifying geologic features such as domes, faults (see figure 11.20), or **anticlines** that may contain oil-bearing deposits.

Similarly, information on volcanoes and impending volcanic eruptions has been provided from thermal infrared sensors capable of detecting small changes in temperature. Thermal infrared sensing techniques are also contributing to investigations in surface and groundwater hydrology. For example, imagery was obtained over the coastline of Hawaii near Hilo where a number of underground fresh water springs discharge into the ocean. The contrasting light and dark images indicated points along the coast where cooler fresh water was intermixing with warmer ocean water (see figure 11.21). Inasmuch as fresh water is in short supply in some parts of Hawaii, this information could be extremely valuable in locating usable sources of groundwater.

Other types of water resource planning and management information can also be obtained from earth-orbiting satellites. Satellite photography can supplement data from stream recorders and monitoring instruments by providing interpretative information between recording station points. Conversely, the stream point data can provide a ground check for calibrating the satellite remote sensors. Data on such items as urban

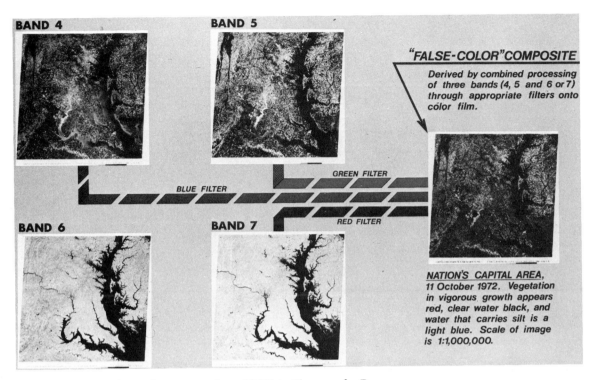

Figure 11.18. *Multiband Exposure from ERTS-1, Chesapeake Bay — Washington, D.C., Area, October 1972. The imagery shown as Bands 4, 5, 6 and 7 are from various parts of the electromagnetic spectrum. Each photo is able to highlight varying geographic features and therefore assist in overall interpretation. When the 3 of 4 images are processed through filters, a color composite shown on the right is produced, further highlighting features not clearly interpretable from the individual images. In the composite, vegetation appears red, clear water black, and silt-laden water light blue.* PHOTO COURTESY: *U.S. Geological Survey, EROS Data Center.*

development in a drainage basin, vegetative indicators of hydrologic conditions, sedimentation patterns, flood potential, and distribution of snow pack are obtainable from space.

Remote sensing can assist investigation and management of coastal resources, including research on current patterns, coastal erosion, and analysis of the effects of groins and jetties on beaches. Early detection of oil spills has been shown feasible using ultraviolet and color photography; thermal infrared imagery can also record temperature differences between oil and ocean water (see figure 11.22).

The recent realization that high altitude aircraft and satellite imagery would be valuable for natural resource studies means that

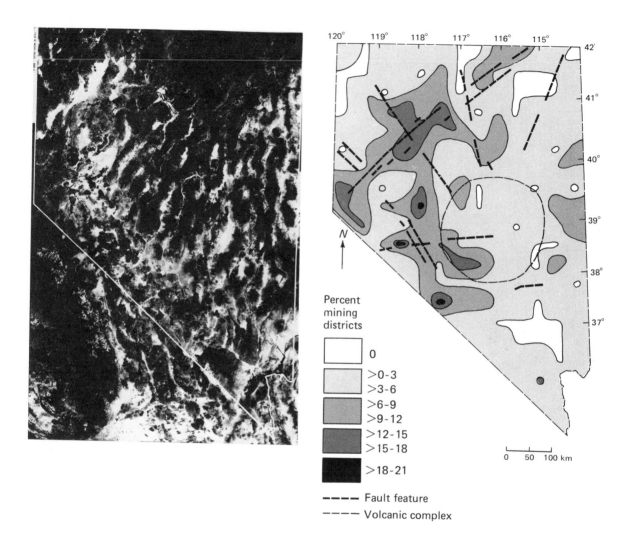

Figure 11.19. *ERTS Imagery of Nevada. This satellite image mosaic shows a fault lineation extending from near Lake Tahoe to northeast Nevada. Volcanic activity along these zones of weakness have produced ore-bearing minerals. Discovery of these faults on remote sensing imagery has led to increased mineral exploration in several existing mining areas. Although undetectable on this particular photo, other types of imagery can distinguish individual rock units, enabling an interpreter to identify the type and potential value of the mineral resource.* PHOTO COURTESY: *U.S. Geological Survey.* MAP REDRAWN FROM: *L. C. Rowan, "Application of Satellites to Geologic Exploration,"* American Scientist *63 (1975): 394.*

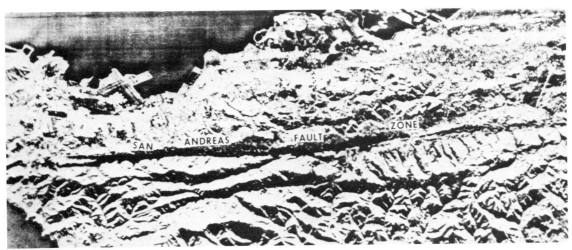

Figure 11.20. *Radar Imagery, San Andreas Fault Zone in California. This side-looking radar image shows remarkable definition of the San Andreas Fault and its related branch faults near San Francisco Bay. Sag ponds, lakes, valleys, near surface moisture and other fault features reflect radar waves away from the receiver and therefore appear black in the picture.* PHOTO COURTESY: *U.S. Geological Survey.*

Figure 11.21. *Infrared Image of Hilo, Hawaii. Dark streaks on this infrared image show the escape of cooler, underground fresh water into the warmer salt water of the ocean. This type of imagery can assist in locating wells to tap groundwater supplies.* PHOTO COURTESY: *U.S. Geological Survey.*

THERMAL INFRARED

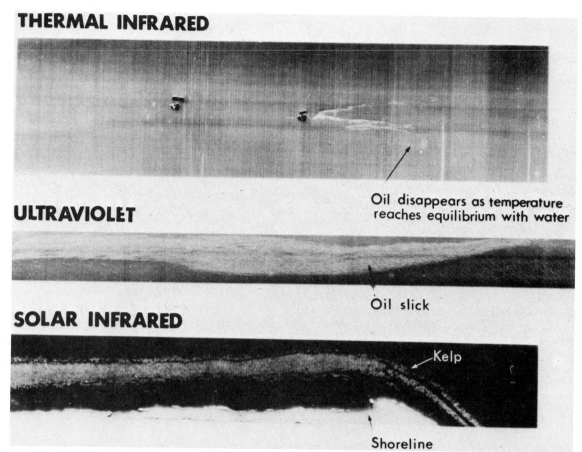

Oil disappears as temperature
reaches equilibrium with water

ULTRAVIOLET

Oil slick

SOLAR INFRARED

Kelp

Shoreline

Figure 11.22. *Oil Slick Monitoring, Santa Barbara, California. Airborne monitoring of this oil slick using three wavelength bands provides information on the source of oil (thermal infrared), distribution of oil (ultraviolet), and identification of kelp which could be contaminated (solar infrared).* PHOTO COURTESY: *U.S. Geological Survey.*

many programs to date are experimental, and additional remote sensing techniques and applications will be forthcoming in the next few years. After the Gemini and Apollo spacecraft missions confirmed the potential of remote sensing from satellites, ERTS-1 was launched in 1972 specifically for the purpose of providing useful information concerning natural resources and environmental quality. ERTS-1 circled the earth fourteen times per day, continuously relaying sensor imagery back to data collection points. The satellite covered every point on the earth's surface every eighteen days. In January 1975, Landsat-1 and Landsat-2 satellites were launched to replace ERTS-1. Interpretation of Landsat photography has demonstrated that surprisingly high image resolution can be achieved — that, in fact, objects smaller than 50 meters (165 feet) in size can be recognized from photographs taken 900 kilometers (400 miles) above the earth.

Remote sensing techniques have tremendous possibilities for the future. One prospect is that earth-orbiting satellites will carry multiband sensing equipment together with a computer. Equipped in this way, the satellites could record resource data for any particular area, produce an immediate printout that would amount to an areawide resource map, and read into the computer programmed instructions for comparison of various resource management techniques. The decision for optimum management of the resource in question would then be telemetered to the ground for whatever action is deemed necessary (Colwell, 1968).

NEW APPROACHES TO ENVIRONMENTAL PLANNING

As described in the previous sections of this chapter, the planning process consists of several stages, which include the following:

1. statement of community goals and objectives

2. collection and interpretation of data

3. formulation of a general plan

4. implementation of the plan through zoning, subdivision regulation, and special ordinances

However, this approach to planning has not been effective in the past in assisting planners and decision makers in making intelligent land use decisions. In many cases, general plans have been ignored, or inconsistently followed, either because they were unrealistic or outdated. Economic considerations, such as land cost and accessibility to transportation, have often decided the use of urbanizing land rather than compatibility with physical-environmental factors. This practice has resulted in the loss or degradation of natural resources as well as the development of portions of communities in hazardous areas.

Land Capability Mapping

Within recent years several techniques have evolved that provide a better method for evaluating alternative land uses, utilizing data on environmental sensitivities and physical constraints and thereby making it possible to incorporate environmental constraints into the planning process. These techniques are variously called "constraint mapping," "suitability mapping," or "land capability analysis"; they involve a recognition of the intrinsic value of ecological systems. McHarg (1969) perfected and popularized a method by which the physical suitability of land can be rapidly evaluated for specific land uses. His technique involves preparation of successive (transparent) acetate overlays, each of which displays an individual physical factor. Factors considered depend upon the regional setting, but often include slope, surface water and groundwater resources, soil limitations, bedrock geology, vegetation, and wildlife (see figure 11.23). These factors are individually evaluated, rated for their value in

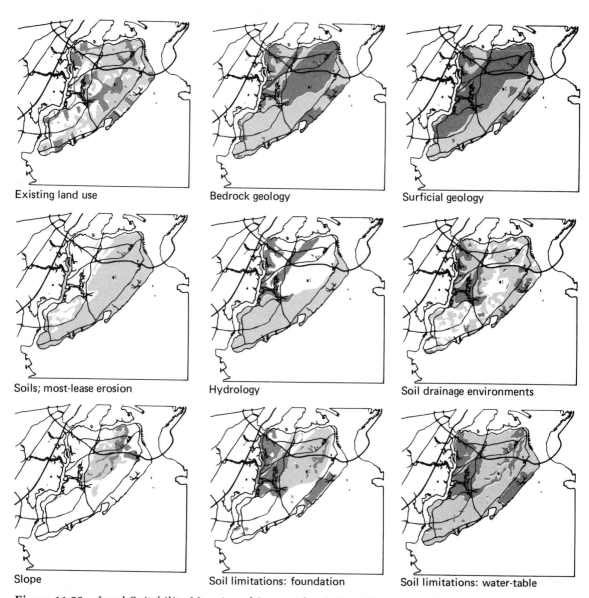

Figure 11.23. *Land Suitability Mapping of Staten Island, New York. Maps show examples from the thirty-two physical, social, and economic factors identified as important in this regional plan. Each factor was evaluated and ranked in terms of suitability for various land uses: resource conservation, passive and active recreation, residential development, commercial and industrial development. The resulting map shadings express pictorially each factor's intrinsic suitability, so that the least acceptable regions are darkest.* REDRAWN FROM: *I. L. McHarg,* Design with Nature *(Garden City, N.Y.: Natural History Press, 1969), pp. 105–11.*

the overall natural ecosystem, and then related to specific land uses. The geographic distributions of these factor levels are then displayed by shading or coloring the acetate overlays so that dark areas indicate regions having greatest environmental significance or constraint. Conversely, optimum areas for development or land use changes are indicated by light area on the composite map, derived by overlaying all the individual acetate factor sheets (see figure 11.24).

A similar technique was used in Kansas where maps of physical-environmental factors were programmed for land use planning purposes. Maps prepared depicted slopes, drainage basins, soil characteristics, bedrock geology, mineral resources, vegetation, land use, and historic and aesthetic factors. Two or more of these maps can be combined into a composite suitability map, which is utilized by developers in the initial stages of the planning for a particular project and by county planners in making recommendations on the project to the planning commission (Kansas Geological Survey, 1968).

The use of suitability maps can be illustrated by a study evaluating sanitary landfill waste disposal sites in DeKalb County, Illinois (Gross, 1970). Important physical factors in this study were identified: the type and thickness of unconsolidated deposits, type of bedrock, and surface and groundwater conditions, with particular consideration given to aquifers used for domestic water supply. In general, it was noted that at least 10 meters of relatively impermeable material must be present between the base of the waste material and the top of a groundwater aquifer for proper operation of the sanitary landfill. The map shown as figure 11.25 is the final product of this study. In DeKalb County the only areas consistently meeting

this requirement were those covered by a clay-rich glacier **till** (shown by symbol G on figure 11.25).

There are, however, several drawbacks to the suitability analysis approaches illustrated above.

1. There is a definite limitation on the number of environmental factors that may be considered in a study. As evidenced in the McHarg approach, that limitation is imposed by the number of acetate overlays that can be manipulated at any one time.

2. Each factor can only be subdivided into a limited number of subcategories for evaluating. This introduces some inaccuracy and generalization into the system.

3. Because data must be transferred from base maps to overlay sheets, if a large number of physical factors are evaluated, the process can be slow and cumbersome.

4. Weighting the relative importance of various factors is difficult, and usually must be done in qualitative, rather than quantitative, terms. In many cases, this oversimplifies actual physical conditions and can render the composite map(s) inaccurate and somewhat inflexible.

5. As new environmental information becomes available, new maps must be prepared, if updating is required (Tilmann et al., 1975). Some of these difficulties can be overcome or reduced by using computers for the collection, storage, analysis, and presentation of environmental information.

This type of analysis, although cumbersome and potentially inaccurate, does have certain advantages such as simplicity and

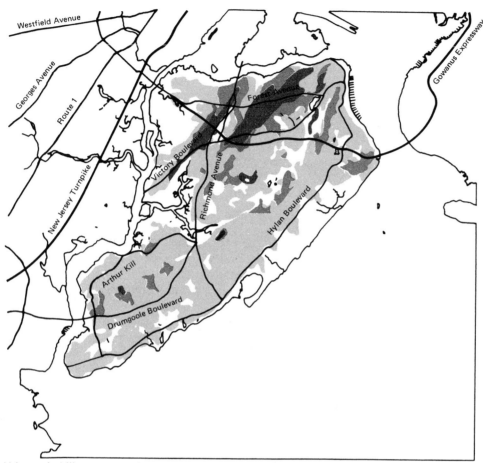

Urban suitability

Figure 11.24. *Composite Map of Staten Island, New York. The composite suitability map for urbanization (residential, commercial, and industrial development) was derived by overlaying individual factor maps exemplified in figure 11.23. A total of fourteen factors were used to produce this map. Optimum areas for development are indicated by light areas, while those that should be avoided because of unsuitable factors or constraints are shown by darker shadings.* REDRAWN FROM: *I. L. McHarg,* Design with Nature *(Garden City, N.Y.: Natural History Press, 1969), p. 113.*

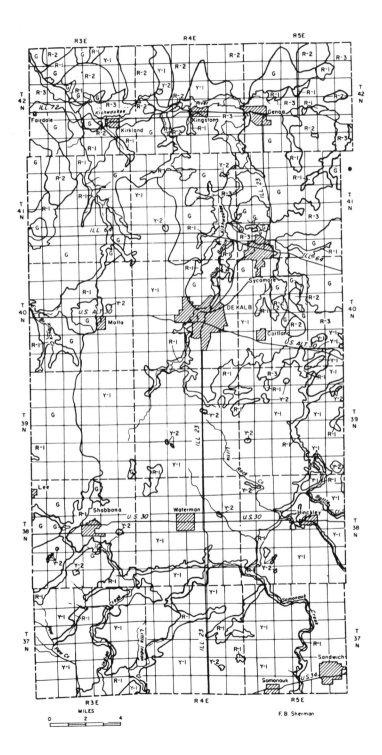

G   Suitable disposal sites probably available; areas of glacial till where there are no data suggesting the occurrence of permeable sand or sand and gravel.

Y-1   Suitable disposal sites may be available; extremely varied material; primarily till with many inclusions of sand or sand and gravel.

Y-2   A few suitable disposal sites probably available; isolated sand or sand and gravel deposits; usually associated with extremely varied till.

R-1   Disposal sites are likely to pose pollution hazards unless special engineering precautions are taken; surficial silt, sand, or sand and gravel underlain by till at less than 20 feet.

R-2   Disposal sites are liable to pose pollution hazards unless special engineering precautions are taken; less than 50 feet of glacial drift over the bedrock surface; surficial material varies from sand to till.

R-3   Disposal sites are liable to pose pollution hazards unless special engineering precautions are taken; upper 20 feet of surficial material consists of silt, sand, or sand and gravel.

Figure 11.25. *Solid Waste Disposal Suitability Map, DeKalb County, Illinois. This map was developed by color-coding relevant physical factors including geology, topography, surface water and groundwater hydrology. Green (favorable conditions), yellow (caution) and red (unfavorable conditions) are represented by the letters G, Y, and R. Shades of each color, represented by the numbers 1, 2, and 3, are used to depict various limitations within each color group. Thus G indicates areas with fewest limitations for solid waste disposal, while R-3 indicates areas containing the most severe problems.* PHOTO COURTESY: *D. L. Gross, "Geology for Planning in DeKalb County, Illinois," in Environmental Geology Notes, N. 33 (Urbana: Illinois State Geological Survey, April 1970), p. 10.*

easy comprehension by user groups. It is a major step toward a thorough evaluation of natural processes for use in regional resource analysis and land use decision making.

Computer Applications

The environmental capability mapping process described above is readily adaptable to the computer, which can assist in storage, retrieval, manipulation, and display mapping of geologic data for assessing suitable areas for specific uses. Existing base maps are used as data sources; however, instead of manually transferring this data to acetate overlays, the information is digitized (programmed into the computer) using standardized grid cells and transferred onto

"data records" with one record for each grid cell. (A data record is that set of cards that contains all the information about one grid cell.) Data files are normally organized on the basis of the geographic location of consecutive grid cells. To retrieve from the computer a specific piece of information about a location, one searches the data record referring to the pertinent location and then finds the desired piece of information. The computer can produce a written printout or display maps for various environmental factors (see figure 11.26). Successive environmental factors can be weighted (see table 11.4) and combined by the computer to produce interpretative maps that can show areas optimally suited for an intended land use project (see figure 11.27). In a similar manner the interpretative maps can be used by planning staffs to prepare general plans,

Table 11.4. *Housing Model Factors, Bretton Woods, New Hampshire*

	5 (High Suitability)	4	3	2	1 (Low Suitability)	0 (Not Suitable)
Soil Erosion	0–4	4–6	6–8	8–10	8–10	10+
Cost of Grading	1–2	2–4	4–6	6–9	9–12	12+
Septic Tank Suitability	Slight		Moderate		Severe	Very severe
Soil Creep	1	2	3		4	5
Vegetation Type & Density	Open	Decidious	Mixed	Mixed & decidious upland	Conifer	Conifer on uplands
Water Constraints						Floodplain rivers, lakes
Slope	0–3%	3–5%	5–10%	10–15%	15–20%	20%
Visual	0–2	2–4	4–6	6–8	6–8	8+

NOTE: This chart illustrates the various factors that were used for the housing model (figure 11.28), and how these factors were associated and weighted.

SOURCE: Environmental Systems Research Institute, "Environmental Site Analysis for Bretton Woods, Mt. Washington," Redlands, Calif., 1972.

The hazard for soil creep is lowest for level 0 and highest for level 4. The variables used were soil texture, rock depth, and slope.

Septic suitability is best for level 1 and worst for level 4. The variables used were slope, landforms, and water table depth.

Data mapped in 5 levels between extreme values of 1.00 and 5.00.

Data mapped in 4 levels between extreme values of 1.00 and 4.00.

Absolute value range applying to each level:

Minimum	1.00	1.80	2.60	3.40	4.20
Maximum	1.80	2.60	3.40	4.20	5.00

Absolute value range applying to each level:

Minimum	1.00	1.75	2.50	3.25
Maximum	1.75	2.50	3.25	4.00

Percentage of total absolute value range applying to each level:

20.00 20.00 20.00 20.00 20.00

Percentage of total absolute value range applying to each level:

25.00   25.00   25.00   25.00

Frequency distribution of data point values in each level:

Frequency distribution of data point values in each level:

Figure 11.26. *Computer Display Maps of Bretton Woods, Near Bethlehem, New Hampshire. This soil creep hazard map was produced by using soil and slope information available for this area. Those areas shown as dark shades are most susceptible to soil creep problems. The septic tank suitability map illustrates where septic tanks can be safely used as a sewage disposal system.* PHOTO COURTESY: Environmental Systems Research Institute, "Environmental Site Analysis for Bretton Woods, Mt. Washington" (Redlands, Calif., 1972), pp. 8–9.

Housing model:
   Level 0—those areas not well suited for housing
      to
   Level 4—those areas well suited for housing
Variables used in this model—erosion hazard, relative costs
for grading, septic suitability, soil creep, vegetation type,
drainage, slope, landforms, geology, soils.

Data mapped in 5 levels between extreme values of 7.00
and 32.00.

Absolute value range applying to each level:
   Minimum   7.00   12.00   17.00   22.00   27.00
   Maximum   12.00   17.00   22.00   27.00   32.00

Percentage of total absolute value range applying to each
level:

          20.00   20.00   20.00   20.00   20.00

Frequency distribution of data point values in each level:

Figure 11.27. *Housing Model, Bretton Woods, New Hampshire.
Using the weighted factors shown in table 11.4, the computer is able
to plot this composite interpretive map. Areas lightly shaded
(Level 0) are poorly suited for housing, while dark areas (Level 4)
are well suited for residential development.* Photo courtesy: *Environmental Systems Research Institute, "Environmental Site
Analysis for Bretton Woods, Mt. Washington" (Redlands, Calif.,
1972), p. 15.*

environmental impact statements, or even
specific environmental performance ordinances.

The weighting of individual environmental factors is an important part of any
computerized system. It requires certain dis-
cretionary judgment by the geologist or
planner. By varying the weightings, different environmental management alternatives
can be tested rapidly and easily by the computer. For instance, if in the sanitary landfill
example described above soil permeability is

critical in the selection of an appropriate
site, then that factor should receive a relatively high weighting to emphasize its importance. However, if it is later discovered
that all areas have high permeability, and
that factor is not a significant problem, other
factors such as depth to groundwater aquifers can be weighted to reflect the new emphasis. After any desired statistical treatment, the factor scores for each grid cell are
simply compiled by the computer. In the
final printout, high sums will indicate po-

tential waste disposal sites, while low sums will show areas that should be avoided. In comprehensive planning studies, the computer can be programmed for factors that should be avoided entirely in devising management policy or choosing site alternatives. For instance, if a local ordinance prohibits waste disposal on floodplains, all floodplain zones can be given a zero score even though other factors may be acceptable (Tilmann et al., 1975).

The computer approach to land capability analysis has definite advantages over manual techniques. The number of environmental factors and variables that this system can store, manipulate, and evaluate is almost limitless. Grid cell data is amenable to various aggregation techniques that can be used to define, test, and portray natural processes. Furthermore, if new data on a certain factor becomes available, the data record file may be updated by simply reentering the new information. The appropriate factors can then be sorted, weighted, and new interpretative maps produced relatively quickly and inexpensively.

However it should be noted that this approach, like any automated system, has certain limitations that should be recognized for proper application of the method. First, the computer outputs, and specifically the interpretative maps, are only as valid as the initial data base. Furthermore, if information on a certain factor is not available or is too general, the computer cannot be used to provide it or increase its validity. The technique is not designed to replace sound professional geological experience and judgment. Second, the computer is a tool that can be used in the development of models for comprehensive land use, environmental impact evaluation, and project site analysis. By itself, the technique does not make decisions, but it can be an important informational contribution to the decision-making process.

ENVIRONMENTAL IMPACT ANALYSIS

National Environmental Policy Act (NEPA)

In 1969, in what is considered by many to be landmark legislation, the National Environmental Policy Act (NEPA) was adopted by Congress. The intent of the NEPA was to bring fundamental reform to all levels of federal decision making concerning the environment. All relevant environmental data had to be considered in the decision-formulating process before a project or improvement was authorized. The NEPA's underlying premise is that policy decisions would be improved and a better balance of environmental and developmental objectives would emerge if the total range of environmental parameters and project alternatives were examined early in the process, so that modifications in the project to accommodate environmental objectives could be readily made.

The National Environmental Policy Act sets forth the requirement that all federal actions or projects that could have a significant effect on the environment must have an environmental impact statement (EIS) prepared. The NEPA also applies to local projects that involve federal expenditures and to private projects that must go through approval procedures in a federal agency. Although each federal agency has developed its own procedures in conformance with the NEPA, every environmental impact statement must have five required components discussed:

1. the environmental impact of the proposed action

2. any adverse environmental effects that cannot be avoided should the project be implemented

3. alternatives to the proposed action

4. the relationships between local short-term uses of the environment and the maintenance and enhancement of long-term productivity

5. any irreversible and irretrievable commitments of resources that would be involved in the proposed action should it be implemented

The NEPA does not specify who should prepare the EIS, but rather indicates that it shall be prepared by the "responsible officer" in each agency. The individual federal agency has the authority to designate the responsible official in its procedures. When a federal regulatory or permit agency requires an EIS, the agency prepares the statement but the individual or organization proposing the action can be asked to provide a preliminary environmental impact assessment. The agency is fully responsible for the content of the report, however, and must review it for adequacy and modify it as required.

State Environmental
Impact Legislation

As of January 1, 1975, thirty-two states had followed the federal lead and enacted legislation or administrative regulations to establish environmental impact assessment in their decision-making processes. Fourteen

other states at this time were considering environmental legislation similar to the NEPA (Burchell and Listokin, 1975). To some extent, the state legislation was modeled on the federal law, but in many jurisdictions there are important differences. For instance, the California legislation — the California Environmental Quality Act (CEQA) adopted in 1970 — indicates two factors that must be contained in an environmental impact report, in addition to the five enumerated in the federal legislation:

1. mitigation measures proposed to minimize the impact

2. discussion of the growth-inducing impact of the proposed action

A second significant development in California occurred in a 1972 court case (*Friends of Mammoth* v. *Mono Company Board of Supervisors*) when the environmental impact of private construction was questioned. This led to the requirement of an environmental impact report for all private projects as well as those conducted by governmental agencies. These two developments — expansion of the EIS to include mitigation and growth inducement and the application of environmental legislation to private development projects requiring some form of permit from a governmental agency — have been incorporated into the EIS requirements in many states and to some extent into recent federal guidelines for NEPA implementation.

Content of an Environmental
Impact Statement

Environmental impact statements (in California and several other states, EISs are called Environmental Impact Reports or EIRs) have ranged from one page statements to several hundred page volumes of prose, statistics, tables, diagrams, and maps. Furthermore, no single format has been established for preparing an EIS. Because the laws are new and environmental review is still in its infancy, agencies and courts are still deciding what an adequate document is, and major changes in content and format can be expected in the future. Therefore, the following discussion is offered as a general guide only. For further information, and certainly before preparing an EIS on a specific project, the reader should obtain the environmental impact guidelines from the appropriate federal, state, or local government agency.

Project Description

Using available maps and photos, the project should be described in relation to its surroundings. The current status and all phases of the project should be discussed. Project objectives should be delineated and compared with any plans or adopted policies that various jurisdictions have for the area.

Description of Existing Environment

The environment in the vicinity of the project should be described as it exists before commencement of the project, from both a local and regional perspective. This description should be given in sufficient detail so that the reader will be able to make an independent evaluation of environmental factors affected by the project. This description, however, need not include environmental aspects that would be unaffected by the project.

Environmental Impact of the Proposed Action

This section must be comprehensive and include the physical, cultural, and economic impacts of the project on the environment. It is necessary that this evaluation include secondary as well as primary impacts, and long-range as well as short-term effects. For example, a road project might cause erosion and sedimentation in a nearby stream, but the residential expansion, made possible by the new road, could cause further sedimentation, contribute to areal flooding with new impervious surfacing, and deplete groundwater supply.

Some of the major geology-related factors that should be considered in an impact evaluation include: contamination or depletion of groundwater supply, reduction of aquifer recharge capability, impact on surface drainage, erosion and siltation, subsidence affecting off-site property, impact on mineral resources, impact on beaches or coastal landforms, and project relation to areas of suspected geological hazard (see EIS format example in appendix E).

Because of the necessity to assess a variety of environmental factors and conditions in this section, methodologies have been developed to ensure a systematic approach.

1. *Checklists.* Checklists are compiled either in topical or questionnaire form. They assume enough knowledge about a project to be able to extract items that apply. Checklists, varying in detail, have been developed by many local, state, and federal agencies; the example shown in appendix E is typical.

2. *Matrix.* The matrix is a visual organization of cause-effect environmental relationships. Usually aspects of the environment

that could be affected by a project are listed in one column, while actions that cause the impact on the environment are noted in the other. Thus a chart is created that identifies environmental concerns that need to be discussed. This approach has been used by Leopold et al. (1971) to assign relative values for magnitude and importance of impact (*see* figure 11.28). This technique is valuable in providing a summary of the EIS as well as an organizational format. Users should recognize the subjectivity of the numerical values and the consequent limitations of such a system. A more comprehensive approach was taken by Sorenson (1971) in using a series of matrices to relate a chain or cycle of cause-effect relationships (see figure 11.29).

3. *Capability mapping.* The techniques used by McHarg and others, explained ear-

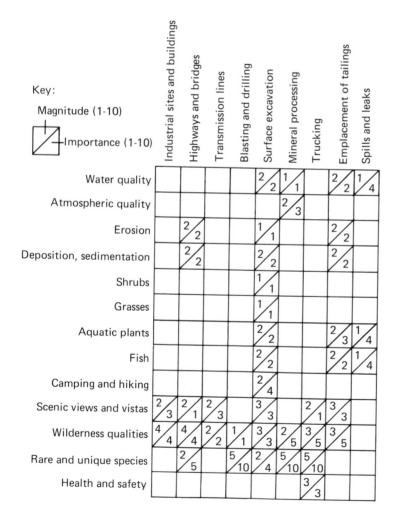

Figure 11.28. *Environmental Impact Matrix for a Phosphate Mining Lease. Activities associated with phosphate mining are listed along the top, and existing environmental characteristics are shown in the vertical column. Where an activity could impact an environmental condition, a numerical evaluation was made in terms of magnitude and importance. For example, in the top column, spills and leaks could impact water quality, but were considered sufficiently rare to warrant a magnitude rating of 1. However, if they occurred they would be moderately important and were therefore given an importance value of 4.* RE-DRAWN FROM: *L. B. Leopold et al., "A Procedure for Evaluating Impact,"* Geological Survey Circular 645 (1971).

lier in this chapter, can also be used to assess environmental impacts. This method is most useful for projects covering large areas, and for environmental assessments made early in the project planning stage, so that locational adjustments can be easily accommodated.

4. *Computer data systems.* Computer mapping and modeling techniques, described on p. 438 in this chapter, are of tremendous assistance in environmental impact evaluation in areas where computer information systems have been developed. They can also serve to catalogue and store data from EISs that have been completed and reviewed. In this way research done for impact statements can become part of a larger body of knowledge, providing a better understanding of environmental interrelationships and improved ability to predict environmental impact in the future.

Unavoidable Adverse Impacts

The purpose of this section is to highlight adverse project impacts mentioned in the previous section, including those that may be reduced in significance but not eliminated. Discussion in this section should be related to the evaluation of alternatives (*see below*). If alternatives exist that can minimize or eliminate adverse effects, the reasons why the project is being proposed should be discussed.

Mitigation Measures Proposed to Minimize Impact

Mitigation measures are project design or construction features that are proposed to reduce adverse project impacts. For example, erosion control techniques to reduce stream siltation can be suggested for construction projects involving some grading. This discussion should state acceptable levels of impact reduction and why such levels are considered acceptable. Where alternative mitigation measures are available, or mitigation measures in themselves have some impact, reasons for the choices made should be given. Mitigation measures should be chosen realistically so that they can be carried out in the project design. In many jurisdictions suggested measures are imposed as conditions of project approval.

Alternatives to the Proposed Action

An important part of an EIS, one which is often inadequately discussed, is alternatives considered to the proposed project. This evaluation should include any possible modifications of the project or portions of it, other types of projects that may meet the same objectives, alternative project locations, and other changes that can reduce impact even if these could be more costly or impede the attainment of project objectives. The specific alternative of "no project" should always be evaluated, along with the impact.

There is often confusion in the distinction between alternatives and mitigation measures. Mitigation measures are specific methods for reducing identified impacts. Alternatives, on the other hand, are other types of projects or a significant variation of the action proposed that may or may not meet the basic objectives of the proponent. For instance, alternatives to a proposed office complex could include single family housing, a community park, or the office complex proposed at a different location.

Figure 11.29. *Environmental Impact Matrix — Human Activities in the Coastal Zone. This matrix illustrates the secondary impacts that result from various human uses in the coastal zone. The "uses" in the upper left corner require various supportive activities or structures, shown as "causal factors" in the middle column. These, in turn, create the adverse impacts used in the last three columns. The "initial condition" shows the primary environmental degradation that may result from the activity. The "consequent condition" and "effect" columns indicate secondary impacts that may result in a cause-effect relationship from the initial condition. The matrix illustrates the complexity of environmental impact relationships and the multitude of problems incurred when People affect a natural system. The numbers preceding the last three columns provide an index for computer treatment of the matrix.* REDRAWN FROM: *J. C. Sorensen, "A Framework for Identification and Control of Resource Degradation in the Multiple Use of the Coastal Zone." Master's thesis, University of California at Berkeley (1971).*

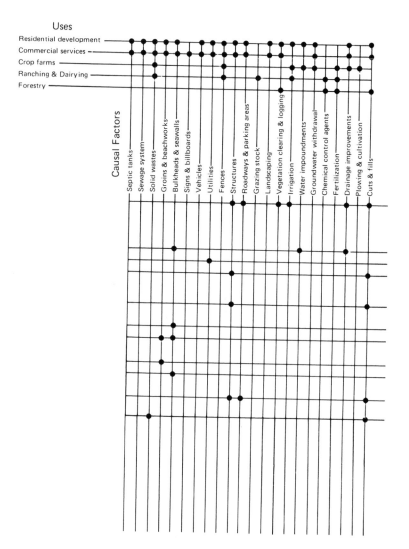

Relationship Between Short-term Uses and Long-term Productivity

In this section a description of the cumulative and long-term effects of the project that adversely affect the environment should be given. The closing of future options and possible risks to future public health and safety should be covered specifically. Any beneficial effects of the project should be weighed to ascertain why it should be carried out now rather than reserving the option for the future. Reference to adopted policies and general plans of local agencies is advisable. A city with a policy of protecting coastal land for the enjoyment of all residents should find it difficult to approve a residential subdivision on a coastal bluff that provides access to a popular beach.

Possible Adverse Impact		
Initial Condition	Consequent Condition	Effect

101.1 Increased fresh water flows into estuary.	101.2.1 Stimulate or increase cliff erosion.	101.2.1.1 Imperil cliff edge structures & utilities.
	101.2.2 Reduce salinity in estuarine waters.	-------------- A 130.1
	101.2.3 Increased frequency & size of floods.	101.2.3.1 Imperil costal flood plain development.
	101.2.4 Increased sheet and gully erosion.	-------------- A 301.1
302.2 Blockage of sand erosion & transport.	302.2.1 Decreased deposition of sand to beaches.	302.2.1.1 Erosion of beaches. A 101.2.1 & C 398.2.1
304.1 Increased sedimentation.	----------- D 304.1	
308.1 Stimulation of landslides.	308.2.1 Sheet & gully erosion into coast drainage.	-------------- A 301.1
	308.2.2 Highly visible areas of earth & vegetation disturbance.	-------------- D 678.1
334.1 Overloaded unconsolidated sediments.	334.2.1 Support failure and collapse of fill.	334.2.1.1 Structural damage to development on fill.
	334.2.2 Increased susceptability to earthquakes.	334.2.2.1 Structural damage to development on fill.
344.1 Increased slope of shore.	344.2.1 High wave run-up & overtopping.	344.2.1.1 Inland flooding.
348.1 Reflection of onshore waves.	348.2.1 Increase backwash across shore.	348.2.1.1 Increased beach erosion. A 302.2.1.1
	348.2.2 Reflected waves into waterways.	348.2.2.1 Small boat navigation hazard.
350.1 Interception of longshore drift.	----------- B 350.1	
354.1 Irregular bulkhead line.	354.2.1 Visual fragmentation of shoreline expanse.	354.2.1.1 Aesthetic displeasure.
	354.2.2 Collection of floatable trash & litter.	354.2.2.1 Aesthetic displeasure. Attraction of pests.
370.1 Dredging or excavation for landfill.	----------- C 320.1, C 373.1, C 324.1, C 326.1, C 328.1, C 375.1	
375.1 Filling of wet and submerged lands.	375.2.1 Change slope of shoreline.	-------------- A 344.1
	375.2.2 Change angle of shoreline.	375.2.2.1 Increase wave erosion on shoreline. A 133.1
		-------------- B 350.1
	375.2.3 Reduce estuarine tidal prism.	-------------- C 109.1
	375.2.4 Remove shoreline buffer strip.	375.2.4.1 Expose shoreline to storm & erosion damage.
	375.2.5 Convert open lands into private holding.	375.2.5.1 Prohibit public access. A 501.2.1, A 660.1
	375.2.6 Decreased oxygenation capacity.	375.2.6.1 Reduced ability to assimilate pollutants.
	375.2.7 Destroy nursery grounds.	375.2.7.1 Reduce nursery dependent populations.
	375.2.8 Destroy waterfowl nesting & feeding area.	375.2.8.1 Reduce waterfowl populations.
	375.2.9 Reduce nutrient supply to coastal system.	375.2.9.1 Decreased primary production. Decreased size and number of species.

Irreversible Environmental Changes

The focus in this section should be on resource utilization and preemption rather than on a regurgitation of impacts from previous sections. Use of nonrenewable resources, or commitments of land that would make removal or nonuse of a constructed facility unlikely, are irreversible. An obvious example of an irreversible commitment is the residential development of areas having an underlying mineral deposit, preempting any recovery of the mineral resources. If commitments of resources that cannot be reversed are planned, an evaluation should be made as to why such consumption is necessary.

Essay

ENVIRONMENTAL IMPACT STATEMENT
GRAND TETON/JACKSON HOLE AIRPORT, WYOMING

A small airport servicing private and commercial propeller aircraft now exists within Grand Teton National Park, near the town of Jackson in Wyoming. To accommodate commercial jet aircraft, and to meet certain air safety requirements, expansion and improvement of the airport was proposed. In July 1973, in order to meet the requirements of NEPA, the U.S. Department of Interior (National Park Service) released a 135-page draft environmental impact statement on the project. This was followed by several months of public hearings, oral testimony, and written comments on the draft EIS. The written documents were incorporated into a final EIS completed in March 1974.

Proposed Project

The basic proposal included the following: extension of the runway from the present 6305 feet to 8000 feet; expansion of the runway width from 100 to 150 feet; construction of a taxiway; construction of a control tower; installation of an instrument landing system and flood lights; expansion of airplane parking aprons; construction of a new automobile parking area and new access road; and, installation of a new sewage system (see figure 11.30).

Environmental Setting

The airport lies within the southern portion of the national park on a glacial outwash valley. The park is well known for its geological and biological features, which contribute to a unique scenic beauty and a steadily increasing number of visitors. It is dominated by towering steep granitic peaks, glacial-scoured canyons, and large lakes at the

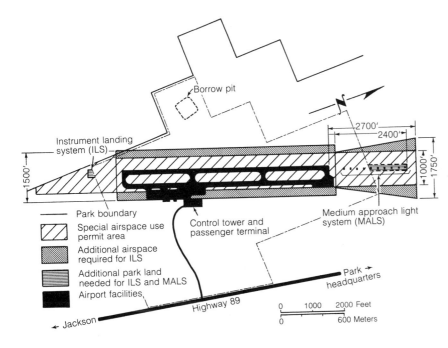

Figure 11.30. *Jackson Hole Airport Grand Teton National Park, Wyoming. This diagram is a project description for the expansion and improvement of the airport.* REDRAWN FROM: *"Draft Environmental Impact Statement, Grand Teton-Jackson Hole, Airport Expansion," U.S. Dept. of Interior, National Park Service (July 1973).*

mountains' base. Grand Teton and surrounding areas contain unusual wildlife species including large populations of elk and moose, pronghorn, bighorn sheep, black bear, bison, eagles and other raptor bird species. Visitor use of the park is varied and consists of such activities as back country hiking, mountain climbing, fishing, wildlife viewing, and rafting on the Snake River. Large portions of the park are now proposed for a wilderness preserve.

The existing airport facilities allow Frontier Airlines to operate Convair turboprop planes providing passenger service to the town of Jackson and Grand Teton National Park. Aircraft take-off and landings occur over southern portions of the park and create some noise for park visitors in these areas. The effects of this noise on wildlife populations has not been examined by the Park Service.

Impacts

1. *Noise.* Expansion of the airport runway allowing use of commercial jet aircraft (Boeing 737) would increase noise levels up to 17 decibels depending upon location and flight patterns. The noise could detract from visitor wilderness experiences and could have an adverse impact upon wildlife. The statement indicates that although little is known about the effects of aircraft noise upon wildlife, significant disruption of wildlife populations within the Park, along the Snake River, and in the National Elk Refuge must be considered a possibility.

2. *Air pollution.* Additional aircraft emissions from airport expansion were considered to be minimal with no foreseeable impact, according to the EIS.

3. *Aesthetic.* Construction of airport facilities such as the control tower and lighting system will result in some increased visual intrusion although the overall impact would be minimal. However, the Park Service also indicated that even minimal intrusions can be significant in a national park setting and will be clearly visible to visitors looking down on the valley from the mountains.

4. *Economic.* Airport improvements will cost $1,377,000, but the increased passenger traffic permitted by the expansion should provide economic benefits to Frontier Airlines and stimulate economic growth in Jackson and for recreational facilities in the surrounding region.

Mitigating Measures

The EIS discusses use of noise control equipment to reduce sound impacts from commercial jet aircraft and limitations on take-off and landing routes to minimize noise impacts on park visitors and wildlife. Construction of the control tower will be low profile and designed to blend with the existing terminal building to mitigate visual impact from the nearby Jackson Hole Highway. Use of the lighting system will be restricted to periods of actual operation at night or during instrument landings.

Alternatives

In considering alternatives, the EIS provides an analysis of a reduced project, no project, and relocation of the airport to another site. One of the most controversial parts of the statement was the suggestion for an alternative airport site. The two chosen for detailed study included a location along the Snake River south of the park, but still in the Jackson Hole area, and a site near Driggs, Idaho, west of Teton Pass (see figure 11.31). Both locations meet FAA navigable airspace requirements, but the EIS concluded the Snake River site has many of the aesthetic and ecological objections present in the site. The Driggs, Idaho, location is thirty-four miles from Jackson and forty-six miles from the park, requiring a greater expenditure of time and money for passengers.

As a result of the analysis provided by the EIS, public testimony, and written comments received, the Department of Interior decided in May 1974, to recommend against extension of the runway length, and therefore against use of the airport by commercial jet aircraft. It also decided that a regional transportation study should be prepared before these issues were reconsidered and any long-range commitments were made regarding the airport. However in 1975, some airport improvements were made including the runway widening, construction of the taxiway, the parking lot expansion, and construction of new sewage treatment facilities.

Despite criticism leveled at the draft EIS by both sides on the airport expansion issue, the statement (including comments received) served its purpose in providing a full analysis of project environmental issues and promoting public participation in the decision process.

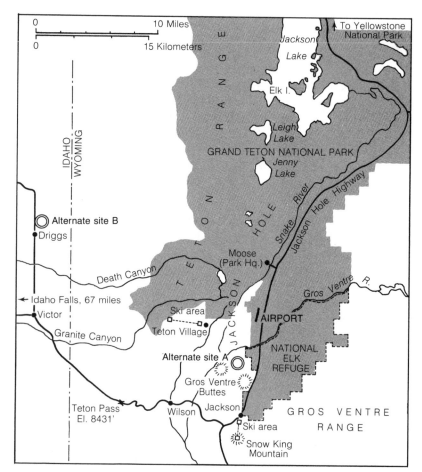

Figure 11.31. *Jackson Hole Airport and Vicinity. The map shows location of the proposed project in relation to Grand Teton National Park, town of Jackson, Wyoming, and the two alternative sites considered along the Snake River and at Driggs, Idaho.* REDRAWN FROM: *C. Stallings, "Flyway to Tourist-Trapdom," Andubon 76, no. 1 (January 1974) p. 107.*

Growth-inducing Impact

Growth here refers to facilitation of population growth, either directly or indirectly, and encouragement or facilitation of other projects that may have an impact on the environment. A major highway project, for example, can facilitate residential development, which in turn can promote growth of community service facilities. Each of these could have separate physical impacts which should be discussed.

The Final EIS

The final Environmental Impact Statement contains, in addition to the items above, a statement of comments from agencies, groups, and individuals in a public hearing and review process, and the responsible agency's reply to these comments. This review process is an important aspect of the EIS procedure in that it encourages coordination with other public agencies at all governmental levels that might ultimately be affected by the project. Additionally, it provides a mechanism for interested private groups and individuals to obtain information and to become involved in decisions that affect the environment. The adequacy of review and public participation depends upon the degree of public notice given a draft EIS, as well as upon the amount of time allowed for review. The report should be available early enough to allow time for a reviewer who questions an analysis of an impact to conduct an independent investigation. Review periods for local agencies vary considerably, but most state and federal agencies allow thirty to forty-five days for review of an EIS (or EIR).

Problems in the Environmental Assessment Procedure

A recently completed analysis of environmental impact statements for U.S. Army Corps of Engineers water projects (Ortolano and Hill, 1972) provides some useful information on the overall quality and usefulness of the statements in evaluating environmental impact as well as some insight into deficiencies shared by many state and federal agencies. In the survey, 234 statements prepared by the corps through August 1971 were examined in detail. In general terms, the majority of the statements were decidedly less than adequate. They did not seem to be written with the view of providing nontechnically oriented readers with the kinds of insights and information necessary for effective participation in the decision-making process. While they did not reflect the careful integration of environmental issues into the project evaluation process, this could not always be expected, considering

that many of the projects were in rather advanced stages of design. Some of their suggestions for improvements that should be helpful in all EIS preparation include the following:

1. *Reduce the level of generality.* A large number of reported impacts are so general as to be unusable, the phrases "elimination of vegetation," "loss of wildlife habitat," "increase in turbidity," and the like are typical.

2. *Identify all significant impacts.* Many statements are incomplete in not discussing all the probable significant impacts, to the extreme of stating that projects would have no significant environmental impact at all.

3. *Suggest alternatives to the proposed action.* Usually insufficient consideration was given to the alternatives to the proposed action; alternatives were usually rejected on economic and technological grounds without a discussion of environmental implication.

Several other considerations in the environmental impact assessment process, not discussed in this study, should also be noted.

1. *Objectivity.* Far too often, reports are slanted toward the project and are not objective evaluations of environmental impacts. An agency or private developer is unlikely to propose a project unless it believed it was beneficial for at least some of the public. Therefore, there is a natural tendency to minimize adverse impacts and to prepare the EIS as a justification or promotion for the project. In the case of private development projects subject to permit authority by a local

planning commission, this problem can be solved by requiring that the agency's planning staff, or an outside consultant responsible to the staff, prepare the report.

2. *Qualification of individuals preparing reports.* Resource evaluation and planning is a dynamic, complex process that should only be attempted by persons qualified in the respective environmental fields. An interdisciplinary approach is required using geologists, hydrologists, biologists, and other individuals trained in various disciplines to evaluate impacts as required by the specific situation encountered in the project. However, in terms of the final product, it is important that the report be integrated — that is, the various environmental concerns are synthesized to produce a systematic, balanced approach covering all impacts in relation to their importance.

3. *Lengthiness.* The increasing size (and cost) of EISs has become a disturbing trend in recent years. But rather than assisting in the analysis of environmental problems involved with a project, the increased size, volume of data, and sometimes irrelevant text have detracted from the process. As a result decision makers are unable to adequately review an EIS, and the public is less able to contribute meaningful comments. This, in turn, defeats the major purposes of environmental impact assessments: providing public participation, communication between governmental entities, and informed environmental decision making. This problem can be partially solved by use of appendices to include supporting data. A second solution is to eliminate or only briefly discuss insignificant issues and reserve the majority of the report for evalua-

tion and presentation of support data on issues identified initially as important.

Despite the problems encountered, the environmental impact process is a valuable tool for assessing geologic resources and hazards associated with developmental projects. Properly used, it provides a method for full public disclosure of geologic issues, as well as an opportunity for integration of these into the planning process. More than anything else, impact assessment can clarify the consequences of public decisions, and can communicate clearly the implications of choosing one action over another.

REFERENCES

References Cited in the Text

Burchell, R. W., and Listokin, D. *The Environmental Impact Handbook.* Rutgers, N.J.: Center for Urban Policy Research, 1975.

California Division of Mines and Geology. *First Report of Governors Earthquake Council.* Sacramento, Calif.: Resources Agency, 1972.

California Government Code Section 65302. In *West's Annotated California Codes Volume 36A.* St. Paul, Minn.: West Publishing Company, 1975. Pocket Supplement, pp. 85–86.

Colwell, Robert. "Remote Sensing of Natural Resources." *Scientific American* 218(1968): 54–69.

Flawn, Peter T. *Environmental Geology.* New York: Harper and Row, 1970.

Goldman, H. B., ed. *Geologic and Engineering Aspects of San Francisco Bay Fill.* California Div. of Mines and Geology Special Report 97. San Francisco: California Div. of Mines and Geology, 1969.

Gross, David. "Geology for Planning in DeKalb

County, Illinois." *Environmental Geology Notes No. 33*. Urbana: Illinois State Geological Survey, 1970.

Hansen, W.R., *Effects of the Earthquake of March 27, 1964 at Anchorage, Alaska*. U.S. Geological Survey Professional Paper 542A, Washington, D.C.: U.S. Government Printing Office, 1965.

Kansas Geological Survey Study Committee. *A Pilot Study of Land Use Planning and Environmental Geology*. Kansas Geological Survey Report 15D, 1968.

Lauchenbruch, A. M. *Some Estimates of the Thermal Effects of a Heated Pipeline on Permafrost*. U.S. Geological Survey Circular 632, 1970.

Leighton, F. B. "Landslides and Hillside Development." In *Engineering Geology in Southern California*. Association of Engineering Geologists Special Publication 1966, pp. 149–90.

Leopold, L. B. et al. *A Procedure for Evaluating Environmental Impact*. U.S. Geological Survey Circular 645, 1971.

Liang, T., and Belcher, D. J. "Airphoto Interpretation." In *Landslides and Engineering Practice*. *Highway Research Board Special Report 29*, NAS-NRC 544. Washington, D.C. U.S. Government Printing Office, 1958, pp. 69–92.

McCulloch, D. S. et al. *A Preliminary Study of the Effects of Water Circulation in the San Francisco Bay Estuary*. U.S. Geological Survey Circular 637-A B, 1970.

McHarg, Ian. *Design with Nature*. New York: Natural History Press, 1969.

Mader, G. G., and Crowder, D. F. "An Experiment in Using Geology for City Planning-the Experience of the Small Community of Portola Valley, California." In Nichols and Cambell, ed. *Environmental Planning and Geology*,

Association of Engineering Geologists Proceedings, Washington, D.C.: U.S. Government Printing Office, 1969, pp. 196–89.

Miller, R. D., and Dobrovolny, E. "Surficial Geology of Anchorage, Alaska and Vicinity, Alaska." *U.S. Geological Survey Bulletin 1093*, 1959.

Nichols, D. R. "Seismic Safety Legislation and Engineering Geology." *Symposium on Geology, Seismicity, and Environmental Impact*. Annual Meeting, Association of Engineering Geologists, 1973, pp. 1–12.

Nichols, D. R., and Buchanan-Banks, I. M. *Seismic Hazards and Land Use Planning*. U.S. Geological Survey Circular 690, 1974.

Ortolano, L., and Hill, W. W. *An Analysis of Environmental Statements for Corps of Engineers' Water Projects*. U.S. Army Engineers Institute for Water Resources Report 72-3. Alexandria, Va. 1972.

Rowan, Lawrence. "Application of Satellites to Geologic Exploration." *American Scientist* 63(1975): 393–403.

Scullin, C. M. "History Development and Administration of Excavation and Grading Codes." In *Engineering Geology in Southern California*. Association of Engineering Geologists Special Publication, 1966, pp. 227–36.

Slossen, J. E., and Hauge, C. J. "The Public and Geology-Related Legislation in California, 1968–1972." *Geology, Seismicity, and Environmental Impact*. Association of Engineering Geologists Special Publication. Los Angeles: University Publishers, October 1973, pp. 23–28.

Sorenson, J. C. "Coastal Planning: The Impact of Use on Environment." Master's Thesis. University of California at Berkeley, 1971.

Tilmann, S. E.; Upchurch, S. B.; and Ryder, Graham. "Land Use Site Reconnaissance by Computer-Assisted Derivative Mapping."

Geological Society of America Bulletin 86(1975): 23–34.

Thurow, C.; Toner, W.; and Erley, D. *Performance Controls for Sensitive Lands: A Practical Guide for Local Administrators.* Washington, D.C.: Office of Research and Development, U.S. Environmental Protection Agency, 1975.

Other Useful References

Adams, Ukton W. *Earth Science Data in Urban and Regional Information Systems: A Review.* U.S. Geological Survey Circular 712, 1975.

Bosselman, F. and Callies D. *The Quiet Revolution in Land Use Control.* Washington, D.C.: Council on Environmental Quality, 1971.

Citizens Advisory Committee on Environmental Quality, Task Force on Land Use and Urban Growth. *The Use of Land: A Citizens Policy Guide to Urban Growth.* New York: Thomas Crowell, 1973.

Goodman, W. I., ed. *Principles and Practice of Urban Planning.* Washington, D.C.: International City Managers Association, 1968.

Heffernan, P. H. and Corwin R., eds. *Environmental Impact Assessment.* San Francisco: Freeman, Cooper & Co., 1975.

Nichols, D. R., and Campbell, C. C., eds. *En-* *vironmental Planning and Geology.* Proceedings of a Symposium on Engineering Geology in the Urban Environment of the Association of Engineering Geologists. Washington, D.C.: U.S. Government Printing Office, 1971.

Pepper, J. E. *An Approach to Environmental Impact Evaluation of Land Use Plans and Policies: The Tahoe Basin Planning Information System.* Berkeley: University of California, 1972.

Pessl, F.; Langer, W. H.; and Ryder, R. B. *Geologic and Hydrologic Maps for Land Use Planning in the Connecticut Valley with Examples from the Folio of the Hartford North Quadrangle, Connecticut.* U.S. Geological Survey Circular 674, 1972.

Rudd, R. D. *Remote Sensing: A Better View.* North Scituate, Mass.: Duxbury Press, 1974.

Sorensen, J. C. and Moss, Mitchell L. *Procedures and Programs to Assist in the Environmental Impact Statement Process.* Berkeley: University of California Press, 1973.

Spangle, W., and Associates; Leighton, F. B., and Associates; and Baxter, McDonald and Company. *Earth Science Information in Land Use Planning: Guidelines for Earth Scientists and Planners.* U.S. Geological Survey Circular 721, 1976.

Way, Douglas. *Terrain Analysis: A Guide to Site Selection Using Aerial Photographic Interpretation.* Stroudsburg, Penna.: Dowden, Hutchinson, and Ross, 1973.

Appendixes

Contents

APPENDIX A
Geologic Time Scale

Era	Period	Epoch	Approximate Age in Millions of Years (beginning of unit)
Cenozoic	Quaternary	Holocene	0.01
		Pleistocene	1.7 to 2.0
	Tertiary	Pliocene	13
		Miocene	25
		Oligocene	36
		Eocene	58
		Paleocene	65
Mesozoic	Cretaceous		135
	Jurassic		181
	Triassic		225
Paleozoic	Permian		280
	Pennsylvanian		325
	Mississippian		345
	Devonian		405
	Silurian		425
	Ordovician		500
	Cambrian		570
Precambrian	No worldwide classification	Approximate age of oldest rocks discovered on earth	4000
		Approximate age of earth and oldest meteorites	4500

APPENDIX B
Metric Conversion Table

Length

1 inch = 2.54 centimeters	1 centimeter = 0.39 inches
1 foot = 0.30 meters	1 meter = 3.28 feet
1 yard = 0.91 meters	1 meter = 1.09 yards
1 mile = 1.61 kilometers	1 kilometer = 0.62 miles

Area

1 square inch = 6.45 square centimeters	1 square centimeter = 0.15 square inches
1 square foot = 0.09 square meters	1 square meter = 11 square feet
1 square yard = 0.84 square meters	1 square meter = 1.20 square yards
1 acre = 0.40 hectares	1 hectare = 2.47 acres
1 square mile = 2.6 square kilometers	1 square kilometer = 0.38 square miles

Volume

1 cubic inch = 16.4 cubic centimeters	1 cubic centimeter = 0.06 cubic inches
1 cubic foot = 0.27 cubic meters	1 cubic meter = 0.37 cubic feet
1 cubic yard = 0.76 cubic meters	1 cubic meter = 0.13 cubic yards
1 cubic mile = 4.19 cubic kilometers	1 cubic kilometer = 0.24 cubic miles
1 gallon = 3.8 liters	1 liter = 0.26 gallons

Weight-Mass

1 ounce = 28.3 grams	1 gram = 0.04 ounces
1 pound = 0.45 kilograms	1 kilogram = 2.20 pounds
1 ton = 0.9 metric tons	1 metric ton = 1.1 tons

Temperature

$F = \frac{9}{5}C + 32$

F = degrees farenheit

$C = (F - 32)\frac{5}{9}$

C = degrees celsius (centigrade)

Other Conversion Factors

1 cubic foot/second (cfs) = 0.03 cubic meters/second = 7.48 gallons/second
1 acre foot = 43,560 cubic feet = 1233 cubic meters = 325,829 gallons

---

APPENDIX C

Aquifer Recharge Ordinance of Volusia County, Florida

Findings of Fact

In adopting this ordinance the County Council of Volusia County hereby finds the following facts to be true:

Volusia County is a unique hydrological entity in that an insignificant portion of the water supply for the entire county is derived from sources outside the county. On the other hand, a certain amount of water leaves Volusia County through rivers and streams and other water courses as well as the Clastic and Floridan aquifers. Volusia County is almost totally dependent upon rainfall within the county for its entire water supply. Water is supplied to users in Volusia County from wells driven into the Floridan Aquifer. As water is used, the aquifer is replenished or "recharged" primarily from rainfall, which ultimately finds its way to the Floridan Aquifer.

The area [known] as the "Potential Recharge Area" is essential to the water quantity and quality of Volusia County for the reasons more specifically stated hereafter. Water recharge occurs throughout Volusia County; however, the Potential Recharge Area has the greatest potential for recharge of the Volusia County water supplies as water is drawn from the Floridan Aquifer. Within the Potential Recharge Area, as the level of water in the Floridan Aquifer is reduced through use, the rejected recharge which currently occurs in the area would be decreased by capture of such water, thus increasing the amount of water available for use.

In those areas of Volusia County currently identified as the best defined areas for recharge to the Floridan Aquifers, the recharge ability cannot be increased. This is due to the fact that these areas, particularly the DeLand Ridge, the Rima Ridge, and the Atlantic Coastal Ridge, generally are accepting the maximum amount of rainfall available. Continued development of these areas will decrease the recharge capability because of less natural area available to absorb rainfall and an increase in the volume and speed of surface water runoffs, which in turn increases the amount of water lost to rivers and streams and due to evapotranspiration.

Control of development within the Potential Recharge Area is imperative for the following reasons:

(1) The previously mentioned ridges are currently the areas of greatest development and development potential in Volusia County, resulting in adverse effects on their future potentials as areas of recharge to the Floridan Aquifer.

(2) The greatest amount of fresh water is available from the Floridan Aquifer within the Potential Recharge Area with a minimum danger of salt water intrusion into the water supplies of Volusia County.

(3) Land within the Potential Recharge Area is currently undeveloped and essentially undevelopable due to the existing surface and subsurface water conditions. Developments for other than limited agricultural and timber production uses could only occur with the use of substantial drainage and significant alterations of the present hydrological and geological conditions. Such alterations would adversely affect the recharge potential of the Potential Recharge Area by decreasing the capability of the area to provide continuous recharge to the Floridan Aquifer.

(4) The natural vegetation within the Potential Recharge Area as well as the interaction of surface water areas with the Clastic and Floridan aquifers provides the superior filtration of waters recharging the Floridan Aquifer, reducing the amount of natural and man-made pollutants reaching the Floridan Aquifer. Such filtration is being lost in the present Potential Recharge Area due to development.

(5) If the Potential Recharge Area is sufficiently controlled and generally preserved in its natural state, it will assure the maintenance of a water recharge system sufficient to supply the projected needs of the entire county for water well beyond the projected population increases through the year 2000. Uncontrolled development of the Potential Recharge Area, combined with the continued development of all other areas of Volusia County, will seriously impair the water supply of Volusia County by significantly reducing recharge capabilities to the Floridan Aquifer as use of water continually increases. Such a situation could ultimately result in increased saltwater intrusion in various areas of the Floridan Aquifer as well as potential future water shortage.

(6) Any uncontrolled drainage or alteration of the natural movement of surface water through construction of roads, fill of land, excavation, and similar alterations would adversely affect the relationship between the surface waters and the recharge of the Floridan Aquifer which makes the Potential Recharge Area the optimum area for potential recharge of the Volusia County water supply system.

While certain types of development will have no adverse effects on the beneficial aspects of the Potential Recharge Area, or may actually enhance positive qualities, uncontrolled development will give rise to substantial destruction of the potential abilities of this area to sustain adequate, quality water supplies for Volusia County for the reasons more specifically set forth heretofore.

Purpose

"The purpose of this ordinance is to protect the water resources of Volusia County, prevent the development or use of land in the Potential Recharge Area in a manner tending adversely to affect the quantity of water within Volusia County or tending to destroy or have a substantially adverse effect on the environment of the county by virtue of pollution of the air, land, or water by foreign substances, including noxious liquids, gases, or solid wastes, or pollution by virtue of the creation of potentially harmful conditions, including the creation of unnecessary injurious heat, noise, or odor, and preserve the aesthetic qualities of the county in order to enhance the overall development of the county and the proper planned promotion of agriculture, tourism, and appropriate residential, commercial, and industrial development within Volusia County. It is intended by this council that this ordinance be interpreted liberally in view of the paramount public interest involved in the preservation of the Potential Recharge Area."

Short Title

This ordinance shall be known and may be cited as "The Potential Water Recharge Area Preservation Ordinance of Volusia County."

Definitions

action. Any application for a permit under this ordinance or any development or use encompassed within the jurisdiction of this ordinance.

development permit or *permit.* Includes any building permit, zoning permit, plat approval or rezoning, certification, variance, or other action having the effect of permitting development as hereinafter defined.

development. The carrying out of any building, agricultural, or mining operation or the making of any material change in the use or appearance of any structure or land, and the dividing of land into two or more parcels. The following activities or uses shall be taken, for the purposes of this ordinance, to involve development as defined herein.

(1) Any construction, reconstruction, alteration of the size, or material change in the external appearance of a structure on land.

(2) Any change in the intensity of use of land, such as an increase in the number of dwelling units in a structure or on land, or a material increase in the number of businesses, manufacturing establishments, offices, and dwelling units, including mobile homes, campers, and recreational vehicles, in a structure or on land.

(3) Any agricultural use of land including, but not limited to, the use of land in horticulture, floriculture, viticulture, forestry, dairy, livestock, poultry, bee keeping, pisciculture, and all forms of farm products and farm production.

(4) The commencement of drilling, except to obtain soil samples, or the commencement of mining, or excavation on a parcel of land.

(5) Demolition of a structure.

(6) Clearing of land as an adjunct of construction for agricultural, private, residential, commercial, or industrial use.

(7) Deposit of refuse, solid or liquid waste, or fill on a parcel of land.

(8) Construction, excavation, or fill operations relating to the creation of any road or street or any drainage canal.

parcel of land. Any quantity of land capable of being described with such definiteness that its location and boundaries may be established, which is designated by its owner or developer as land to be used or developed as a unit, or which has been used or developed as a unit.

person. An individual, corporation, governmental agency, business trust, estate, trust, partnership, association, two or more persons having a joint or common interest, or any other legal entity.

potential recharge area. That area of real property located in Volusia County as described and incorporated herein as part of this ordinance.

planning department. The Planning Department of Volusia County or the planning staff that has been designated by the county council as a planning staff for Volusia County.

Prohibition

Except as otherwise provided [below], in the Potential Recharge Area as herein defined . . . , no person may erect any permanent structure, including the placement of mobile homes, nor shall any person engage in the development of land, whether for residential purposes or otherwise, for any commercial, agricultural, or industrial pursuit, whether temporary or permanent, unless such person first obtain a use permit from the county council in the manner set forth in this ordinance.

Within six months from the effective date of this ordinance all persons engaged in any activities set forth above shall submit to the Development Coordination Department a statement indicating the nature and extent of activities being carried on. Such statement shall include the information required [below].

Application for Permit

All applications for a use permit shall be made to the Department of Development Coordination in the manner and form prescribed by said department. The application shall include a description of the proposed action, use, or development, including information and technical data adequate to allow for a careful assessment of the application in light of the guidelines set forth [below]. Where relevant, maps and other information shall be provided upon request from any department or agency examining the application.

Upon receipt of an application for a use permit pursuant to this ordinance, the Department of Development Coordination shall submit copies of such application to the Department of Environmental Control, the Planning Department, the Department of Public Works, and such other county, state, or federal departments or agencies it deems should be advised of the proposed action for review and comment.

Within 90 days after receipt of an application, the aforementioned departments shall submit written reports to the Director of the Department of Development Coordination, which report shall include recommendations from each department. In preparing such reports each department shall take into account the guidelines set forth below. All departments and agencies may cooperate and submit a single, comprehensive report.

Upon receipt of the reports, the Director of Development Coordination shall forthwith place the matter on the agenda for county council consideration, and shall notify the applicant when the matter will be considered by the council.

Whenever the county council determines that an application is one which would generate substantial public interest or could cause a significant change in the Potential Recharge Area, the council may call for a public hearing with at least 15 days public notice, published once in a newspaper of general circulation in the county.

Each department may request the assistance of any other department or agency of any local government, the government of Volusia County, or any state or federal department or agency. The department may also require such additional information from the applicant as it deems is reasonably necessary in order to furnish the county council with a complete report.

After July 1973, whenever an application is submitted which, but for the fact that it affects only Volusia County, would be a development of regional impact pursuant to the criteria established in accordance with Chapter 380, Florida Statutes, such application shall be accompanied by an environmental impact statement which shall be prepared in accordance with the criteria established by the National Council on Environmental Quality pursuant to the National Environmental Policy Act of 1969 (Public Law 91-190, January 1, 1970). All such applications shall be subject to a public hearing. . . .

Guidelines

In determining whether a permit should be granted, the county council shall apply the following guidelines:

(A) The probable impact of the proposed action on the environment, including impact on ecological systems such as wildlife, fish, and marine life. Both primary and secondary significant consequences for the environment should be examined. For example, the implications, if any, of the action on population distribution or concentration should be estimated and an assessment made of the effect of any possible changes in population pattern upon the Potential Recharge Area, including land use, water, and public services in the Potential Recharge Area.

(B) Any probable adverse environmental effects which cannot be avoided (such as water or air pollution, undesirable land-use patterns, increased water runoff, damage to life systems, urban congestion, threat to health, or other consequences adverse to the goals and the purposes of this ordinance).

(C) Alternatives to the proposed action, use, or development. In considering alternatives, the council shall attempt to assure that the purposes of this act are complied with and that the action or use proposed is best fitted to meet these purposes. A rigorous exploration and objective evaluation of alternative action that might avoid some or all of the adverse environmental effects is essential. Efficient analysis of such alternatives and their costs and impact on the environment of the Potential Recharge Area should be examined in order not to foreclose prematurely options which might have fewer detrimental effects.

(D) The relationship between local, short-term uses of man's environment, and the maintenance and enhancement of the long-term productivity of the Potential Recharge Area. This will require the county council to assess the actions

for cumulative and long-term effects from the prospective that each generation is the trustee of the environment for succeeding generations.

(E) Any irreversible and irretrievable commitments of resources, particularly water, which would be involved in the proposed action, use, or development, should it be implemented. The county council should identify the extent to which the action curtails the range of beneficial uses of the environment of the Potential Recharge Area. Compatibility of the proposed use with existing uses in the Potential Recharge Area, and compatibility with the zoning and land-use planning for all lands which would be affected by the proposed use.

(F) The effects and compatibility of the use or development or action with regard to the matters set forth in the findings of fact [above].

Grant of Permit, Alteration of Use

No permit shall be issued unless and until the county council has examined the application in light of the foregoing criteria and determined that the use, activity, or development as proposed would not adversely affect the environment of the Potential Water Recharge Area, would be in harmony with the purposes of this ordinance, and would not have any adverse effect with regards to the findings in fact set forth [above].

In granting a use permit, the county council may prescribe appropriate conditions and safeguards in conformity with this ordinance and, where applicable, may prescribe a reasonable time within which the action permitted is required to be begun, completed, or terminated. Violation of such conditions, safeguards, or time limits, when made a part of the terms under which the use permit is granted, shall be deemed a violation of this ordinance.

The permit issued shall specify the use permitted and any conditions set forth by the council to assure the intent and standards of this ordinance are complied with. Where applicable, plans and specifications shall be attached to and made a part of the use permit. The permit shall state that the granting of the use permit under this ordinance does not waive the requirements of any other county, state, or federal law, ordinance, or regulation.

Whenever any person has obtained a permit and thereafter desires to alter the use in any way from the use as proposed and authorized, such person shall make application for a new permit.

Issuance of Development Permits

No development permit, as defined herein, shall be issued within the Potential Recharge Area until such time as the applicant therefor has secured a use permit pursuant to the provisions of this ordinance.

Exceptions

The following uses are excepted from the provisions of this ordinance:

Any parcel of land may be used for private recreational purposes, such as hiking, camping, hunting, or fishing, without a use permit, when no change in the zoning classification of the property, conditional use, or special exception is required for such use.

Any agricultural use as defined in this ordinance shall be permitted without a use permit, when no roads or drainage canals or ditches are constructed subsequent to the effective data of this ordinance which would have the effect of permanently impounding, obstructing or diverting surface or subsurface waters. Nothing in this section shall be construed as prohibiting the construction of irrigation ditches, temporary canals, plowing of land, and similar uses which are ordinarily a normal part of agricultural operations unless undertaken for the sole or predominant purpose of impounding or obstructing surface waters, nor shall this section by construed as prohibiting the construction of temporary roads and drainage canals incidental thereto, which roads are constructed solely for inspecting, harvesting, or planting forestry or agricultural crops, when such roads are ordinary and incidental to a forestry or agricultural operation.

Any use which was in existence as of the effective date of this ordinance shall be permitted to continue indefinitely without a use permit, provided that no new roads or drainage canals or ditches shall be constructed as accessory to such use except in accordance with the provisions of this ordinance, and provided further that the requirements [above] of this ordinance are complied with. Nothing in this section shall be construed to permit the construction of a residential structure or the placement of a mobile home on an individual parcel of land upon the grounds that said parcel was purchased for the purpose of development, as defined by this ordinance, prior to the effective date of this ordinance, without meeting the requirements for a permit as provided in this ordinance. Nothing herein, however, shall be construed to deny or prohibit the addition to an existing structure or its appurtenances or any accessory uses, provided that a notice of intent to expand such structure, appurtenance, or accessory use is filed with the Development Coordination Department.

No part of this ordinance shall be construed to prevent the doing of any act necessary to prevent the harm to or destruction of real or personal property as a result of a present emergency such as fire, infestation by insects or other pests, or flood hazards resulting from heavy rains or hurricanes, when the property is in imminent peril and the necessity of obtaining a permit is impractical and would cause undue hardship in the protection of the property.

Applicability to Incorporated Municipalities

This ordinance shall be applicable throughout the area described [herein], including any incorporated towns or municipalities whether now in existence or hereafter created.

Penalties and Enforcement

Any person, whether an owner, lessee, principal, agent, employee, or otherwise, who violates this ordinance or causes or participates in a violation shall be guilty of a misdemeanor and shall be punished by a fine not to exceed $500 or by imprisonment in the county jail for a period not to exceed 60 days, or by both such fine and imprisonment. Each day upon which a person is in violation of this ordinance shall constitute a separate offense hereunder.

In addition to any other remedies, whether civil or criminal, the violation of this ordinance may be restrained by injunction, including a mandatory injunction, and otherwise abated in any manner provided by law. It is hereby declared by this county council that the violation of this ordinance is a hazard to public health, safety, and the general welfare, and is therefore a public nuisance.

Severance Provisions

If any part of this ordinance is held to be unconstitutional it shall be construed to have been the legislative intent to pass this ordinance without such unconstitutional portion, and the remainder of this ordinance shall be deemed and held to be valid as if such portion had not been included herein. If this ordinance, or any provision hereof, is held to be inapplicable to any person, group of persons, property, kind of property, circumstances, or set of circumstances, such holding shall not affect the applicability hereof to any other persons, property, or circumstances.

APPENDIX D

Storm Runoff Control Ordinance of Naperville, Illinois

Be it ordained by the City Council of the City of Naperville, Du Page and Will counties, Illinois, in the exercise of its home rule powers, . . . that Chapter 12, "Plumbing, Sewers and Water," of the Municipal Code of Naperville of 1960 as amended, be, and the same is hereby amended . . . to read as follows:

Definitions

The following definitions shall be applicable:

storm water runoff. Water that results from precipitation which is not absorbed by the soil or plant material.

natural drainage. Channels formed by the existing surface topography of the earth prior to changes made by unnatural causes.

excess storm water. That portion of storm water runoff which exceeds the transportation capacity of storm sewers or natural drainage channels serving a specific watershed.

by-pass channel. A channel formed in the topography of the earth's surface to carry storm water runoff through a specific area.

storm water runoff release rate. The rate at which storm water runoff is released from dominant to servient land.

storm water storage area. Areas designated to store excess storm water.

tributary watershed. All of the area that contributes storm water runoff to a given point.

recognized agency. An agency or governmental unit that has statistically and consistently examined local and climatic and geologic conditions and maintained records as they apply to storm water runoff, e.g., Metropolitan Sanitary District of Greater Chicago, U.S. Weather Bureau, University of Illinois Engineering Experiment Station, Illinois State Water Survey, etc.

dry bottom storm water storage area. A facility that is designed to be normally dry and contains water only when excess storm water runoff occurs.

wet bottom storm water storage area. A facility

that is designed to be maintained as free water surface or pond.

control structure. A structure designed to control the volume of storm water runoff that passes through it during a specific length of time.

positive gravity outlet. A term used to describe the drainage of an area by means of natural gravity so that it lowers the free water surface to a point below the existing grade or invert of storm drains within the area.

groundwater recharge. Replenishment of existing natural underground water supplies.

safe storm drainage capacity. A term used to describe the quantity of storm water runoff that can be transported by a channel or conduit without having the water surface rise above the level of the earth's surface over the conduit, or adjacent to the waterway.

Regulations

It is not the intent of this article to take areas out of use for the sole purpose of storing excess storm water. Nor is it the purpose of this article to restrict land use or increase development costs. The basic purpose of this article is to eliminate the storage or transportation of excess storm water in or through habitable structures. The use of "natural" paths of storm water runoff to form the "bypass" channel and the restriction of this channel to form storage areas is encouraged. Since political and ownership boundaries often make the use of "natural" drainage patterns difficult, the earthmoving that is accomplished to create the maximum land usage should also be planned to provide a "bypass" channel for storm water that will not create a diversion of storm water drainage or radically change the watershed boundaries. The drainage scheme presented by those who wish to develop property in the City of Naperville should be planned to accomplish all of the following storm water controls without major loss of land use.

The controlled release and storage of excess storm water runoff shall be required in combination for all commercial and industrial developments and for residential developments that contain an area in excess of two-and-one-half acres.

The controlled release rate of storm water runoff from all development described [above] shall not exceed the existing "safe" storm drainage capacity of the natural downstream outlet channel or storm sewer system. The release rate shall be an average value computed as a direct ratio of the tributary watershed area. This value shall not exceed an average runoff rate of 0.15 inches per hour which is compatible with the "safe" capacity of the West Branch of the Du Page River and the Des Plaines River. The rate at which storm water runoff is delivered to a designated storm water storage area shall be unrestricted.

A "natural" or surface channel system shall be designed with adequate capacity to convey through the development the storm water runoff from all tributary upstream areas. This "bypass" channel shall be designed to carry the peak rate of runoff from a 100-year storm, assuming all storm sewers are blocked and that the upstream areas are fully developed and have been saturated with antecedent rainfall. No habitable structures shall be constructed within this floodway; however, streets and parking or playground areas and utility easements shall be considered compatible uses.

Design of this floodway system shall also take into consideration control of storm water velocity to prevent erosion or other damage to the facility which will restrict its primary use. Depths of flow shall be kept to a minimum and retention of channel configurations shall be totally under municipal control. In the event that the area within this "bypass" channel is reshaped or restricted for use as a floodway, the municipality will cause to have any restrictions removed at the expense of the party or parties causing said restriction.

Should the development contain an existing "natural" waterway this land configuration shall be preserved as part of the "bypass" channel system. Construction of a "low-flow" system of storm sewers to carry the minor storm runoff and re-shaping of the channel with a maximum of six horizontal to one vertical side slopes and bottom of a width adequate to facilitate maintenance and carry the flood runoff without eroding velocities shall be included in the plans for land development.

The required volume for storm water detention shall be calculated on the basis of the runoff from a 100-year frequency rainfall of any duration as published by a recognized agency. This volume of storage shall be provided for the fully developed watershed that is tributary to the area designated for detention purposes. The storm water release rate shall be considered when calculating the storm water storage capacity and the control structure designed to maintain a relatively uniform flow rate regardless of the depth of storm water in the storage area.

Dry bottom storm water storage areas shall be designed to serve a secondary purpose for recreation, open space, or other types of uses that will not be adversely affected by occasional or intermittent flooding. A method of carrying the low flow through these areas shall be provided, in addition to a system of drains, and both shall be provided with a positive gravity outlet to a natural channel or storm sewer with adequate capacity. . . .

The combination of storage of the water from a 100-year storm and the design release rate shall not result in a storage duration in excess of 72 hours. Maximum depth of planned storm water storage shall not exceed four feet, unless the existing natural ground contours and other conditions lend to greater storage depth, which shall be approved by the municipality. Minimum grades for turf areas shall be two percent and maximum slopes shall be 10 percent (10 units horizontally to one vertically). Storage area side slopes shall be kept as close to the natural land contours as prac-

tical, and a 10 percent slope or less shall be used wherever possible. If slopes greater than 10 percent are necessary to meet storage requirements or area restrictions, approval shall be obtained from the municipality and suitable erosion control provided, in addition to the protection required to insure public health, safety, and welfare.

Outlet control structures shall be designed as simply as possible and shall require little or no attention for proper operation. Each storm water storage area shall be provided with a method of emergency overflow in the event that a storm in excess of the 100-year frequency storm occurs. This emergency overflow facility shall be designed to function without attention and shall become part of the "natural" or surface channel system. . . . Hydraulic calculations shall be submitted to substantiate all design features.

Both outlet control structures and emergency overflow facilities shall be designed and constructed to protect fully the public health, safety, and welfare. Storm water runoff velocities shall be kept at a minimum and turbulent condition at an outfall control structure will not be permitted without complete protection for the public safety. The use of restrictive fences shall be kept to a minimum and used only as a last resort when no other method is feasible.

Wet bottom storm water storage areas shall be designed with all of the items required for dry bottom storm water storage areas, except that a low-flow conduit and a system of drains with a positive gravity outlet shall be eliminated. However, the following additional conditions shall be complied with:

(1) Water surface area shall not exceed one-tenth of the tributary drainage area.

(2) Shoreline protection shall be provided to prevent erosion from wave action.

(3) Minimum normal water depth shall be four feet. If fish are to be used to keep the pond clean, a minimum of one-fourth of the pond area shall be a minimum of 10-feet deep.

(4) Facilities shall be available, if possible, to

allow the pond level to be lowered by gravity flow for cleaning purposes and shoreline maintenance.

(5) Control structures for storm water release shall be designed to operate at full capacity with only a minor increase in the water surface level. Hydraulic calculations shall be submitted to substantiate all design features.

(6) Aeration facilities to prevent pond stagnation shall be provided. Design calculations to substantiate the effectiveness of these aeration facilities shall be submitted with final engineering plans. Agreements for the perpetual operation and maintenance of aeration facilities shall be prepared to the satisfaction of the municipality.

(7) In the event that the water surface of the pond is to be raised for purposes of storing water for irrigation or in anticipation of the evapotranspiration demands of dry weather, the volume remaining for storage of excess storm water runoff shall still be sufficient to contain the 100-year storm runoff.

Paved surfaces that are to serve as storm water storage areas shall have minimum grades of one percent and shall be restricted to storage depths of one foot maximum. Rooftop storage shall be designed with permanent-type control inlets and parapet walls to contain runoff on the rooftop. Emergency overflow areas shall be provided to insure that the weight of water stored will not exceed the structural capacity of the roof. Release rates and storage volume requirements for paved storage areas remain the same as outlined [above]. If a portion of an area within a storm water storage area is to be paved for parking or recreational purposes, the paved surface shall be placed at the highest possible elevation within the storage area. Maximum parking lot grades shall not exceed normal design parameters of three to five percent.

Where developments form only a portion of a watershed or contain portions of several watersheds, the requirement for providing storage shall be based upon the proportion of the area being developed, as compared to the total watershed tributary to the storage area. Compensating storage will be acceptable whenever it is justified and feasible. As a watershed is developed with a series of storm water storage facilities, due consideration will be given for calculation of the allowable release rate and capacity of the "natural" or surface channel system. . . .

Plans, specifications, and all calculations for storm water runoff control as required hereunder shall be submitted to the City of Naperville for review and approval prior to the approval of a final plat, in the case of subdivisions and planned unit developments, or issuance of a building permit, in the case of commercial or industrial construction.

Where development of a property presents the threat of flooding or damage by flash runoff to downstream residents, the facilities for storm water runoff control shall be constructed prior to any earthmoving or drainage construction on the project site.

The construction of the storm water control system shall be accomplished as part of the cost of land development. If the amount of storage capacity can be increased to provide benefit to the municipality, negotiations for public participation in the cost of development may be feasible.

The ability to retain and maximize the groundwater recharge capacity of the area being developed is encouraged. Design of the storm water runoff control system shall give consideration to providing ground water recharge to compensate for the reduction in the percolation that occurs when the ground surface is paved and roofed over. Specific design calculations and details shall be provided with the final plans and specifications presented for municipal review. The use of natural gravel deposits for the lower portions of storm runoff storage areas, the flattening of drainage slopes, and the retention of existing topography are examples of possible recharge methods.

During the construction phases of land development, facilities shall be provided to prevent the erosion and washing away of the earth. Silting of downstream areas can be prevented through the strategic use of stilling basins, sodding of runoff channels, and by limiting the period of time during which the earth is stripped of vegetation.

Final engineering plans shall show complete details for all of the items covered in this ordinance and shall be submitted for review and approved prior to the start of construction.

Penalties

Any person, firm, or corporation violating any provision of this article shall be fined not more than $500 for each offense, and a separate offense shall be deemed committed on each day a violation continues.

APPENDIX E

Environmental Impact Report: Format and Checklist

As defined by State guidelines, EIRs are "informational documents" presenting a statement of the facts. EIRs should be based, as much as possible, on available scientific and technical data. Sources must be referenced and material substantiating claims made within the body of the EIR must be included in the appendixes.

Well-written EIRs are balanced (environmental concerns viewed in conjunction with social and economic considerations), objective and thorough. Avoid statements of opinion (on your part) and value judgments. You are the reporter presenting the facts. The facts, if complete, will supply sufficient information for the reader to make his/her own judgment about the proposed project.

If, in your research and consultations, a source expresses an opinion you believe should be included in the EIR, use a direct quote. You may also paraphrase by writing, "In the opinion of . . ." or "So-and-so, a registered whatever, believes. . . ." The conversation, letter, document, or whatever you are quoting, must be referenced.

For clarity and completeness, tests or analyses often need to be made. If you are not qualified to do this kind of work, the appropriate person should be hired to do the necessary testing. The applicant must be informed of the need for the work and the extra cost.

The California Environmental Quality Act (CEQA) delineates four specific sections for the draft EIR: description of the project; description of the environmental setting; discussion of environmental impacts; and identification of organizations and persons consulted. These state-mandated requirements have been included in this checklist.

The EIR section describing impacts must discuss seven concerns specified by CEQA:

1. the environmental impacts (direct and indirect) of the proposed action;

2. *any* adverse environmental effects *that cannot be avoided* if the proposal is implemented;

3. mitigation measures, including energy conservation measures, proposed to minimize the impact;

4. alternatives to the proposed action;

5. the relationship between local short-term uses of man's environment and the maintenance and enhancement of long-term productivity;

6. any irreversible environmental changes that would be involved in the proposed action should it be implemented;

7. the growth-inducing impact of the proposed action.

Any environmental conditions that are not affected or impacted by the proposed project do not need to be discussed. However, the environmental impact section should mention which conditions are not impacted and, in concise terms, explain why they were not discussed. This discussion should follow the order of the EIR format.

A final EIR contains: the draft EIR or a revision of that draft; comments and recommendations received that pertain to the draft EIR; a list of persons, organizations and public agencies that commented on the draft EIR, and the Environmental Planning Division's response to "significant environmental points raised in the review and consultation process."

Please remember that this checklist is *not* all-inclusive and does not cover all potential environmental conditions, impacts and mitigation measures. The mitigation measures are only recommendations and are not applicable in all cases.

EIRs should be typed with single-spacing, reproduced on both sides of the page where possible, and all pages should be numbered.

EIR Format

  I.   Summary
 II.   Table of contents
III.   Project description
 IV.   Environmental considerations
       A. Geologic and soil
           1. Existing geologic conditions
           2. Geologic impacts
           3. Unavoidable adverse geologic impacts that will occur if project is implemented
           4. Mitigation of geologic impacts
           5. Irreversible geologic impacts

Follow this same format for the remaining categories (in order):

       B. Hydrologic
       C. Atmospheric
       D. Sonic
       E. Biotic
       F. Energy
       G. Cultural-aesthetic-scientific
       H. Socioeconomic

Describe the following according to specifications given in Article V of the Guidelines:

  V.   Growth-inducing impact of the proposed action
 VI.   The relationship between local short-term uses of man's environment and the maintenance and enhancement of long-term productivity
VII.   Alternatives to the proposed action

VIII. Persons and organizations consulted in EIR preparation
 IX.   Identification and qualifications of persons, firm or agency preparing the EIR
  X.   Bibliographical references
 XI.   Appendixes

Checklist

The following checklist is designed to specify most environmental conditions and impacts that might be covered in an EIR:

  I.   Project and site description
       A. Maps
           1. 7½ minute USGS map with project area location, boundaries and major physical features
           2. A regional (county) map showing location of project
       B. Project location (address and nearest intersection/landmark)
       C. Assessor's parcel number
       D. Adjacent structures or physical features
       E. Applicant (name, address)
       F. Type of project
       G. Objectives of project
       H. Current zoning of project site, and any zoning changes necessitated by project
       I. Current land use of project site and any land use changes necessitated by project
       J. General Plan and other pertinent plan designations of project area; adopted policies relating to project area — e.g., mountain development standards
       K. Any projects implemented or proposed that depend on, or have a direct relation to this project
       L. Physical description of project, for example:
           1. Size of structures to be constructed
           2. Number and height of buildings
           3. Type of buildings and materials used

4. Number, size, and price of dwelling units, if any
5. Number of parking spaces, if any
6. Construction timetable, if available
7. Water, sewage, access, and solid waste disposal considerations
8. Resources to be extracted: trees, minerals, etc.

II. Environmental considerations
   A. Geologic and soils
      1. Existing geologic conditions
         a) soil type, rating, and grouping
         b) agricultural and/or timber potential of soil
         c) site elevation
         d) slope
         e) present, and/or potential soil erosion
         f) seismic hazards on or near site:
            fault zone, ground shaking
            tsunamis, seiches
            landslides
            liquefaction and associated ground failures
         g) soil stability
            subsidence
            shrink-swell potential
            bearing capacity
            landsliding
            surface creep
      2. Geologic and soil impacts
         a) soil alteration
            erosion
         b) seismic safety
         c) major land displacement, grading
         d) inducement of landslides, or effects upon project from landsliding
         e) coastal cliff erosion
      3. Mitigation of soil and geologic impacts
         a) erosion control procedures
            alternate land use for project site
            installation of energy dissipat-

ers and drop inlets
terracing
only grade areas of immediate construction
stockpiling and replacement of surface soils
immediate revegetation and mulching of disturbed surfaces with native species (refer to Environmental Health's pamphlet about grading and building sites)
retain and protect existing vegetation
vegetate strips adjacent to waterways
diversion of water from erodible areas to well-vegetated areas
setback from riparian corridor
revegetate/landscape with native plant species

b) seismic safety
alternate land use for project site
fault zone:
   appropriate setbacks from known traces (50′ single-family residences, 100′ critical facilities, high occupancy structure, etc.)
landslides:
   locate septic tank and leach field as far from home as possible neither directly up- nor downslope from structures
   minimize watering on landslide material, revegetate with native species
divert drainage
construct retaining structure
excavate landslide deposits
install drainage and dewatering devices, e.g., hydrogers

liquefaction:

  construction techniques such as mat or floating foundation, drilled pier or caisson-pile foundations (set in firm bedrock) may reduce liquefaction hazard

general:

  if applicable —

    insure careful design and construction to resist severe ground shaking

    avoid construction on fill unless engineered to withstand severe seismic shaking

    site-specific investigation by soils engineer, geologist or engineering geologist

    all large objects (refrigerators, water heaters, large furniture, etc.) should be fastened to walls or floor to prevent movement or overturning

    plan for procedures for both during and after an earthquake

4. Unavoidable adverse impacts that will occur if project is implemented
5. Irreversible impacts

B. Hydrologic
1. Existing hydrologic conditions
  a) watershed in which project located
  b) surface and groundwaters
    water sources on site
    water sources receiving water from site
    water sources supplying water to site
    water quality
    surface water capacity (channel and floodplain)
    evidence of ponding and/or gullying (drainage) on site
  c) water table, aquifers and springs
  d) soils of low infiltration capacity
  e) percolation rates
  f) tsunami or flood hazards
  g) source of water for project
  h) estimated average daily consumption of water for project
  i) does construction of project involve dredging, tunneling or trenching?
  j) cubic foot per second run-off at given storm intensity ($Q = CIA$)
2. Hydrologic impacts
  a) water quality affected during construction and operation of project by:
    inorganic and organic substances
    stream bank alteration
    siltation potential
  b) water sources affected by project (on site, downstream)
    areas sensitive to change in water quality or quantity
    water-dependent recreation areas
    public water supplies
    effects on watershed
  c) effects from flood or tsunami hazard
  d) effects upon groundwater aquifers and aquifer recharge areas:
    depletion from consumption; percolation of treated waste; and limited recharge due to impervious cover
  e) effects upon surface runoff, drainage, increased flow
  f) depletion of water sources

g) local water supplies affected by water diversion

h) irrigation

i) channelization

j) salt water intrusion

k) change in water volume configuration (Q = CIA)

3. Mitigation of hydrologic impacts

a) alternate land use for project site

b) evaluation and control of waste product disposal — ensure compliance with existing local, state and federal discharge requirements

c) leach lines below impervious layers

d) minimize water consumption and/or waste discharge by limiting construction

e) limit impervious surfacing

f) regular sweeping and clearing of impervious surface

g) minimize use of fertilizer and pesticides:
use of native species
nonchemical and biodegradable methods
slow-release fertilizers
test soil to determine allowable amounts of fertilizers
consult with agricultural advisor for proper method of applying fertilizers and pesticides

h) improved irrigation practices

i) check dams and debris dams to prevent siltation of water sources and allow for recharge of underground water sources

j) cleanable traps to prevent grease and/or sediment entry into water sources

k) regulation of water, diversions, dredging, filling, spoil, disposal, etc., to minimize impacts on fish and wildlife

l) power generation facilities:
treatment and coding of thermal discharge
effective fish screens on intake pipes
recycle cooling waters
proper location and design of discharge outfalls

m) structural measures to reduce flood frequency, e.g., dams, levees, channel enlargement, bypasses

n) prohibit septic systems in floodplain

o) prohibit storage of chemical materials in floodplain, or allow when in flood-proof structures

p) elevate structures above expected flood level

q) native species to minimize water loss

r) special appliances to limit private water consumption

s) control of drainage from spoil areas

4. Unavoidable adverse impacts that will occur if project is implemented

5. Irreversible impacts

C. Atmospheric

1. Existing atmospheric conditions

a) airshed or basin in which project located (particularly downwind)

b) local climate (micro, macro, daily, seasonal)
circulation patterns
prevailing winds
storm exposure
precipitation
fog patterns
temperature
odor
pollution assimilation capacity

c) quality of air that will receive emissions from project (gases, chemicals, particulates, clarity)

d) downwind and/or airshed conditions that may be impacted

    public facilities

    recreation areas

    vegetation and wildlife areas particularly sensitive to pollution

    residential areas

2. Atmospheric impacts

  a) air quality degradation due to traffic generation, industry

  b) increased local temperature

  c) vegetation or crop damage

  d) noxious or displeasing odors and/or fumes

  e) dust and/or particulate matter

  f) block or decrease wind circulation

  g) cumulative effects

  h) air quality affected by agricultural operations, burning, or chemical sprays

3. Mitigation of atmospheric impacts

  a) air emissions of project meet federal, state and local air pollution control district atmospheric quality standards

  b) implementation of best available pollution prevention and control technology:

    odor control devices, such as air scrubbers, enclosed digestors

    dust particulates control through gas control regulation, e.g., electrostatic precipitators

  c) eliminate open burning of solid waste

  d) minimize project's parking spaces

e) decrease number of parking spaces and encourage public transit

f) provide facilities for alternate modes of transportation, such as pedestrian walkways and bicycle paths

4. Unavoidable adverse impacts that will occur if project is implemented

5. Irreversible impacts

D. Sonic

1. Existing sonic conditions

  a) current sources of noise emitted from project site

  b) off-site sources of noise detectable on site

  c) proximity of construction to residences or business

  d) describe any potential noise levels from project in terms of decibels, time of noise, duration and types of noise and vibration

2. Sonic impacts

  a) new sources of noise caused by construction and operation of project

E. Biotic

1. Existing biotic conditions

  a) inventory and location of dominant plant species

  b) inventory of resident and migratory animal species

  c) vegetation of high fire potential on or near site

  d) on-site areas of low revegetation potential

  e) generally undisturbed or unique plant and animal communities on or near site

  f) rare and endangered plant and animal species on or near site

  g) animal resting, feeding, nesting, breeding area

h) riparian habitat and other wetlands on or near project site

i) biotic community of scientific
and/or educational value

j) wildlife migration corridors

k) economic and pest species

l) agriculture and timber productivity

2. Biologic impacts

a) clearing involving use of herbicides, defoliants, blasting,
cutting, bulldozing or burning

b) limited of revegetation potential causing erosion and/or siltation and aesthetic impacts

c) introduction of exotic flora and
fauna

d) compaction of soil, damage to
native root system

e) alteration of vegetative cover

f) effect on habitat of rare and
endangered species

g) effect on highly productive
habitats supporting sport or
commercially valued species

h) modification and interruption
of habitat for nesting, feeding,
breeding, and migration

i) predator control

j) effect upon stability of ecosystems

k) disturbance or removal of undisturbed, unique, or rare and
endangered plant or animal
species

l) alteration or destruction of
riparian habitats

m) increased forest fire potential,
or location in an area of known
hazard

n) removal of trees

o) impacts of clearing using herbicides, defoliants, blasting,
cutting, bulldozing or burning

p) forestry impacts — list

3. Mitigation of biotic impacts

a) site planning to avoid locations of rare or endangered
plants

b) careful evaluation and control
of uses made of less developed
portion of property

c) control of access to and activities within area supporting
biotic communities of concern

d) site planning to avoid or
minimize disturbance of riparian vegetation, wildlife,
key habitat areas, etc.

e) potential acquisition and protection of adjacent resource
areas by industrial concerns

f) forestry:
selective cutting only (recommended % cut)
clear posting of roads being
used by logging trucks, list-
ing of time schedule of normal use and marking of all
logging truck highway
crossings
replant skid trails, temporary
roads (list species); reforestation

g) minimize use of pesticides

h) careful placement of tailing,
spoil and overburden

i) avoid riprapping

j) site planning of channel alterations, spoil disposal areas, etc.,
to minimize impact on key
habitat areas

k) undercrossings, speed signs
and planting of vegetation not
attractive to wildlife to minimize road kills

l) landscape with native plant
species and/or special fire re-
sistent ones

m) minimize impacts on migration corridors

n) avoidance of construction near

waterways and reparian zones

4. Unavoidable adverse impacts that will occur if project is implemented

5. Irreversible impacts

F. Energy

1. Existing energy conditions on site

   a) existing resources

   b) resources that would be diminished by project implementation, e.g., fossil fuels, mineral ores

   c) finite resources

   water

   oil

   coal

   tar sand

   d) alternative sources (geothermal, solar)

2. Energy impacts

   a) depletion of existing and finite resources

   b) inefficient, wasteful, or unnecessary consumption of resources

   c) secondary effects of energy consumption (air, water quality degradation)

   d) impacts from use of atomic fuels background reduction, acceptable levels; impacts from disposal

3. Mitigation of energy impacts

   a) insulation and other protection from heat loss of heat gain to conserve fuel used to heat or cool buildings and mobile homes

   b) use of resource-conserving forms of energy such as solar energy for water and space heating, wind for operating pumps, and falling water for generating electricity

   c) energy-efficient building design including such features as

orientation of structures to summer and winter sunlight to absorb winter solar heat and reflect or avoid summer solar heat

d) measures to reduce energy consumption in transportation such as:

   providing access to alternative means of transportation for people, such as bus lines, mass transit, bicycle lanes, and pedestrian facilities

   use of small cars rather than large cars where possible

   use of alternative means of shipping that would allow for energy savings

e) efficient lighting practices including use of indirect natural light, use of efficient lighting fixtures, establishment of reasonable lighting criteria to prevent overillumination, and minimum use of architectural or display lighting

f) energy conserving construction practices

g) utility rate structures that discourage unnecessary energy consumption

h) use of human or animal power where such use is feasible

i) waste heat recovery

j) recycling and use of recycled materials

k) careful site planning for drilling sites, access roads, etc.

l) maintain resource areas in open space or passive recreation area

m) minimize consumptive water use in industrial processes

n) implementation of best available pollution and accident prevention technology

o) only security lighting of project during hours when closed
4. Unavoidable adverse impacts that will occur if project is implemented
5. Irreversible impacts

G. Cultural, scientific, and aesthetic
1. Existing cultural, scientific, and aesthetic conditions
    a) unique or scenic landforms:
        hillslopes
        caves
        waterfalls
        stratigraphy
    b) views
    c) culturally significant features or scenic areas that will be seen in juxtaposition to project
    d) public recreation areas, park lands or residential areas that will have a view of the project
    e) commercial attractions directly benefiting from their view, that will have a view of the project
    f) wildlife, domestic stock within viewshed that would be particularly sensitive to project activities
    g) visual qualities of project
    h) historical, archaeological, or architectural resources on, within or near project site (an archaeological reconnaissance by a qualified archaeologist should be completed and placed in an appendix)
2. Cultural, scientific, and aesthetic impacts
    a) alteration or destruction of:
        unique or scenic landforms
        viewshed
        present site visual qualities
        historic or archaeological resource

b) effects on culturally significant features or scenic areas near project site
c) visual effect on public recreation areas, park lands or residential areas within view of project site
d) effects on commercial attractions near project site and benefiting from current view
e) effects of project activities on wildlife or domestic stock on or near project site
f) effect upon archaeological, historical or architectural resource(s)
g) visual effects of grading and landform alteration

H. Socioeconomic
1. Existing socioeconomic conditions
    a) adjacent and vicinity land use
    b) agricultural and open space uses in the vicinity
    c) residences, businesses and public facilities which may need to be relocated
    d) public service facilities: sewer, water, fire
    e) tax base and assessment
    f) employment characteristics and opportunities
    g) transportation modes and networks adjacent and in vicinity of project
    h) access conditions
    i) housing needs and present composition, e.g., vacancy rates
        special consideration given low- and moderate-income housing needs in area
    j) population characteristics
        income, age, education, family size, and ethnic background
        distinct socioeconomic groups

k) community characteristics
l) community and/or neighborhood identity, environment, and/or interaction
m) public health and safety conditions
n) existing environmental degradation

2. Socioeconomic impacts
   a) change in surrounding land use caused by new infrastructure needed for project
   b) impacts on housing
   c) preemption of current and alternative land use for project site
   d) relocation of uses preempted from project site, or denial of future use of site
   e) alteration of natural or present character of area
   f) displacement and relocation impacts on people, residences, businesses, and/or public facilities
   g) permanent loss of open space, agricultural lands, and/or resource areas
   h) public service facility requirements for project and surrounding area
   i) impacts on public service facilities
   j) temporary and permanent effects upon income distribution, employment, and/or tax revenues
   k) changes in tax base and assessment for project site and surrounding area
   l) employment opportunities for various socioeconomic groups
   m) impacts on access or modes of transportation, e.g., proposed bike, riding, and hiking routes

— Trails Council, state, county
   n) impacts on social affiliation and interaction of neighborhood
   o) impacts on public safety and/or health
   p) visual and/or noise impacts on surrounding area
   q) convenience to public services
   r) impacts on privacy and/or amenable climate of surrounding area
   s) degradation of recreational or resource areas due to overuse

3. Mitigation of socioeconomic impacts
   a) alternate site
   b) site planning and screening to reduce visual and noise impacts on surrounding area
   c) landscape all highly visible structures
   d) subdued signs made from natural materials
   e) subdued glare-free and indirect lighting
   f) undergrounding of utilities
   g) cluster development to provide greenbelts of large open space areas
   h) recreational or resource areas: restrictions on intensity, location and type of activities allowed, control of access to resource areas and unauthorized access
      spark arrestors and legally specified sound baffles on off-road vehicles
      fire prevention and control programs
      strict enforcement and supervision of federal, state and local fire protection regulations and codes, includ-

ing: adequate access, water facilities, vegetation clearance around structures, etc.

i) structural engineering of buildings to allow full and equal access for the elderly and the visually and physically handicapped

j) fencing of plants and corporation yards with scenic landscaping

k) build out of flood zone

l) different housing to adequately meet community's housing needs or housing mix to meet variety socioeconomic needs

m) land use in conformance with general plan elements and/or present character of surroundings

n) developer pay all or part of relocation and utility costs that are created by project

o) developer to pay all or part of any new public service facilities that are necessitated by project

p) requirement that developer use local labor to maximum extent possible

q) fence and post warning signs around excavated areas and pits

r) fence land fills to reduce blowing debris, cover daily

4. Unavoidable adverse impacts that will occur if project is implemented

5. Irreversible impacts

Glossary

Accretion: The gradual addition of new land to old by the deposition of sediment carried by water.

Acre-foot: Volume of liquid required to cover 1 acre to a depth of 1 foot.

Active fault: A fault that has moved in the last 10,000 years (within Holocene time).

Adobe soils: Calcareous clay and silt deposits found in the semiarid basins of southwestern North America and in Mexico. The material is used extensively for making sun-dried brick.

Adsorption: Adhesion of molecules of a gas, liquid, or dissolved substance to the surface of a solid body with which it is in contact.

Aerobic bacteria: Any bacteria requiring free oxygen for the breakdown of materials.

Aftershock: An earthquake that follows a larger earthquake and originates at or near the focus of the larger earthquake. Generally, major earthquakes are followed by a large number of aftershocks, which may last many days or even months.

Aggregate: Sand, gravel, or other mineral material that is used with cement to form a mortar or concrete.

Amplitude: The elevation of the crest of a wave above the adjacent troughs.

Andesite: A fine-grained extrusive igneous rock composed of plagioclase feldspars and ferromagnesian silicates.

Anticline: An upfold in which each limb dips away from the axis. Oldest rocks are in the center of the fold.

Aquiclude: An impermeable geologic formation or layer that is capable of slowly absorbing and containing water, but that is not able to transmit it to supply a spring or well.

Aquifer: A water-bearing layer of permeable rock, sand, or gravel. *See also* Confined aquifer.

Arroyo: A flat-floored, steep-walled channel of an intermittent stream typical of semiarid climates.

Artesian well: A well containing water under sufficient pressure to cause it to rise above the top of the aquifer, where it is first encountered. In some cases the water may flow to the surface without pumping.

Artificial recharge: Deliberate recharge of groundwater by artificial means using, for example, injection wells or infiltration ponds.

Bank storage: Groundwater that is stored in the bank of a stream during high runoff or a flood, or in the geologic strata adjacent to a reservoir after filling. Much of this water will be released after the flood or when the reservoir is emptied.

Barrier island: A low, sandy island near the shore and parallel to it.

Base flow: That portion of stream flow coming from groundwater discharge. It may vary from a negligible portion of total flow during periods of high surface runoff to the total flow during drought periods.

Bedding plane: The surface between layers of sedimentary rocks that separates deposits of distinctive character.

Bedload: Sediment and rock debris transported along the bottom of a stream by the moving water.

Benthic organism: Organisms that live at the bottom of the ocean or other body of water.

Bentonitic clay: A clay formed from decomposition of volcanic ash containing a large portion of montmorillonite. The clay commonly has great ability to absorb water and to swell accordingly.

Biochemical oxygen demand: The amount of dissolved oxygen required to meet the metabolic needs of microorganisms and to break down organic matter contained in waste water. Abbreviated BOD.

Biosphere: Zone of life about the earth.

Block glide: Failure or downslope movement of earth materials along a nearly planar surface, such as a bedding plane, joint, or fault.

Breakwater: A structure protecting a harbor, shore area, inlet, or basin from waves.

Brecciated rock: Angular, coarse fragments of rock.

Calcareous: Pertaining to rocks that contain calcium carbonate ($CaCO_3$).

Caldera: A roughly circular steep-sided volcanic basin with a diameter at least three or four times its depth. Commonly at the summit of a volcano. *See also* Crater.

Carbonate rocks: Sedimentary rocks composed chiefly of carbonate minerals, most commonly calcite (calcium carbonate).

Carbonation: A process of chemical weathering by which minerals containing lime, soda, and potash are changed to carbonates by the action of carbonic acid in water or air.

Cementation: The process by which a binding agent is precipitated into the spaces between individual particles of an unconsolidated deposit. The most common cementing agents are calcite, dolomite, and quartz.

Chert: Very fine-grained sedimentary rock composed of microcrystalline silica, similar in appearance to flint.

Chiton: A small mollusk enclosed in a platy shell.

Chromatography: A method of qualitative analysis in which a solution is tested by applying it to a treated porous paper and identifications are made on the basis of the nature and location of resulting color spots.

Clastic: Consisting of fragments of rocks that have been moved individually from their places of origin.

Clay: The term carries with it three implications: (1) a natural material with plastic properties; (2) a composition of particles with very fine grain sizes (.005 millimeters and smaller); and (3) a composition of crystalline fragments of minerals that are essentially hydrous aluminium silicates or occasionally hydrous magnesium silicates. The most common clay minerals belong to the kaolinite, illite, and montmorillonite groups.

Cleavage: The capacity of a mineral to break along weak plane surfaces as determined by the crystal structure.

Compressional wave: The most rapidly moving seismic wave produced by an earthquake, its motion is of a push-pull nature. Also called a primary or P wave.

Confined aquifer: Groundwater that is confined under pressure greater than atmospheric by overlying, relatively impermeable strata. When encountered by a well, water in a confined aquifer will often rise about the level of the water table, and may rise to the surface without pumping. Also known as artesian water.

Conglomerate: A sedimentary rock composed of rounded pebbles or larger particles within a finer matrix.

Connate water: Water trapped in the interstices (pore spaces or voids) of rock at the time of its deposition. Such water is usually of poor quality due to its salt content.

Continental shelf: The gently sloping, relatively shallow ocean area between the shoreline and the steeply descending continental slope that extends to approximately a 135 meter depth surrounding a continent.

Crater: A roughly circular steep-sided volcanic basin with a diameter less than three times its depth. Commonly at the summit of a volcano. *See also* Caldera.

Creep (fault): An imperceptibly slow, more or less continuous movement resulting from long application of stress along an active fault. *See also* Creep (mass movement).

Creep (mass movement): A slow, downslope movement of soil and other surficial earth material that is sufficient to produce permanent deformation but too small to produce shear failure, as in a landslide.

Crystalline rock: A general term for igneous and metamorphic rocks consisting of intergrown minerals as opposed to sedimentary rocks.

Debris avalanche: The rapid downslope movement of soil and rock on steep slopes that may be caused by complete saturation through protracted rains or by seismic shaking.

Debris flow: The relatively rapid but viscous flow of mud and other surficial material downslope.

Derivative maps: Maps that are derived by combining, overlaying, or interpreting several environmental factors. For example, a simple landslide susceptibility map may be derived by interpreting geology, slope, and rainfall. Also called interpretative maps.

Dilatancy: Inelastic increase in volume of a rock that begins after stress is applied. Several phenomena related to dilatancy can be monitored to forecast the ultimate rupture of the rock, causing an earthquake.

Dimension stone: Stone that is quarried or cut in accordance with required specifications.

Dolomite: A mineral composed of the carbonates of calcium and magnesium. Also used as a rock name for formations composed largely of the mineral dolomite.

Drawdown: The lowering of the water table caused by pumping (or artesian flow).

Earth flow: Downslope movement of water-saturated soil or other colluvial material in a manner similar to that of a viscous fluid.

Ephemeral: A temporary or short-lived condition. Characteristic of some beaches, lakes, or streams that change rapidly.

Epicenter: The point on the earth's surface directly above the focus of an earthquake.

Erosion: The breakdown and transportation of earth materials at the surface. Agents of erosion include water, wind, ice, and gravity.

Estuary: The lower portion of a river valley where river water and tidal currents meet and intermix.

Expansive clays: Clays that, because of their molecular structure, shrink and swell by taking up and releasing water.

Fault: A fracture or fracture zone along which there has been vertical or horizontal displacement of the sides relative to one another. The displacement may extend for several inches or for many miles.

Fault gouge: Finely ground material occurring between walls of a fault.

Field Act (1933): California legislation that gave the State Division of Architecture authority for approving the design and supervising the construction of public schools. The act resulted in more adequate earthquake standards in building codes for public schools and established severe penalties for violations.

Fluvial erosion: Erosion caused by running water.

Focal point: The source within the earth of a given set of earthquake waves.

Foliation: A layering in some rocks produced by parallel alignment of minerals. Foliation is important in that it can affect the strength and hydrologic properties of rocks.

Foreshock: An earthquake that precedes a larger earthquake within a fairly short time interval (several days or weeks) and that originates at or near the focus of the larger earthquake.

Foreshore: Lower shore zone between ordinary low and high water levels.

Fumarole: A vent for volcanic steam and gases.

Gaining stream: A stream that intersects the water table and receives flow from the groundwater.

Geodimeter: An electronic-optical instrument that measures distance on the basis of the velocity of light.

Geomorphic: Of, or pertaining to, the form of the earth's surface.

Geothermal energy: Energy derived from heat stored within the earth.

Glacial drift: A general term applied to sedimentary material transported and deposited by glacial ice.

Glacial rebound: The slow regional uplift of large

land areas, such as Scandinavia, as a result of the melting and removal of the glacial ice load of the Pleistocene ice ages.

Glacial till: Unstratified and unsorted glacial material deposited directly by glacier ice.

Gneiss: A coarse-grained metamorphic rock with alternating bands of granular and schistose minerals.

Granite: A coarse-grained igneous rock dominated by light-colored minerals (quartz, orthoclase).

Groin: A device, usually constructed perpendicular to the coast, for building or protecting a shoreline by trapping littoral drift or for protecting a beach by retarding loss of beach materials.

Ground acceleration: A measurement of the intensity of ground motion that occurs as seismic waves pass beneath an area, commonly used as a factor in building design.

Gypsum: A mineral composed of hydrated calcium sulfate ($CaSO_4$ $2H_2O$), which may have internal planes of weakness.

Half-life: The time required for one-half of a given material to undergo chemical breakdown. Usually refers to radioactive atoms undergoing radioactive decay.

Humus: Decomposed or partially decomposed organic matter in the soil.

Hydration: The chemical combination of a substance with water.

Hydrocompaction: Loosely consolidated material that is compressed or compacted with the addition of water.

Hydrograph: A graph of stream discharge or stage with time.

Hydrologic cycle: The water cycle, beginning with the evaporation of water from the oceans and including all the processes whereby water returns to the sea again.

Hydrolysis: The chemical process whereby water breaks down and reacts with other compounds.

Hydromulching: A method of erosion control in which a mixture of plant seed, fertilizer, and mulch is applied to potentially erodible soil, usually by a mechanical sprayer.

Igneous rock: Rock composed primarily of silicate minerals, formed by cooling and solidification of molten magma.

Inactive fault: A fault in which there is no indication of movement within Holocene time (last 10,000 years), and where there is no reason to expect a recurrence of movement. Sometimes, a fault that has moved during Quaternary time (last 3 million years) and has displayed some potential for renewed movement is classified as potentially active.

Infiltration: Downward movement of water into soil and rock.

Infiltration capacity: The maximum rate at which the soil, when in a given condition, can absorb precipitation.

Intensity: A subjective measure of the force of an earthquake at a particular place as determined by its effects on persons, structures, and earth materials. *See also* Magnitude.

Intermediate regional flood: A 100-year flood calculated using recent flood recurrence intervals and regional hydrological characteristics, when 100 years of data are not available for a watershed.

Ion: A charged atom formed by the addition or loss of one or more electrons.

Ion exchange: Reversible exchange of ions in a crystal accomplished without destruction of the crystal structure or disturbance of electrical neutrality.

Isohyet: A line drawn between points of equal precipitation.

Jetty: A structure extending into a body of water to direct and confine a stream or tidal flow to a selected channel. Jetties are built in pairs at the mouth of a river or at the entrance to a bay to help protect or stabilize a channel.

Juvenile water: Water that is derived from the interior of the earth and has not previously existed as atmospheric or surface water.

Kaolinite: A common clay mineral composed of hydrous aluminium silicate.

Karst: A type of limestone topography characterized by closed depressions (sinkholes), caves, and subsurface streams.

Lava flow: A stream or river of molten material derived from a volcanic eruption.

Leachate: Referring to solid waste, it is the liquid that has passed through the solid waste in a refuse disposal site and has accumulated various contaminants.

Leach field: A subsurface drainage system that disposes of liquid waste from a septic tank through percolation into the soil.

Limestone: A sedimentary rock composed largely of the mineral calcite ($CaCO_3$) that has been formed by organic or inorganic processes.

Limpet: A shellfish consisting of a single cone-shaped shell and a thick fleshy foot used to cling to rocks.

Liquefaction: Earthquake-induced deformation whereby saturated, loose, granular materials (sand, silt) are transformed from a stable state into a fluidlike state in which the solid particles are in suspension, similar to quicksand. Loss of strength and ground failure often results.

Lithosphere: The outermost, rigid layer of the earth, commonly 100 km thick, consisting of the crust and upper mantle.

Littoral drift: The movement of beach sand along the coast due to wave action.

Loess: Homogeneous, unconsolidated deposits of silt, deposited primarily by wind.

Long-period structures: A taller building (more than 2 stories) characterized by a relatively long wave period. During an earthquake, a damaging resonance commonly develops where the fundamental wave period of the building coincides with the natural period of the ground. Therefore tall buildings on loosely consolidated ground are particularly vulnerable.

Longshore current: A current generated by waves approaching the coast at an angle, which then move parallel and adjacent to the shoreline. Often called littoral current.

Losing stream: A stream that recharges groundwater by giving up flow.

Magma: Molten fluids generated within the earth from which igneous rocks are derived by crystallization or other processes of consolidation.

Magnitude: A measure of energy released by an earthquake. The rating of a given earthquake is the logarithm of the maximum P wave amplitude recorded on a seismogram 100 kilometers (62 miles) from the epicenter. See *also* Intensity.

Mantle: The intermediate zone of the earth. Surrounded by the earth's crust, it rests on the core at a depth of about 2900 kilometers.

Mass wasting: A general term for a variety of processes by which large masses of rock or earth material are moved downslope by gravity, either slowly or quickly.

Meteoric water: Water that occurs in or is derived from the atmosphere.

Microearthquake: A small earthquake having a magnitude of 2.0 or less on the Richter scale.

"Mining": Referring to groundwater withdrawal where the rate of withdrawal is greater than natural recharge.

Montmorillonite: A group of clay minerals composed of aluminium silicates that are characterized by swelling when water is present.

Moraine: A general term applied to landforms composed of sediment (till) deposited by glaciers along their margins and at their terminus.

Multispectral photography: Photography that is taken in several or more bands of the electromagnetic spectrum (visible, infrared, thermal infrared, for example).

Nuée ardente: A highly heated mass of gas-charged volcanic material that, when ejected from a volcano, flows as an avalanche by virtue of its extreme mobility.

Oil shale: Shale containing such a proportion of hydrocarbons as to be capable of yielding oil on slow distillation.

Overburden: Surficial materials that overlie an extractable ore or coal deposit, or that simply overlie bedrock.

Overdraft: Continued lowering of the groundwater table due to excessive withdrawal.

Oxidation: Process of combining with oxygen. Many minerals and most metals oxidize to some extent when exposed to air or water.

Perched beach: A beach formed by placing coarse sand on top of an existing beach to reduce offshore sand losses.

Perched water table: A small groundwater body separated from the main groundwater basin below by a relatively impermeable stratum of small aerial extent. Wells tapping perched water tables yield only temporary or small quantities of water.

Permafrost: Permanently frozen ground occurring mainly in high latitudes and locally at high elevations.

Permeability: The ability of a rock or earth material to transmit fluids.

pH: A term used to describe the relative acidity or alkalinity of a solution.

Pholad: A type of boring clam normally found in the intertidal zone.

Placer deposit: A mass of gravel, sand, or similar material containing gold, platinum, tin, or other valuable minerals that have been derived from rocks or veins by erosion.

Plate tectonics: Deformation, on a global scale, involving differential movement of major crustal blocks.

Porosity: A measure of the volume of open space, or pores, in a rock or in soil.

Primary treatment: Removal of floating solids and suspended solids, both fine and coarse, from raw sewage by a settling process.

Progradation: A seaward advance of the shoreline resulting from the near-shore deposition of sediments brought to the sea by rivers.

Pyrite: A mineral, iron sulfide (FeS_2), commonly occurring in coal deposits. Colloquially called fool's gold.

Pyrolysis: A process by which solid waste is chemically converted into natural gas or oil by use of heat.

Quick clay: A water-bearing clay that readily liquifies when disturbed.

Rating curve: A graph showing the relation between elevation of the stream water surface, or stage, and stream discharge.

Recurrence interval: The average time period between storms of a certain size, or between flood events of a certain magnitude.

Remote sensing: Acquiring information about a feature by a recording device that is not in physical or intimate contact with the feature under study. The technique uses such sensors as cameras and radar systems flown in aircraft and spacecraft.

Rip current: A seaward-moving current that returns water carried landward by waves.

Riprap: A foundation or wall of broken rock placed irregularly so as to protect embankments or shorelines from erosion by running water or breaking waves.

Rockfall: A form of mass movement resulting from the relatively free fall of a newly detached segment of bedrock from a cliff or steep slope.

Room and pillar method: A mining technique requiring that thick columns of coal or other material are left in place to support the overlying roof.

Runoff: That portion of rainfall that flows over the land surface as slope wash and in stream channels.

Runout zone: In reference to avalanches, this is the area at the base of a slope where the snow and associated avalanche material comes to a halt.

Safe yield: Amount of groundwater that can be extracted from an aquifer without producing an overdraft or negative effects.

Sandstone: A sedimentary rock formed of individual sand-sized grains predominately composed of quartz.

Saturation: A condition where all the interstices of rock or soil are filled with water or another liquid.

Scarp: A cliff or steep slope formed by a fault, generally by one side moving up relative to the other.

Schist: A metamorphic rock dominated by fibrous or platy minerals. It exhibits cleavage in which grains and flakes are clearly visible ("schistose cleavage").

Schistosity: A type of foliation (parallel lamination) that occurs in coarser grained metamorphic rocks. This structure is generally the result of the parallel arrangement of platy mineral grains.

Scoria: A dark, cindery crust on the surface of a lava flow.

Scour: The erosive action of streams or waves in excavating and carrying away material from the bed and banks of a stream or beach.

Sea stack: Near-shore coastal island of resistant rock separated from the mainland as a result of wave erosion.

Seawall: A structure built along a portion of a coast to prevent erosion and other damage by wave action. Seawalls are usually more massive structures than other types of barriers.

Secondary effluent: The discharge of sewage that has had secondary treatment.

Secondary treatment: A biooxidation process in which up to 90 percent of the organic component of sewage is removed. Types of secondary treatment processes include: trickling filter, activated sludge, and oxidation ponds.

Sedimentary rock: Rock formed from accumulations of sediment, which may consist of rock fragments or particles of various sizes (conglomerate, sandstone, shale), the remains or products of animals or plants (certain limestones, coal), the product of chemical action or of evaporation (salt, gypsum), or mixtures of these materials.

Seismograph: An instrument for recording earthquake waves and other vibrations.

Serpentine: A hydrous magnesium silicate usually grayish-green in color that may contain planes of weakness.

Shale: Fine-grained sedimentary rock composed of clay and silt-sized particles. Shales are distinguished from mudstones by thin-bedded stratification.

Shear strength: The internal resistance that tends to prevent adjacent parts of a solid from "shearing" or sliding past one another parallel to the plane of contact. It is measured by the maximum shear stress that can be sustained without failure.

Shear stress: A stress causing adjacent parts of a solid to slide past one another parallel to the plane of contact.

Shear wave (S-wave): A distortional, transverse, or secondary wave. See also Transverse wave.

Shoal: A ridge of sand just below the surface of the sea or a river.

Shrink-swell ratio: Amount of heave caused by the swelling of expansive clays.

Silt: A sediment in which most of the particles are between $\frac{1}{16}$ and $\frac{1}{256}$ mm. in diameter.

Siltation: The process by which silt and similar sized particles are transported and deposited by streams.

Sinkhole: A topographic depression developed by solution of limestone, dolomite, or rock salt.

Slickenside: Polished and striated (scratched) surface that results from friction along a fault plane.

Slump: Downward movement of earth materials, either in mass or as several subsidiary units, characterized by rotational motion.

Soil mantle: Loose, unconsolidated surficial deposits overlying bedrock. Also called regolith.

Spalling: Rock that breaks off in relatively thin layers parallel to a surface.

Spring tide: A tide occurring at or shortly after the new and full moon; normally the highest tides of the month.

Standard project flood: A major flood that can be expected to occur from a severe combination of meteorological and hydrologic conditions. Such a flood might have a recurrence interval of several hundred years or more and is used as design criteria for major flood control projects.

Stream gauging: The measurement of stream flow.

Strike-slip fault: A fault in which movement is principally horizontal.

Subduction: In plate tectonics, the process by which one crustal block descends beneath another into the mantle.

Subduction zone: In plate tectonics, a linear zone where two plates or blocks are pushing together and subduction is taking place. Oceanic trenches, lines of volcanoes, and intense faulting and folding occur in such zones.

Submarine canyon: A deep, steep-sided submarine valley commonly crossing the continental shelf and slope.

Suspended sediment: Fine-grained sediment that remains in suspension in water for a considerable time without contact with the bottom.

Synergism: The combined action or effect of several pollutants or variables which may be much more severe than any of the individual variables.

Tailings: Those portions of washed ore that are regarded as too poor to be used or treated further.

Talus: A collection of fallen disintegrated rock material that has formed a slope at the foot of a steeper declivity.

Tectonic: Rock structure or external land forms resulting from deformation of the earth's crust.

Tephra: A collective term for all pyroclastic materials ejected from a volcano during an eruption. Includes volcanic dust, ash, cinders, lapilli, pumice, and larger particles.

Theodolite: A very accurate surveying instrument used for measuring horizontal and vertical angles.

Thrust fault: A steeply inclined fault in which the block above the fault has moved relatively upward or over the block below the fault.

Till: *See Glacial till.*

Tiltmeter: An instrument for measuring the tilt of the ground. Used to indicate the swelling of a volcano during periods of rising magma preceding an eruption. Also used for observing surface disturbances in an attempt to predict earthquakes.

Tips: Coal waste deposits.

Trace: The intersection of a fault and the earth's surface. Two or more parallel fault traces normally constitute a fault zone.

Track: Downslope path of avalanche material.

Transform fault: A lateral fault forming the edge of a tectonic plate. The fault ends suddenly at a point where the movement is transformed into a different structure, such as an oceanic ridge.

Transpiration: The process by which water vapor escapes from a plant and enters the atmosphere.

Transverse wave: Wave in which the motion of the particles, or the entity that vibrates, is perpendicular to the direction of progression of the wave train.

Tsunami: A large ocean wave generated by earthquake activity, or rarely, by volcanic eruption. Also called seismic sea wave.

Turbidity current: A current in which a mass of turbid or muddy water moves downslope relative to the surrounding water because of its greater density.

Unconfined aquifer: Groundwater that is not overlain by impermeable material.

Water table: The upper surface of the zone of groundwater saturation, where all the pore spaces are filled with water.

Wave period: The time for a wave to advance the distance between two successive wave crests (a wavelength).

Wave refraction: The process by which the direction of a train of waves moving into shallow water is changed.

Weathering: The mechanical and chemical breakdown of rocks as they are exposed to water, air, living matter, and temperature changes.

Index